Green Energy and Technology

Climate change, environmental impact and the limited natural resources urge scientific research and novel technical solutions. The monograph series Green Energy and Technology serves as a publishing platform for scientific and technological approaches to "green"—i.e. environmentally friendly and sustainable—technologies. While a focus lies on energy and power supply, it also covers "green" solutions in industrial engineering and engineering design. Green Energy and Technology addresses researchers, advanced students, technical consultants as well as decision makers in industries and politics. Hence, the level of presentation spans from instructional to highly technical.
Indexed in Scopus. **Indexed in Ei Compendex**.

More information about this series at https://link.springer.com/bookseries/8059

Augustine O. Ayeni
Olagoke Oladokun • Oyinkepreye David Orodu
Editors

Advanced Manufacturing in Biological, Petroleum, and Nanotechnology Processing

Application Tools for Design, Operation, Cost Management, and Environmental Remediation

Springer

Editors
Augustine O. Ayeni
Department of Chemical Engineering
Covenant University
Ota, Nigeria

Olagoke Oladokun
Department of Chemical Engineering
Covenant University
Ota, Nigeria

Oyinkepreye David Orodu
Department of Chemical Engineering
Covenant University
Ota, Nigeria

ISSN 1865-3529 ISSN 1865-3537 (electronic)
Green Energy and Technology
ISBN 978-3-030-95819-0 ISBN 978-3-030-95820-6 (eBook)
https://doi.org/10.1007/978-3-030-95820-6

This Springer imprint is published by the registered company Springer Nature Switzerland AG
The registered company address is: Gewerbestrasse 11, 6330 Cham, Switzerland

Preface

The 1st International Conference on Energy and Biochemical Engineering (ICEBE2021) was successfully held in Ota, Nigeria, on September 27–29, 2021. ICEBE2021 provided a platform for knowledge exchange and transfer among academic scientists and engineers, industry researcher, scholars, and students to share experiences and research results, thereby ensuring adequate dissemination of information for national and international development.

This book, *Advanced Manufacturing in Biological, Petroleum, and Nanotechnology Processing: Application Tools for Design, Operation, Cost Management, and Environmental Remediation*, emerged from papers submitted to the ICEBE2021 from universities and industries across the world. All accepted papers were subjected to strict peer-reviewing by expert referees. Papers have been selected for this volume based on quality and relevance to the conference. This book has been organized into three sections exploring advances in the design and application of process systems, pollution control and management, and process modelling and simulation, and targets readers in industrial processes, smart control systems, environmental pollution, control and monitoring, biopolymer production, fluids and coolants, machining operations, and biomedical waste minimization and eradication. The book will reveal novel advanced manufacturing of products,process systems,modelling, and simulation aimed at introducing future process designs to new application tools for optimal process design, operation, and cost management. Various studies documented in this book can provide new platforms for developing further research efforts toward advanced manufacturing processes.

The editors believe that this book will enable readers to view broadly the new advances in sustainable manufacturing, cutting across biological, petroleum, and nanotechnology processes.

ICEBE2021 would like to thank all authors for their contributions, which made this book possible. Our appreciation also goes to reviewers that participated in the peer-review process before the papers were accepted. The contributions of the reviewers greatly enhanced the quality of the chapters in this book. Finally, we appreciate the various committees of ICEBE2021 that actively participated in making the event a huge success.

Ota, Nigeria

Augustine O. Ayeni
Olagoke Oladokun
Oyinkepreye David Orodu

Contents

Part I Design and Application of Process Systems

Nanotechnology: Applications, Challenges, and Prospects 3
C. N. Egwu, R. Babalola, T. H. Udoh, and O. O. Esio

Development of Energy Efficient Processes and Products from Renewable and Nonrenewable Resources in Nigeria 17
O. J. Oyebode, A. T. Adeniyi, U. S. Gekwu, K. O. Olowe, and A. O. Coker

An Overview Application of Natural Oil as a Sustainable Plasticizer in Production of Biopolymers 29
F. Iriaye, A. A. Abioye, O. O. Yusuf, M. E. Emetere, S. O. Ongbali, A. A. Noiki, S. A. Afolalu, and F. Ademuyiwa

Manufacturing of Brake Pad Using Aluminium Silicon Carbide Reinforced with Alumina for Automobile Industry 37
P. O. Babalola, T. A. Okunuga, A. O. Inegbenebor, O. Kilanko, and M. O. Udo

Overview of Nanofluid Applications and Its Sustainability 45
F. Iriaye, A. A. Noiki, O. O. Yusuf, S. A. Afolalu, and M. E. Egbe

Influence of Cutting Fluid and Parameters on Machining and Cooling Techniques in Recent Technology 55
F. Ademuyiwa, S. A. Afolalu, O. O. Yusuf, and M. E. Emetere

Nanofluid: A Sustainable Alternative Coolant for Metalworking and Machining Operations 75
M. Udo, A. A. Noiki, O. O. Yusuf, M. E. Emetere, S. A. Afolalu, and S. O. Ongbali

The Essence of Intermetallic Phases in AA6061/Clay Composites 85
N. E. Udoye, O. S. I. Fayomi, and A. O. Inegbenebor

A Short Overview on the Role of Nanotechnology in Different Sectors of Energy System 99
O. Agboola, A. O. Ayeni, O. S. I. Fayomi, O. Oladokun, A. A. Ayoola, E. D. Babatunde, V. E. Efeovbokhan, O. A. Odunlami, A. Adeniyi, and E. R. Sadiku

Experimental Study of Enhanced Oil Recovery Potential of Nanoparticle (Silicon Dioxide) Coated with Guar Gum 117
O. O. Olabode, O. Okafor, B. Oni, P. Alonge-Niyi, and V. Abraham

Part II Pollution Control and Management

Public-Private Partnership: A Veritable Tool for Handling Environmental Pollution and Infrastructural Degeneration in Nigeria 131
O. J. Oyebode

Smart Toilets and Toilet Gadgets in Sustainable Smart Cities: An Overview of Personal Health Monitoring 143
C. C. Mbonu, O. Kilanko, M. B. Kilanko, and P. O. Babalola

Biomedical Engineering Education: Equipment, Prospect and Challenges for Environmental Healthcare in Nigeria 157
O. J. Oyebode

Inhibition Performance of Admixed Grapefruit and Lemongrass Oil Extracts on Low Carbon Steel in Weak Acid Formulation 165
R. T. Loto, P. Okpaleke, and U. Udoh

Corrosion Inhibitive Behaviour of *Moringa Oleifera* in Acidic Medium 175
A. A. Ayoola, S. C. Okwuonu, B. M. Durodola, E. E. Alagbe, O. Oladokun, O. Agboola, and R. Babalola

Phosphating Technique: A Reliable Approach for Corrosion Resistance of A36 Mild Steel 185
S. N. Ezekiel, A. A. Ayoola, B. M. Durodola, O. Odunlami, and O. A. Oyeniyi

Adoption of Environmental Engineering Strategies for Public Health and Sustainable Development 197
O. J. Oyebode

Adsorptive Performance of Immobilized Activated Carbon for Effective Removal of Dibenzothiophene from a Synthetic Petroleum Distillate in a Packed-Bed Column 211
O. O. Sadare and M. O. Daramola

Overview of Nano-agro-composite Additives for Wastewater and Effluent Treatment 223
S. A. Afolalu, O. Okwilagwe, O. O. Yusuf, O. R. Oloyede, S. O. Banjo, and F. Ademuyiwa

Sustainable Waste Management Towards Circular Economy in Nigerian Context: Challenges, Prospects and Way Forward 237
O. J. Oyebode

Biomedical Waste Management Practices in Sub-Saharan Africa: Insights of Its Impacts and Strategies for Its Mitigation 249
O. J. Oyebode, A. O. Coker, M. K. C. Sridhar, and N. M. Ogarekpe

Bioplastic: A Sustainable Remedy to Manage Environmental Waste 257
S. A. Afolalu, O. O. Yusuf, M. E. Emetere, S. O. Ongbali, and F. Ademuyiwa

A Short Review on Land/Soil Pollution: The Pollutants and the Treatment Techniques 267
F. B. Elehinafe, O. G. Olomukoro, A. O. Ayeni, and O. B. Okedere

The Effect of Air-Fuel Ratio on Tailpipe Exhaust Emission Measurement of Motorcars 277
O. A. Odunlami, F. A. Akeredolu, K. O. Oderinde, A. A. Ayoola, A. A. Busari, and V. E. Efeovbokhan

Part III Process Modelling and Simulation

Design of Water Retaining Structures and Application of Environmental Engineering for Sustainable Environment 287
O. J. Oyebode

Development of Robust Fractional-Order Controllers to Provide Effective Economic Load Dispatch Management in Interconnected Area Networks 301
J. Okoronkwo, A. Mati, Y. Jibril, G. Olarinoye, and A. S. Abubakar

Digital Technology and Sustainable Manufacturing: The Nexus 315
A. Noiki, F. Ademuyiwa, S. A. Afolalu, M. B. Edun, O. O. Yusuf, and M. E. Emetere

Gasific: A Python Tool for Biomass Gasification Process 335
O. Oladokun, B. Nyakuma, W. S. Luing, E. Oladimeji, O. Abatan, A. O. Ayeni, O. Agboola, A. A. Ayoola, M. O. Ojewunmi, V. E. Efeovbokhan, L. Olagoke-Oladokun, and O. A. Odunlami

Index 355

Part I
Design and Application of Process Systems

Nanotechnology: Applications, Challenges, and Prospects

C. N. Egwu, R. Babalola, T. H. Udoh, and O. O. Esio

1 Introduction

Nanotechnology is one of the thriving progressive advances in recent times. It encompasses many technologies which are carried out on the scale of nanometer (having one or more dimensions of the order of 100 nm or less) with a wide application in different disciplines of biological science and engineering. Many other technologies have come into existence directly from a particular scientific discipline, unlike nanotechnology. The panel of the British Royal Society and the Royal Academy of Engineering characterized nanotechnology "as the plan, portrayal, creation and use of structures, gadgets and frameworks by controlling shape and size at the nanometer scale" (SCENIHR, 2006).

Nanotechnology can be applied in various areas such as security, transportation, electronics, cosmetics, fuel cells, renewable energy, zeolite synthesis, etc. This study seeks to review the applications of nanotechnology in five different areas, namely, textile, water treatment, food preservation, agricultural production, and medicine & healthcare. The challenges and prospects of the application of nanotechnology in these areas are also presented.

Reducing the size (and shape) of some materials has shown to greatly affect their properties. For example, when gold particle size is reduced from 10 nm to 2 nm, its melting point reduces from 527 °C to 327 °C. Similarly, when the particle size of calcium selenide (CaSe) powder is reduced, its color changes from red to yellow. Studies have shown that addition of nanoparticles of aluminum or nickel to rocket fuel doubles its heat of combustion (Qhatan, 2017).

C. N. Egwu (✉) · R. Babalola · T. H. Udoh · O. O. Esio
Department of Chemical/Petrochemical Engineering, Akwa Ibom State University, Mkpat-Enin, Nigeria
e-mail: rasheedbabalola@aksu.edu.ng; tinuolaudoh@aksu.edu.ng; esioboho@aksu.edu.ng

A. O. Ayeni et al. (eds.), *Advanced Manufacturing in Biological, Petroleum, and Nanotechnology Processing*, Green Energy and Technology,
https://doi.org/10.1007/978-3-030-95820-6_1

2 Materials and Methods Used in Nanomaterial Synthesis

Nanomaterials are synthesized from different materials such as graphite/carbon, cadmium sulfide, kaolin, oxides of copper, silver, gold, zinc, germanium, etc. Methods deployed in the synthesis of nanomaterials are shown (Table 1). Additionally, examples of some materials of application in nanomaterial synthesis are shown (Table 2).

Microemulsion, spark discharge, inert gas condensation, sputtering, infiltration, spray pyrolysis etc. are other methods.

2.1 Nanotechnology Application in Textiles

Textiles are raw materials for production of yarn. The textile industry is recently combining with other areas of science and engineering to make products with extraordinary properties and uses in areas of textile finishing, medicals, sportswear, conductive textiles, military/security attire, fashion/lifestyles, textile-based sensors, etc. (Fig. 1). Textiles and electronics can be synchronized to yield extraordinary properties (Hassan et al., 2019).

A lot of non-textile materials are now being incorporated into textiles as coatings and conductive materials in fibers. These advancements have made it possible to achieve smartness in textile by manipulation of nanoparticles, nanofilms, and nanocoatings. Additionally, with nanotechnology, enhanced electrical conductivity,

Table 1 Some methods of synthesizing nanomaterials

Method	Description
Ultrasound	Nanomaterials are formed when ultrasonic radiation creates cavitation which generates temperature and pressure for reactions at extreme settings
Hydrothermal technique	Nanomaterials are synthesized in an airtight but heated solution at prevailing temperature and pressure
Microwave-assisted method	Excellent control during synthesis especially for temperature, proper agitation, etc. with high produce and less sideway products
Laser ablation	Beams are used to remove solid (or liquid) particles in a top bottom approach. The intensity of the beam is under control
Template synthesis	Here, Green Chemistry approach is employed; the nanoparticle synthesized has uniform void space because the template/skeleton is designed thus. Nanotube and nanowire can be made using the method
Coprecipitation	This technique takes into cognizance the real-time process of particle development, nucleation, and cluster progressions
Sol-gel	In this chemical procedure, a combination of liquid-solid phase (gel-like) is made from solution (sol). This method is applied in dental/medi-care
Biological synthesis	Here nanotechnology and biotechnology approaches are combined

Source: Ajay et al. (2018)

Table 2 Some nanomaterials required for finishing in textile industry

Finishing	Nanomaterials used
Protection from UV	ZnO, TiO_2
Improved staining and fade reduction	Nanoporous hydrocarbon, carbon black, SiO_2 matrix
Moisture absorbency	TiO_2
Self-cleaning properties and water repellency	TiO_2, fluoroacrylate, CNT, SiO_2 matrix
Medicinal products or fragrances	Montmorillonite (nanoclay), SiO_2 (as matrix)
Antistatic and conductive	Carbon nanotubes (CNTs), carbon black, copper, polypyrrole, polyaniline
Durability	Metal oxides of Al_2O_3, SiO_2, ZnO, CNT, polybutylacrylate
Antibacterial	Chitosan, Ag, SiO_2 matrix, ZnO, and TiO_2
Fire proofing	Boroxosiloxane, CNT, montmorillonite (nanoclay), Sb_3O_2

Source: Hassan et al. (2019)

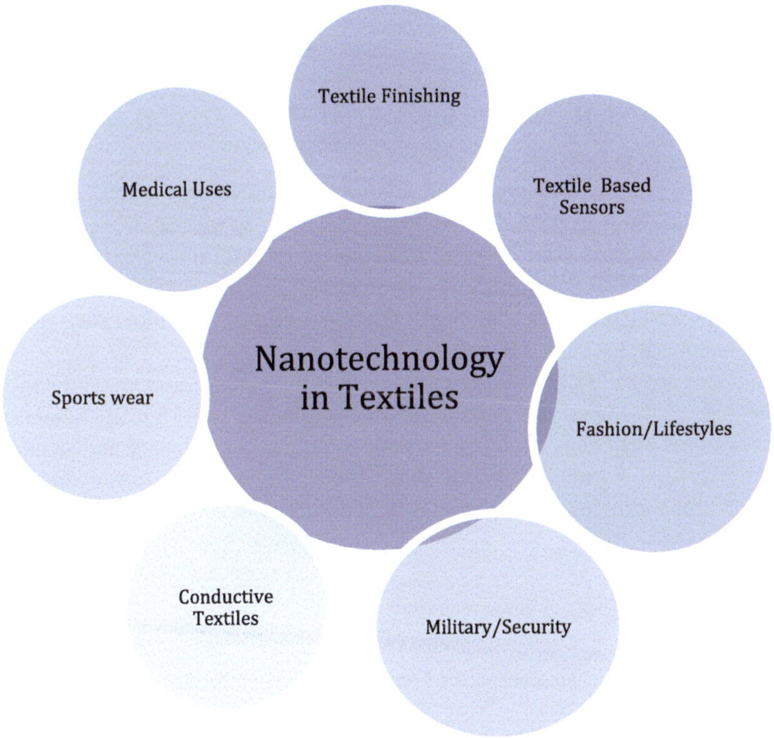

Fig. 1 Nanotechnology in some segments of textile. (Hassan et al., 2019)

high mechanical strength, and thermostability are now easy to achieve. Smart textiles are also finding applications in the area of designs associated with interior parts. Smart furniture are also being delivered in interior decorations by the blend of hardware materials. These features however make the interior products unique and expensive.

Semiconductor ceramics and metallic oxide nano-completions can be applied on materials to get explicit properties. These include fire retardance, capacity to repulse water and oil, warm resistivity and antimicrobial properties, and so on, as illustrated in Table 2.

Engineered nanomaterials (ENMs) containing Ag and Ag^+ nanoparticles (NP) are used as antimicrobial channels with adequate transport properties (Gopal et al., 2006). Covering the materials consistently enhances the durability. Silver nanoparticles covered on cotton filaments through the cushion dry fix technique demonstrated high washing durability, protecting the antibacterial properties and antifungal parasitic properties against numerous microbes even after numerous washes. These properties of silver nanoparticles make the utilization of nanoparticles on materials advantageous (Balakumaran et al., 2016).

2.2 *Application of Nanotechnology in Water Treatment*

Nanotechnology study is a promising significant innovation for the treatment of wastewater which is essential to humanity. These advances are conservative, dependable, quick, and solid to treat wastewaters by wiping out explicit kinds of toxins from water. Application of nanotechnology for potable water supply encapsulates water filtration, chemical injection, and utilizing tools such as nanosensors, nanoparticles, and catalysts.

Nanoclays are naturally existing particles used in water treatment, and nanoclay particles are considered as nanomaterials originating geologically. Nanoclays display various structures which incorporate tetrahedral silicates and octahedral aluminum layers. Various forms of these muds rely upon its arrangement and formation of the structures (Lubomira & Valentin, 2005).

Table 3 Water-related issues of developing countries

Statistics	Realities
3.4 million	In evolving nations, people die from water-associated ailment annually
63 million	In evolving nations like Bangladesh, India, and Nepal, people agonize with arsenic contamination
6 km	Women from Africa and Asian continent walk to draw water
80%	Water-associated bereavement in children between age 0 and 14 years
40%	Water-associated bereavement are due to diarrhea

Source: Yuan and Wu (2007)

Shi and Zhu (2011) explored the use of palladium-graphene nanocomposite and ion fluid as a sensor for chlorophenols. The arrangement of the nanocomposite by a sono-electrochemical course and the conceivable development component was proposed. Scanning electron microscopy (SEM), X-ray diffractometry (XRD), etc. were utilized for the portrayal of the structure and morphology of the nanocomposites. The experimental outcomes indicated that palladium nanospheres are made up of few atoms of palladium (Pd) nanoparticles and are consistently appended to graphene sheets. The properties were explored, and the differential pulse voltammetry (DPV) showed that the Pd-graphene nanocomposite had a high activity for chlorophenol oxidation. In this, 2-chlorophenol was chosen as the model particle. The outcomes indicate that graphene assumed a significant role in the creation of the chlorophenols' sensor. The combined nanocomposite for the most part described by enormous electrochemical active surface prompted a brilliant electrocatalytic activity which can additionally upgrade the synergist action of palladium graphene for chlorophenols seen in a few auxiliary tests from a wastewater treatment plant (Shi & Zhu, 2011).

2.3 *Nanotechnology in Food Preservation*

Nano-based "keen" and "dynamic" food packaging gives a lot of benefits over the typical packaging techniques by guaranteeing better packaging material with improved properties. For instance, properties like mechanical strength, antimicrobial layers, and nanosensing for microbe discovery consequently make purchasers aware of the status of food (Singh, 2018). The food sector is heavily leveraging on the nanotechnology-obtained food packaging materials and its different methods of usage (Duncan, 2011).

In this application, imbuing nanomaterials to develop the wrapping properties is made, for example, in adaptability, temperature steadiness, dampness, and strength of these materials, joining nanoparticles with antimicrobial or oxygen rummaging properties, and "keen" food packaging with nanosensors, and observing and reporting the state of the food and decomposable polymer nanomaterial composites.

Generally, the creation of nanoparticles can be performed by both the "top-down" and "base-up" procedures and nanocapsules are not exempted. For the previous procedure (top-down), the nanonization is accomplished by the utilization of energy, while for the latter, the conglomeration of particles, monomers, ions, or atoms is controlled physiochemically to frame the nanocapsule.

Food added substances like benzoic, citric, and ascorbic acids, dietary enhancements, and useful food ingredients (vitamins A and E, lipoic acid, soybean isoflavones, carotene, lutein, omega-3 unsaturated fats, and coenzyme Q10) are being made with the aid of nanotechnology (Mohammadi et al., 2015).

Furthermore, nanomicelle-based bearers for nutraceuticals and dietary enhancements have been created: nanocochleates (50 nm in size), an example of nanomicelle, in view of a phosphatidylserine bearer got from soya bean, are, for the most part, viewed as safe for consumption. They are acquired by adding calcium ions to little phosphatidylserine vesicles. The nanocochleates' framework is professed to be utilized as insurance for micronutrients and antioxidants from termination during production and storage (Aschberger et al., 2011).

2.4 Application of Nanotechnology in Agricultural Production

Continual population growth places a demand on the need for improved productivity in agricultural production. Sadly, this is not so in many parts of the world, which presupposes that there is a need to upgrade the food production capacity. It is accepted that applying nanotechnology in agricultural production can increase yield, limit the rate of disease attack, and subsequently provide food for the teeming population. Some applications of nanotechnology in agricultural production are (i) nanoform zeolites for the gradual issuance and efficient delivery of water and fertilizers for plants, drugs for livestock, nanocapsules, and herbicide conveyance, (ii) nanosensors for soil texture and for plant health investigation and nanosensors used for pest detection, (iii) nanomagnets for displacement of soil toxins, and (iv) nanoparticles for synthesis of new products of pesticides, insecticides, and insect repellents (Srivastava et al., 2016).

Several novel discoveries are being made in the agriculture sector. For example, the development of nanoseed with built-in pesticide effect and nanoencapsulation technique which has the potential to change dietary configuration, flavor, etc. is tailored to meet consumer needs and physiological requirements. Also, the nanotechnological intersection in horticulture centers around three useful enhancements, namely, expanded adequacy (with high dissolvability, stability, and viability), controlled delivery (due to specific upgrades), and directed conveyance of compost, plant development controllers, and biocides such as fungicides, herbicides, and pesticides (Srivastava et al., 2016).

Nanotechnology-based agriculture products currently undergoing development include the following:

- Nanoclay capsule containing biocontrol specialist and development upgrading chemicals that are intended for the arrival of dynamic fixing ingredients. Models are pyrethroids such as cyhalothrin and cypermethrin and others like artemisia arborescence essential oil (Rai et al., 2009).
- Nano-definitions (nanodispersions/nanoemulsions) of herbicide intended to remove the seed covering of weeds and prevent weed germination.
- Nanotechnology-empowered gadgets such as independent nanosensors linked to a global positioning system (GPS) framework have been utilized for continuous checking of soil conditions and growth of crops. Exactness cultivating, with the assistance of shrewd sensors, can upgrade efficiency in agriculture by giving precise data, which will at last assist farmers and land administrators in proper decision-making.

- Nanoscale gadgets could be utilized for early sickness diagnose and to assist ranchers in making preventive decisions.

Nanotechnology is helping in the development of manure and insecticide delivery systems which would be able to respond to climatic changes. "Gutbuster" is a nanomicrocapsule created and licensed by Syngenta that transports pesticides orchestrated to unveil in an alkaline medium of a bug's stomach. Such nano-exemplification procedures not just give in-assembled pesticides to crops but also adjusted insecticidal crops and guarantees in-fabricated changes to control the delivery, hence ensuing accessibility of pesticides (Klaine et al., 2008). These CNTs had the ability to enter the thick seed coat and enhance water take-up inside the seeds, a process that was able to influence seed germination and growth of tomato seedlings positively.

The physical and chemical properties of nanozeolites, alongside their bounty in sedimentary deposits and rocks with volcanic material, have been instrumental in numerous agricultural uses. Due to their high porosity and high cation trade attributes, nanozeolites are viewed as a significant additive in agricultural and ecological design as they will in general expand the yield and improve the proficiency of supplements/conveying manures, bug sprays, fungicides, herbicides, and catching weighty metals (Klaine et al., 2008; Ehsan, 2019). David (2007) showed that studies on zeolite synthesis from natural sources like kaolin have progressively delivered excellent nanoporous qualities of nanozeolites. Some of these qualities include modest molecule size, a more significant level of surface conductivity, retention ability, optimum silica to alumina ratio, crystallinity, etc. Increase in the internal surface area and the pore sizes and high cationic exchange properties are other great qualities of nanozeolites that enhance their application in nanotechnology (Adeoye et al., 2017). Among the regular zeolites, clinoptilolite is generally plentiful in soils and usually utilized in agricultural tasks and fills in as a soil reformer, and it has contributed to improving nitrogen take-up in soils. Clinoptilolite, from the heulandites group of normal zeolites, has a high bond with ammonium (Ehsan, 2019).

2.5 *Nanotechnology Applications in Medicine and Healthcare*

Today, there are several complex diseases posing problems to mankind. Some of which are dreaded diseases as well as different kinds of serious inflammatory or infectious diseases (including HIV), and most recently the coronavirus (COVID-19). Ricardo and Lino (2010) in their study showed that nanotechnology in medicine and healthcare using nanomaterials can be applied for treatment of various ailments. Nanomaterials have significantly affected their associations with biomolecules and cells due to their unconventional size, shape, chemical synthesis, surface structure, charge, solvency, agglomeration, and so on. For instance, nanoparticles can be used to create extraordinary pictures of tumor destinations. Single-walled carbon nanotubes have been applied as high-proficiency carriers for biomolecules into target cells (Nikalje, 2015).

Nanomedicine is quite a new field of science and technology. By interrelating with biological molecules at nanoscale, nanotechnology application in medicine can be very broad. Nanotechnology has made the communication of nanodevices with each biomolecules to be unraveled both in the extracellular intermediate (in vitro) and inside the human cells (in vivo). Working with nanoscale has made it possible to explore some properties that differ from those seen at microscale such as the volume and surface ratio. Two types of nanomedication that have been effectively tried in mice and are anticipating human preliminaries are the utilization of gold nanoshells to help analyze and fix cancer and the utilization of liposome as immunization adjuvant and as medium for drug transportation (Boisseau & Loubaton, 2011).

The short description of pharmaceutical nanosystem is shown in Fig. 2. Pharmaceutical nanotechnology is separated into two elementary types of nanotools, namely, nanomaterials and nanodevices. These materials can be subclassified into nanocrystalline and nanostructured materials. Nanostructure consists of nanoparticles, dendrimers, micelles, drug conjugates, nano-electromechanical systems, microelectromechanical system, metallic nanoparticles, etc.

Similarly, drug detoxification is another way that nano medicine has been used effectively in rats. The nanodesigned materials are being produced for viably treating sicknesses, infections, and diseases. With the progression of nanotechnology,

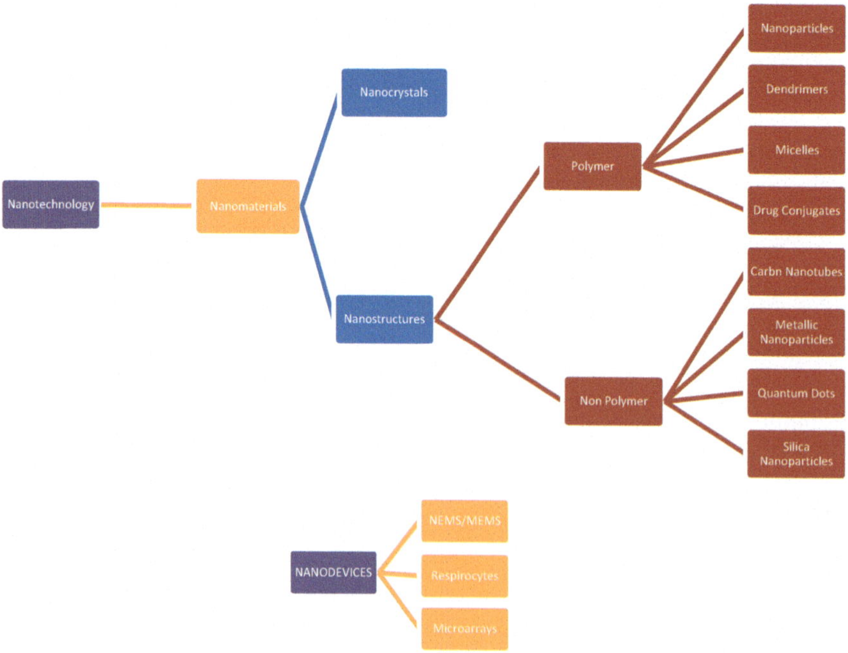

Fig. 2 Pharmaceutical nanosystem. (Nikalje, 2015)

self-collected biocompatible nanogadgets are being made that will distinguish the malignant cells and naturally assess the infection, fix it, and plan reports (Nikalje, 2015). Cadmium selenide nanoparticles as quantum specks are utilized in the location of malignant tumors because they gleam when exposed to bright light. The specialist infuses these quantum dabs into malignant growth tumors and can observe the sparkling tumor; in this way, the tumor can undoubtedly be eliminated (Sahoo et al., 2008).

Nanoparticles are utilized in disease photodynamic treatment, wherein the molecule is embedded within the tumor in the body and is enlightened with photograph light from an external perspective. The molecule retains light and gets warmed because of energy from the light. High-energy oxygen atoms are created because of the light which synthetically responds with and pulverizes tumor cells, without responding with other body cells. Photodynamic treatment has acquired significance as a noninvasive method for managing tumors (Ahmed et al., 2012). Nanofill composite resins are accepted to offer fantastic wear opposition, strength, and extreme style because of their remarkable polishability and shine maintenance. In usable dentistry, nanofillers comprise circular silicon dioxide (SiO_2) particles with a normal size ranging from 5 to 40 nm (Sivaramakrishnan & Neelakantan, 2014).

In optometry, an epic nanoscale scattered eye balm (NDEO) for dealing with serious evaporative dry eye has been effectively evolved (Zhang et al., 2014). The excipients employed as semi-strong lipids were petrolatum and lanolin, as utilized in ordinary eye salve, which was combined with medium-chain triglycerides (MCT) as a fluid lipid; the two stages were then scattered in polyvinylpyrrolidone answer for structure nanodispersions. The curative impacts of NDEO were assessed and exhibited helpful improvement, showing a pattern of a positive relationship with higher centralizations of salve framework in the NDEO reestablished the ordinary corneal and conjunctival morphology conforming its effectiveness for ophthalmic uses (Sahoo et al., 2008).

3 Challenges of the Applications of Nanotechnology

There are several challenges associated with the applications of nanotechnology. One of them is toxicity. To assess the poisonous nature associated with the design of nanomaterials, various techniques are defined. Norms have been made to control the harmfulness of nanomaterials, for example, International Electrotechnical Commission (IEC)/TC113 and International Organization for Standardization (ISO)/TC229. Nanoparticles from engineered or other nanomaterials can penetrate the body by ingestion or by dermal contact. Toxicity of nanoparticles is subject not only to their properties and the course of the passageway into the body fixation and span of openness to nanoparticles but also on individual weakness and condition of the living being. Results of oral course examined gave the indications of harm with moderately high dosages of nanosilver or nanoTiO_2 (Aschberger et al., 2011). Ag, Au, Fe; Ti, Fe, Co-Zn-Fe, etc. have been discovered to be poisonous against soil bacteria (Dinesh et al., 2012). The silver nanoparticle was discovered to be harmful to *E. coli* and *Staphylococcus aureus* (Rai et al., 2009) and *B. subtilis* (Yuan & Wu, 2007). Biosynthesis of silver nanoparticles by organisms has strong action against contagious and bacterial strains like *Aspergillus niger*, *Staphylococcus* sp., *E.*

In food preservation, most of the nanocapsule creation methods are acted upon in a dissolvable media. It is notable that the presence of solvents involves various hindrances, for example, the danger of microbial pollution, expanded expenses, and physicochemical instability (Singh, 2018). Considering this issue, scientists are attempting to utilize these nanoparticles as agents for the conveyance of the compound composts and pesticides to crops (DeRosa et al., 2010).

Tools for routine estimations of free NPs in different media are not sufficient. The accessible facilities to explore the properties of NPs in the climate are not satisfactory. Likewise, almost nothing is thought about the physiology of nanoparticles. The examinations directed in vitro obviously show that ENPs are dangerous to both micro and macrofaunae.

4 Prospects in the Applications of Nanotechnology

There is a great future for various industries in the applications of nanotechnology.

For better understanding and usage of nanotechnology, biology researchers, material experts, physicists, and engineers should collaborate. Nanosystems (nanometer-scale frameworks) additionally need modernized and progressed gear, that way, design and applied sciences can share their abilities. This collaboration is conceivable when all the fields of science share their abilities but keep their natural personalities (Rempel, 2007).

Other than turning toward apparel needs, textile businesses can begin to zero in on non-garment regions of materials such as specialized material. The essential job of specialized material will be to give specialized capacity instead of the usual enrichment attributes (Hassan et al., 2019).

By joining together nanotechnology and biotechnology, an intense new apparatus would be created that can control qualities and even engineer new products. For instance, nanobiotechnologies empower nanoparticles, nanofibers, and nanocapsules to convey unfamiliar DNA and synthetic compounds that adjust qualities (Torney et al., 2007). Notwithstanding the reengineering of existing plants, novel plant assortments might be created to utilize what is manufactured, creating another part of techno-science that draws on the procedures of hereditary design, nanotechnology, and informatics (Lyon et al. 2006). Again, present methodologies will require some modifications in order to deal with hazardous effects associated with nanotechnology.

Obviously, there is an open door for nanotechnology to profoundly affect energy, economy, and the environment, by improving compost items. New possibilities for coordinating nanotechnologies into manures ought to be investigated, mindful of any likely danger to the environment or to human well-being. Because of the little size of nanoparticles, it can get into the fringe tissues, and hence can be taken up in the body in the long run. However, nanoparticles can be utilized effectively and productively during appropriation. Further examination should be possible on nanotoxins with the goal that its clinical uses can be expanded and improved (Suresh et al., 2010).

Advances in nanozeolite science require multidisciplinary abilities (materials' amalgamation and handling, nitty-gritty comprehension of their substance, physical, warm, and optical properties). The designing of gadgets and frameworks dependent on zeolite nanocrystals will keep on including ever-more prominent communications and coordinated efforts between scientific experts, physicists, materials researchers, and engineers (Ehsan, 2019).

5 Conclusion

There is a brilliant future in nanotechnology, by its converging with different advancements and the ensuing rise of intricate and creative mixture of innovations. Nano-innovation is now utilized to control hereditary material, and nanomaterials are now being assembled utilizing organic parts. The capacity of nanotechnology to design matter at the tiniest scale is altering regions and industries including agriculture, clothing/textile, medicine, water treatment, catalyst synthesis, cognitive science, information technology, biotechnology, etc., and this is gradually giving rise to new and interlinking frontiers. Further research on nanotechnology will be of great value for the overall benefit of human existence. Regenerative medication, undifferentiated organism exploration, and nutraceuticals are among the main areas that will be positively altered by nanotechnology advancements. Finally, it is important that standards be followed in the manufacture of nanomaterials and its applications and disposal management.

Conflicts of Interest The authors report no conflicts of interest.

References

Adeoye, J. B., Omoleye, J. A., Ojewumi, M. E., & Babalola, R. (2017). Synthesis of zeolite Y from kaolin using novel method of dealumination. *International Journal of Applied Engineering Research, 12*, 755–760.

Ahmed, R. Z., Patil, G., & Zaheer, Z. (2012). Nanosponges – A completely new nanohorizon: Pharmaceutical applications and recent advances. *Drug Development and Industrial Pharmacy, 39*, 1–10. https://doi.org/10.3109/03639045.2012.69461

Ajay, V. R., Krishnan, K., Abitha, V. K., & Sabu, T. (2018). Methods for synthesis of nanoparticles and fabrication of nanocomposite. *Synthesis of Inorganic Nanomaterials*, 121–139. https://doi.org/10.1016/B978-0-08-101975-7.00005-1

Aschberger, K., Micheletti, C., Sokull-Klüttgen, B., & Christensen, F. M. (2011). Analysis of currently available data for characterizing the risk of engineered nanomaterials to the environment and human health – Lessons learned from four case studies. *Environmental International, 37*, 1143–1156. https://doi.org/10.1016/j.envint.201102.005

Balakumaran, M. D., Ramachandran, R., Jagadeeswari, S., & Kalaichelvan, P. T. (2016). In vitro biological properties and characterization of nanosilver coated cotton fabrics – An application for antimicrobial textile finishing. *International Biodeterioration & Biodegradation, 107*, 48–55.

Boisseau, P., & Loubaton, B. (2011). Nanomedicine, nanotechnology in medicine. *Comptes Rendus Physique, 12*, 620–636. https://doi.org/10.1016/j.crhy.2011.06.001

David, T. W. (2007). Zeolites – Earliest solids state acids. *AU Journal of Technology, 11*(1), 36–41.

DeRosa, M. C., Monreal, C., Schnitzer, M., Walsh, R., & Sultan, Y. (2010). Nanotechnology in fertilizers. *Nature Nanotechnology, 5*, 91.

Dinesh, R., Anandaraj, M., Srinivasan, V., & Hamza, S. (2012). Engineered nanoparticles in the soil and their potential implications to microbial activity. *Geoderma*, 19–27. https://doi.org/10.1016/j.geoderma.2011.12.018

Duncan, T. V. (2011). Applications of nanotechnology in food packaging and food safety: Barrier materials, antimicrobial and sensors. *Journal of Colloid and Interface Science, 363*, 1–24. https://doi.org/10.1016/j.jcis.2011.07.017

Ehsan, K. (2019). Nanozeolites: Synthesized, properties, applications. *Journal of Sol-Gel Science and Technology, 91*, 415–429. https://doi.org/10.1007/s10971-019-05012-4

Gopal, R., Kaur, S., Ma, Z., Chan, C., Ramakrishna, S., & Matsuura, T. (2006). Electrospun nanofibrous filtration membrane. *Journal of Membrane Science, 1*, 581–586. https://doi.org/10.1016/j.memsci.2006.04.026

Hassan, B. S., Islam, G. M. N., & Haque, A. N. (2019). Applications of nanotechnology in textiles: A review. *Advanced Research in Textile Engineering, 4*, 1038.

Klaine, S. J., Alvarez, P. J. J., Batley, G. E., Fernandes, T. F., Handy, R. D., Lyon, D. Y., Mahendr, S. A., Mclaughlin, M. J., & Lead, J. R. (2008). Nanomaterials in the environment: Behavior, fate, bioavailability, and effects. *Environmental Toxicology and Chemistry, 27*, 1825–1851.

Lubomira, T., & Valentin, P. V. (2005). Nanozeolites: Synthesis, crystallization mechanism, and applications. *Chemistry of Materials, 17*, 2494–2513. https://doi.org/10.1021/cm047908z

Lyon, D. Y., Adams, L. K., Falkner, J. C., & Alvarez, P. J. (2006). Antibacterial activity of fullerene water suspensions: Effects of preparation method and particle size. *Environmental Science & Technology, 40*, 4360–4366. https://doi.org/10.1021/es0603655

Mohammadi, A., Hashemi, M., & Hosseini, S.M. (2015). Nanoencapsulation of Zataria multiflora essential oil preparation and characterization with enhanced antifungal activity for controlling Botrytis cinerea, the causal agent of gray mould disease. *Innovative Food Science and Emerging Technologies*, 1–34. https://doi.org/10.1016/j.ifset.2014.12.011

Nikalje, A. P. (2015). Nanotechnology and its applications in medicine. *Medicinal Chemistry, 5*, 081–089. https://doi.org/10.4172/2161-0444.1000247

Qhatan, A. S. (2017). Nanotechnology: Theory, application and explanation of the most important devices used. *Researchgate*, 1–50. https://doi.org/10.13140/RG.2.2.35847.91045.

Rai, M., Yadav, A., & Gade, A. (2009). Silver nanoparticles as a new generation of antimicrobials. *Biotechnology Advances, 27*, 76–83. https://doi.org/10.1016/j.biotechadv.2008.09.002

Rempel, A. A. (2007). Nanotechnologies: Properties and applications of nanostructured. *MaterialsRussian Chemical Reviews, 76*, 435–461. https://doi.org/10.1070/RC2007v076n05ABEH003674

Ricardo, P. N., & Lino, F. (2010). Stem cell research meets nanotechnology. *Revista Da Sociedade Portuguessa D Bioquimica,CanalBQ, 7*, 38–46.

Sahoo, S. K., Dilnawaz, F., & Krishnakumar, S. (2008). Nanotechnology in ocular drug delivery. *Drug Discovery Today, 13*, 144–151. https://doi.org/10.1016/j.drudis.2007.10021

SCENIHR. (2006). The appropriateness of existing methodologies to assess the potential risks associated with engineered and adventitious products of nanotechnologies, 1–79. https://Ec.Europa.Eu/Health/Scientific_Committees/Opinions_Layman/En/Nanotechnologies/L-3/1-Introduction.html. Accessed 25 Jan 2021.

Shi, J. J., & Zhu, J. J. (2011). *Sonoelectrochemical fabrication of Pd-graphene nanocomposite and its application in the determination of chlorophenols Electrochim acta, 56*, 6008–6013. https://doi.org/10.1016/j.electacta.2011.04.099

Singh, P. (2018). Nanotechnology in food preservation. *Food Science Research Journal, 9*, 435–441. https://doi.org/10.15740/HAS/FSRJ/9.2/435-441

Sivaramakrishnan, S. M., & Neelakantan, P. (2014). Nanotechnology in dentistry – What does the future hold in store? *Dentistry, 4*, 2. https://doi.org/10.4172/2161-1122.1000198

Srivastava, M., Sonika, P., Mohammad, S., Anuradha, S., Vipul, K., Shubha, T., Manoj, K. M., Yatindra, K. S., Ved, R., & Srivastava, D. K. (2016). Nanotechnology. *Journal of Pure and Applied Microbiology, 10*, 1055–1060.

Suresh, A. K., Pelletier, O., Wang, W., Moon, J. W., Gu, B., Mortensen, N. P., Allison, O. P., Joy, D. C., Phelps, T. J., & Ooktycz, M. J. (2010). Silver nanocrystallites: Biofabrication using shewanella oneidensis, and an evaluation of their comparative toxicity on gram-negative and gram-positive bacteria. *Environmental Science Technology, 44*, 5210–5215. https://doi.org/10.1021/es903684r

Torney, F., Trewyn, B., Lin, V., & Wang, K. (2007). Mesoporous Silica nanoparticles deliver DNA and chemicals into plants. *Nature Nanotechnology, 2*, 295–300.

Yuan, G., & Wu, L. (2007). Allophane nanoclay for the removal of phosphorus in water and wastewater. *Science and Technology of Advanced Materials, 8*, 60–62. https://doi.org/10.1016/j.stam.2006.09.002

Zhang, W., Yan, W., Benjamin, T. K., Chang, L., Gang, W., & Weiyue, L. (2014). A novel nanoscale dispersed eye ointment for the treatment of dry eye disease. *Nanotechnology, 25*, 1–10.

Development of Energy Efficient Processes and Products from Renewable and Nonrenewable Resources in Nigeria

O. J. Oyebode, A. T. Adeniyi, U. S. Gekwu, K. O. Olowe, and A. O. Coker

1 Introduction

Environmental process engineering needs to be applied for public health and sustainable development. Investigations of the effect of environmental processes on public health and sustainable development were carried out. Development of energy efficient processes and products from renewable and nonrenewable resources is very important for public health, wealth, and better environment. While nearly all products are recyclable, energy sources are not. Thus, to properly understand the issue at hand, we must first define and identify the various types and sources of energy, along with their uses. It exists in various forms from various sources. As it is commonly stated, it is also known as the first law of thermodynamics. This indicates that energy when in use or being generated is not created but transformed to the desired form and when expended, is transformed from its useful form to a byproduct or unwanted form. This brings us to the various types of energy like potential, kinetic, heat, chemical energy, and so on.

O. J. Oyebode (✉) · K. O. Olowe
Civil and Environmental Engineering Department, Afe Babalola University Ado-Ekiti (ABUAD), Ado-Ekiti, Ekiti State, Nigeria
email: oloweko@abuad.edu.ng

A. T. Adeniyi · U. S. Gekwu
Chemical and Petroleum Engineering Department, Afe Babalola University Ado-Ekiti (ABUAD), Ado-Ekiti, Ekiti State, Nigeria
email: adeniyia@abuad.edu.ng; udeagbarasg@abuad.edu.ng

A. O. Coker
Department of Civil Engineering, University of Ibadan, Ibadan, Nigeria

A. O. Ayeni et al. (eds.), *Advanced Manufacturing in Biological, Petroleum, and Nanotechnology Processing*, Green Energy and Technology,
https://doi.org/10.1007/978-3-030-95820-6_2

1.1 Technical Fundamentals of Process Engineering

Technical fundamentals of process engineering are very vital for efficiency and effective performance. Technical fundamentals of process engineering include balancing of energy, materials selection, investigation of thermodynamics of chemical equilibrium and physical equilibrium, transport phenomenon, control, kinetic reactions processes, and dynamics reaction processes. The safety and environmental processes investigations are very essential. There is need for maintenance of biodiversity policies within all important industries to carry out engineering works. Environmental process engineers (EPE) are involved in academics and research work for development of preventive management strategies and technological processes for reduction of emissions from residential buildings and engineering industries with sustainability in view.

2 Environmental Engineering and Management of Processes

Environmental engineers are very important in all nations of the world. We must therefore continue to keep the need for sustainable development in the front burner in all we do. This is because the future of our countries may be irretrievably compromised if we do not pay adequate attention to issues of sustainable development.

Pollution studies are the way out among various management, technological, technical, ecosystems, and natural aspects of our immediate environment. Other construction materials were assessed for environmentally friendly material provision to check the utilization of engineering assets and renewable energy (Oyebode, 2018a). Infrastructure refers to structures that sustain our society, such as highways, water projects, sewerage systems, electrical connections, telecommunications, and many others (Oyebode, 2018b). Urban and rural populations of the world are increasing rapidly where the rate of urbanization is more than strategies for planning and infrastructure development of wastewater (Oyebode, 2015).

United Nations members are compelled to abide by all notable instructions and viable recommendations of the organization (Afionis et al., 2017; Foot, 2007). Renewable energy has great positive impact and enormous significant effect on sustainable development both in developing and developed countries (Guney, 2019).

Developing countries are responsible for most of carbon emission, and this is negatively affecting the health and well-being of the populace. This issue this should be checked and critically examined. Development of sophisticated technology and engineering interventions will go a long way for tackling environmental problems related to processes and pollution of air, soil, and water (Dooley & Gupta, 2017).

Disposal methods and recycling systems should be handled and carefully selected based on location because health, well-being, and safety are very crucial in our environment. Engineering landfill can be designed and carefully constructed to handle electronic wastes, scrap metals, paper, organic wastes, and plastic wastes (Abdul-Rahman & Wright, 2014). There is a decreasing trend in the electricity generation from fossil fuel, and there is increasing use of solar energy and other renewable energy sources. Policy makers and decision-makers have to check carbon emission in our environment as it affects carbon tax in many countries (Benavente, 2016). There is an urgent need for reduction of food wastes from farmlands, markets, homes, and industries and in all major stages of food production, transportation, and consumption. Forty percent of food wastes is at postharvest stage and processing activities (Mopera, 2016). Given Nigeria's location and resources, a number of options become available for the sourcing of renewable energy and products due to the climate and weather naturally available to Nigeria. Figure 1 indicates the map of Nigeria.

In the world today, there are so many things one can talk about concerning environmental pollution. Almost everyone is involved in it; we pollute our environment in so many ways that the scientists have predicted the issue of global warming (Lomborg, 2003). The wastes from our houses, public buildings, and industries that are not disposed of carefully cause soil pollution and air pollution. The chemical wastes thrown into the rivers, ponds, or the ocean are killing the organisms in the water, polluting our water system and air (Praveen et al., 2017). Climate issues can cause market failure in most nations on earth. This will create major problems and result in imminent economic crisis (Stern & Stiglitz, 2021).

Solar energy becomes a viable source of renewable energy, along with wind in the northern parts of the country. Hydroelectric power is easier to attain in the southern region of the country. By diversifying the sources of power in the country, electricity can be more efficiently generated and distributed (Stambouli et al., 2012). This target will help with achieving and sustaining processes of disposing of nonreusable products and reduce the reliance on nonrenewable sources of energy like petrol (Akenji, 2014). Figure 2 gave a graphical representation of recycling.

Fig. 1 Map of Nigeria. (Source: NGSA, 2006)

Fig. 2 Graphical representation of recycling. (Source: Oyebode, 2015)

3 Pollution Reduction and Sustainable Environment Methods

Methodology adopted includes literature survey and other secondary data for process engineering. The part of process engineering that affects public health requires special attention and can be developed for pollution reduction and sustainable environment. It has been concluded that environmental processes have adverse effects on public health and sustainable development. The environmental process engineer has a great task to play in the sustainability of our society. Adequate precautions should be deployed for public health and sustainable development. Most types of energy are named after their source of generation, e.g., renewable form of energy obtained from the radiation of the sun. However, petroleum fuel is considered a nonrenewable form of energy because it is obtained from fossil fuel. Figure 3 presented a plastic waste dump site in Nigeria.

Pollution affects our environment and environmental processes. Figure 4 shows installed biowaste purifier in a process industry, while Fig. 5 shows process engineers at work. Figure 6 presented food waste on dumping site, Figure 7 pollution prevention and control chart, and Fig. 8 stages of water treatment.

Fig. 3 A plastic waste dump site in Nigeria. (Source: Field Study 2021)

Fig. 4 An installed biowaste purifier in a process industry. (Source: Cusenza et al., 2021)

Fig. 5 Process engineers at work. (Source: Oyebode 2018a)

Fig. 6 Food waste on dumping site. (Source: Oyebode, 2015)

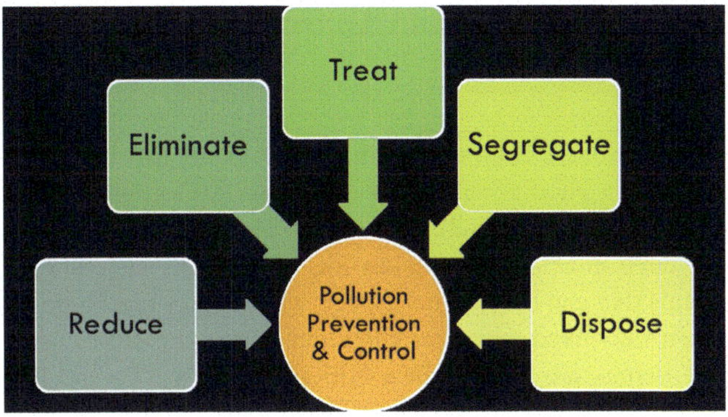

Fig. 7 A pollution prevention and control chart. (Source: Rooney, 1993)

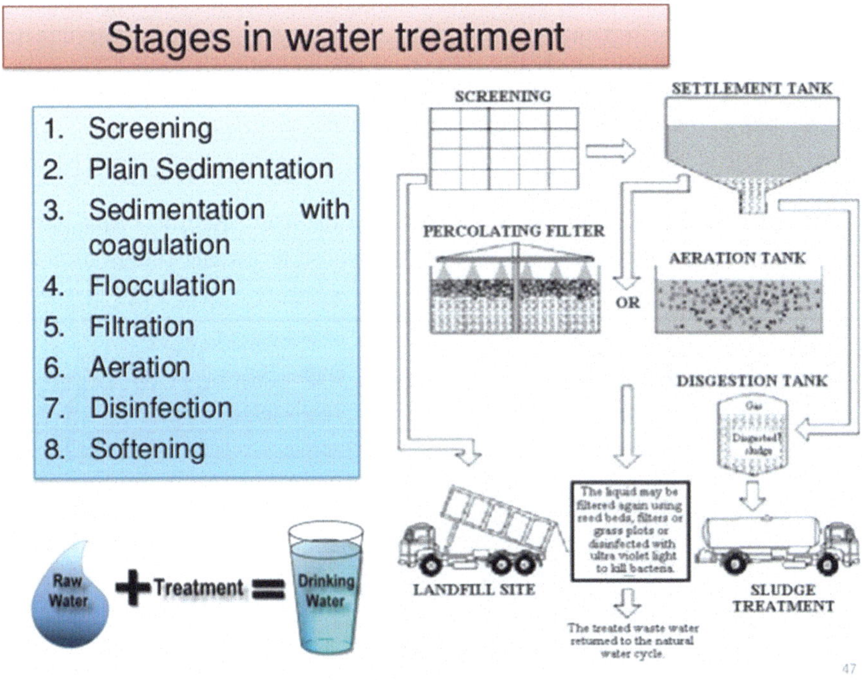

Fig. 8 Stages of water treatment. (Source: Oyebode, 2015)

4 Environmental Process and Approach Using Science and Technology

The environmental process engineer has the task of being involved with the society or the environment; a good environment can bring forth good health in the same way a bad environment can be bad for the body. The environmental engineer helps in keeping our environment clean (fieldwork) and also creates new technological tools or chemicals that can help in this aspect; what good would it be if science and technology wasn't involved when we know that they are there to make man's life easier? Human beings are more than a billion in this world, and the number of people that would be environmental process engineers in the future will be more than enough to enforce the health rules in our society for our benefits; science can be applied in so many ways for the benefits of environmental engineers and so it can be made sure that the problem of pollution would be eradicated with the proper approach using science and technology.

While Nigeria is not in any degree the leading nation in terms of energy generation or management, the country is still blessed with access to resources that can be managed and optimized to truly live up to their title of "Africa's Giant." There are already several small energy generation companies in Nigeria like Afam VI Power Station (IPP), Aba Power Station (IPP), Calabar Power Station and others using steam as well as fossil fuel. Figure 9 presented energy consumption pattern by sector, while Fig. 10 presented recycling of plastic waste.

However, there is hope as the development of the new city "EKO Atlantic" is designed to more efficiently dispose of waste and use renewable energy from the

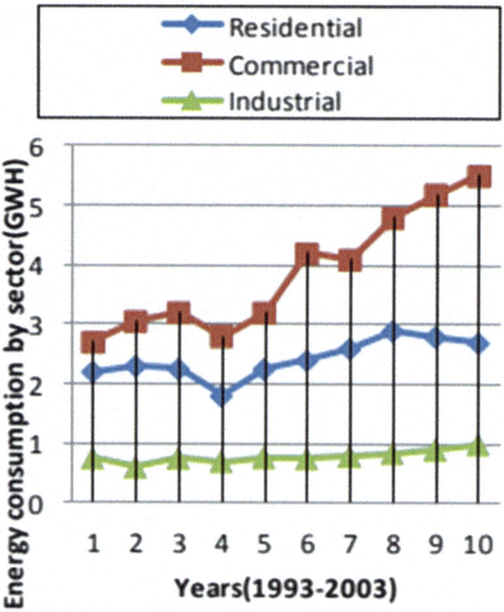

Fig. 9 Energy consumption pattern by sector. (Source: Devi et al., 2009)

Fig. 10 Recycling of plastic waste. (Oyebode 2018a)

ocean. It is also expected to run on its own power grid. The rest of the nation might lag behind. They are far from stagnant as private companies and individuals set out to make the nation greener for the betterment of mankind or just to make their wallets bigger. Other individuals are also helping with the issue, using the materials for artwork and accessories to be admired.

4.1 Challenges Facing Public Health and Sustainable Development

Scheduling and optimization of resources is important for cost reduction and proper allocation of systems and engineering resources to give effective delivery within project completion time. Challenges facing public health and sustainable development in the development of processes and products are:

(i) Global warming, natural hazards, and rise in global instability
(ii) Inadequate planning of projects
(iii) Lack of viable waste management system and workplace innovation
(iv) Lack of accountability and stakeholder involvement in some governmental and private projects
(v) Lack of good governance and priority for public health
(vi) Problems emanating from land use and loss of biodiversity
(vii) Low priority for occupational safety and public health
(viii) Lack of viable evaluation and control of hazards
(ix) Corruption, bribery related issues and lack of governmental commitment to some communities

4.2 Consideration for Efficient Processes and Products

The following must be taken into consideration in order to have efficient processes and products from industries, residential buildings, and institutions:

(i) Drive for quality production of products according to regulations and professional standards
(ii) Engagement of environmental engineers and environmental health professionals in capital projects, localization of sensitive industries, and construction works
(iii) Appropriate testing of materials, products, and processes by certified authorities
(iv) Punishment of any actions and processes that can affect public health, environment, and sustainable development
(v) Establishment of policies and enforcement of laws and order for hygienic environment
(vi) Special counseling, public awareness, proper education, collaboration, advocacy in communities, and effective decision-making by experts
(vii) Special engineering supports, innovative design, and adoption of effective management skills
(viii) Outcome-based engineering education for engineering, environmental health, sciences, and all courses in tertiary institution
(ix) Employment of safety managers in communities, private establishments, industries, institutions, and construction sites
(x) Enforcement of strict legislations for efficient processes, public health, and green environment
(xi) Interventions through legal frameworks, environmental health, and safety
(xii) Systematic and adequate human resource management
(xiii) Inadequate safety experts and job hazard analysis in institutions and establishments
(xiv) Robust waste management and periodic health investigation of populace
(xv) Utilization of environmental health system management

5 Conclusions

Development of energy efficient processes and products from renewable and nonrenewable resources requires adequate and special attention in Nigeria. Life cycle management of engineering materials is critical to health and wealth of developed and developing countries. Notable interventions and challenges facing public health and sustainable development were highlighted. Detailed knowledge of process engineering and the work of a process engineer for the benefit of both the industry and the public are addressed. This would solve two problems at once, as well as create jobs for the unemployed mass. An alternative solution would be to switch over to biodegradable materials. While not having the same number of positives, it will reduce the number of negatives faced by Nigeria in terms of waste management and resource distribution. Safety, green environment, health standards, appropriate regulations, environmental laws, and public health should be given adequate priority in residential buildings, communities, industries, and hospitals and in all public places for environmental health, efficient processes, and sustainable development. There is need to plan, strengthen, and enforce existing laws, implementation strategies, statistical data management, maintenance activities, and responsibilities of regulatory bodies. The environmental process engineer has the task of being involved with the society or the environment; a good environment can bring forth good health in the same way a bad environment can be bad for the body. Processes such as incentivizing the proper sorting of trash are the necessary first step in the development of products and processes from renewable and nonrenewable resources.

6 Recommendations

Recommendations include implementation of effective policy, adequate funding, resource efficiency plan, and involvement of stakeholders at all levels of processes and production. When it comes to products, the materials which they are made of determine what will happen to them after manufacturing and use. Notable engineering interventions and sophisticated equipment should be adopted and applied for development of energy efficient processes and products from renewable and nonrenewable resources in Nigeria. Products made of materials like plastic take decades to decompose. With the available manpower and resources in the nation, it would be advisable to build companies that turn waste into energy. Dealing with the sources of energy and disposal of waste is a vital approach to address most environmental issues. Socioeconomic development and energy and resource efficiency can be achieved by renewable and nonrenewable resources.

References

Abdul-Rahman, F., & Wright, S. E. (2014). *Reduce, reuse, recycle: Alternatives for waste management*.

Afionis, S., Sakai, M., Scott, K., Barrett, J., & Gouldson, A. (2017). Consumption-based carbon accounting: Does it have a future? *Wiley Interdisciplinary Reviews: Climate Change, 8*(1), e438.

Akenji, L. (2014). Consumer scapegoatism and limits to green consumerism. *Journal of Cleaner Production, 63*, 13–23.

Benavente, J. M. G. (2016). Impact of a carbon tax on the Chilean economy: A computable general equilibrium analysis. *Energy Economics, 57*, 106–127.

Cusenza, M. A., Longo, S., Guarino, F., & Cellura, M. (2021). Energy and environmental assessment of residual bio-wastes management strategies. *Journal of Cleaner Production, 285*, 124815.

Devi, R., Singh, V., Dahiya, R. P., & Kumar, A. (2009). Energy consumption pattern of a decentralized community in northern Haryana. *Renewable and Sustainable Energy Reviews, 13*(1), 194–200.

Dooley, K., & Gupta, A. (2017). Governing by expertise: The contested politics of (accounting for) land-based mitigation in a new climate agreement. *International Environmental Agreements: Politics, Law and Economics, 17*(4), 483–500.

Foot, R. (2007). The United Nations, counter terrorism, and human rights: Institutional adaptation and embedded ideas. *Human Rights Quarterly*, 489–514.

Guney, T. (2019). Renewable energy, non-renewable energy and sustainable development. *International Journal of Sustainable Development and World Ecology, 26*(5), 389–397.

Lomborg, B. (2003). *The skeptical environmentalist: Measuring the real state of the world* (Vol. 1). Cambridge University Press.

Mopera, L. E. (2016). Food loss in the food value chain: The Philippine agriculture scenario. *Journal of Developments in Sustainable Agriculture, 11*(1), 8–16.

NGSA. (2006). *The geological map of Nigeria*. Published by Nigeria Geological Survey Agency (NGSA), Abuja.

Oyebode, O. J. (2015). Effective management of wastewater for environment, health and wealth in Nigeria. *International Journal of Scientific and Engineering Research, 6*(7), 1028–1059.

Oyebode, O. J. (2018a). Green building: Imperative panacea for environmental sustainability and life cycle construction in Nigeria. *World Journal of Research and Review, 7*(3), 262584.

Oyebode, O. J. (2018b). Evaluation of civil engineering infrastructures for economy and sustainable construction in Nigeria. In *Proceedings of the 16th international conference and annual general meeting*.

Praveen, K., Ganguly, S., & Wakchaure, R. (2017). Environmental pollution and safety measures international issues and its global impact. *Global Progress in Development of Sustainable Environment*, 40–65.

Rooney, C. (1993). Economics of pollution prevention= how Wmte reduction pays. *Pollution Prevention Review, 261*.

Stambouli, A. B., Khiat, Z., Flazi, S., & Kitamura, Y. (2012). A review on the renewable energy development in Algeria: Current perspective, energy scenario and sustainability issues. *Renewable and sustainable energy reviews, 16*(7), 4445–4460.

Stern, N., & Stiglitz, J. E. (2021). *The social cost of carbon, risk, distribution, market failures: An alternative approach* (No. w28472). National Bureau of Economic Research.

An Overview Application of Natural Oil as a Sustainable Plasticizer in Production of Biopolymers

F. Iriaye, A. A. Abioye, O. O. Yusuf, M. E. Emetere, S. O. Ongbali, A. A. Noiki, S. A. Afolalu, and F. Ademuyiwa

1 Introduction

Plasticizer is a chemical agent which when introduced in plastic as an additive alters its properties and makes it softer and more flexible. It must be thoroughly mixed with the plastic over heat till the plastic dissolves in it to cause any significant changes in the polymer matrix (Godwin, 2011). It serves as a polymer additive which tends to reduce the forces between molecules of the polymer chains, resulting in a flexible and softened polymeric matrix. It directly supports the stretching of the polymer and aids in excellent processing by reducing the softening points, melting points, and the viscosity of the polymer (Beninba & Massardier, 2010) A low glass transition temperature, elastic modulus, flexibility in low-temperature conditions, low tensile strength, and high elongation are properties that a properly plasticized product should possess (Krauskopf, 2003). The continuous growth of the plastic industry has brought about a need for development of plasticizers. This has augmented the expectation in the

F. Iriaye (✉) · A. A. Abioye · S. O. Ongbali · S. A. Afolalu · F. Ademuyiwa
Department of Mechanical Engineering, Covenant University, Ota, Nigeria
e-mail: famous.iriaye@covenantuniversity.edu.ng; abiodun.abioye@covenantuniversity.edu.ng; samson.ongbali@covenantuniversity.edu.ng; adeniran.afolalu@abuad.edu.ng; faith.olumuyiwa@covenantuniversity.edu.ng

O. O. Yusuf
Department of Microbiology, Obafemi Awolowo University, Ile-Ife, Nigeria

M. E. Emetere
Department of Physics, Covenant University, Ota, Nigeria
e-mail: moses.emetere@covenantuniversity.edu.ng

A. A. Noiki
Department of Mechanical Engineering, Ogun State Institute of Technology, Igbesa, Ogun State, Nigeria

A. O. Ayeni et al. (eds.), *Advanced Manufacturing in Biological, Petroleum, and Nanotechnology Processing*, Green Energy and Technology,
https://doi.org/10.1007/978-3-030-95820-6_3

demand of plasticizers to reach. The rapidly growing plasticizer demand will reach 9.75 million tonnes in 2024 (Wei et al., 2019). However, the environmental concern and finite supply reduced their area of application on a large scale, thus resulting in diversification to bio-based sustainable oils (Hassan et al., 2020).

There are about 30,000 materials which have been studied as plasticizers with only about 50 commercially used currently (Malveda, 2015). The most common plasticizers in use are esters, glycerides, polyhydric alcohols, and oils, some of which are glycerol, triacetin, phthalates, and mineral oils. Certain plasticizers, with phthalates topping the list, have been found to be toxic and harmful to humans and agricultural produce, thus creating a need for more eco-friendly plasticizers.

1.1 Plasticization

This method simply means softening and increasing the flexibility of a polymer by introducing a plasticizer. This is also achieved by subjecting the polymer to a melting process and softening by the action of shear forces, heat, and pressures. Plasticizers applied in the process of plasticization alter the mechanical and thermal features of a polymer, whereby lowering its rigidity at room temperature (Johnson et al., 1991). Plasticization can be primarily internal or external.

Internal plasticization: This involves copolymerization, that is, the chemical modifications of the polymer (Johnson et al., 1991). External plasticization: This is achieved by using plasticizers as an additive, which could result in the loss of its original structure through extraction (Johnson et al., 1991). Plasticizers can also be classified based on the solubility of the polymer. These can be either primary or secondary. Primary plasticizers have their polymer soluble in them at a high concentration of the polymer, and they are much compatible with the polymer, while secondary plasticizers do not have their polymer soluble in them, and they have minimal gelation capacity and their compatibility with the polymer is low (Cadogan & Howick, 1996). Plasticizers, made from biopolymer-based films, can also be classified into soluble and insoluble (Pedersen et al., 2008). The film formed from the aqueous dispersion of polymeric plasticizer is based on the quantity and the type of plasticizer (Vieira et al., 2011).

1.2 The Steps Involved in the Plasticization of a Typical Polymer

To obtain a homogenous mixture of polymer and plasticizer, the following steps are observed as illustrated below:

Step I: Mixture of the plasticizer with polymer
Step II: Absorption of the plasticizer and the enlargement of the polymer particles (Johnson et al., 1991)
Step III: Discharge of the polar groups in the polymer (Johnson et al., 1991)

Step IV: The plasticizer's polar group interacting with the polymer's polar group (Cadogan & Howick, 1996).
Step V: Re-establishing the structure of the polymer (Johnson et al., 1991)

Step II is dependent on the viscosity of the plasticizer, the branching degree of the plasticizer, the polymer pore size, the particle size, and the free volume of the polymer. Steps III and IV are dependent on the polarity of the molecules and the volume and the molecular weight of the plasticizers with the exception if there is no retainment of the plasticizers in the end products because this makes it unusable (Cuq et al., 1997). There is higher demand in plasticizers as there is an inclination in the plastic industry and there is discovery in the means for eco-friendly plastic production. Recently, there are variety of choices in selecting appropriate plasticizers with characteristics that suit specific interests. Although there is a query on possibility of toxic effect on plasticizers especially phthalates on human health, there is a diversification to alternate type of plasticizers with low migration levels and low toxicity. Therefore, eco-friendly or natural-based plasticizers are being sought for and applications are being studied (Abioye et al., 2019).

2 Eco-friendly Plasticizers

The materials used for the formation of eco-friendly plasticizers are polysaccharides, lipids, and protein compounds (Choi & Park, 2004). Polysaccharides exhibit excellent film-forming properties and efficient barriers against oils and lipids, although with poor moisture barriers. Proteins also have good mechanical and thermal properties. The unique structure of proteins confers on them a wide range of useful qualities specifically superior intermolecular binding forces (Baltacioğlu & Balköse, 1999). Fatty acids added as plasticizers have shown to have maximal clarity, high elongation, low modulus, and low tensile strength. However, linoleic acid showed more advantages than oleic acid in minimizing water absorption of sheets (Whyuningtiyas & Suryanto, 2017). In design considerations gave characteristics of an eco-friendly plasticizer, some of which are low leachability, energy efficiency, and nontoxicity, all with respect to design parameters of compatibility, efficiency, and performance.

On the journey to the improved development bio-based plastics, material research has seen it as sensible to use bio-based plasticizers as constituent components of products to increase biodegradability (Hassan et al., 2013). Although it is not possible to completely eliminate the use of conventional plasticizers that are used in synthetic plastics, they might not be applicable for all biodegradable plastics, so it is imperative that further studies be carried out to facilitate their replacement. Generally, water is seen as a plasticizer but is less preferred as the products become brittle when at equilibrium with the ambient humidity of the particular environment. In recent times, the use of plasticizers which have natural, low-toxic materials as their bases has grown, with interest in epoxidized triglycerides, such as crude palm oil, linseed oil, and fatty acid esters alongside glycerol (Raju et al., 2015).

2.1 The Utilization of Natural-Based Oil as Plasticizers

In recent times, petroleum products have been used as plasticizers, but depletion of fossil fuels as a result of its broad use has led to discovery of alternate products for the synthesis of plasticizers which are from plants and animals in order to satisfy the requirements of sustainable development and environmental safety. The usage of plant oils as green plasticizers has drawn a great interest in the plastics and rubber industries due to its renewability, low cost, and nontoxicity specifically to public health. Among the variety of plant oils, palm oil (PMO) has been explored more because of its availability and its minimal cost which relies on the saturation level of the PMO type. Highly unsaturated PMO are costly because they are used as additives in food. The other types of PMO which are the low degree unsaturated PMO are economical and are often used as plasticizers in the plastic industry. PMO have shown to exhibit higher molecular weight and free fatty acids as co-activators and are most preferred to conventional synthetic esters. PMO have excellent characteristics, such as antioxidant property, and drive functional ingredients in elastomer composites (.Abioye et al., 2019; Makhtar et al., 2013; Ratnam et al., 2006). Several authors have studied crude palm oil as plasticizer and its effect on polymeric materials. Kamarudin et al. (2019) reported the effects of crude palm oil (CPO) as plasticizer on the morphological, mechanical, and physical properties of blown films made from polypropylene. There was a direct relation between the increase in CPO content and the melt flow rate, elongation at break, and the impact and tear strengths, as opposed to the density and tensile strength.

The CPO was seen to act as a lubricant, create ester linkages within the polymer matrices, and reduce interchain reactions. The study reported a direct relationship between the strengthening of the intermolecular hydrogen bonds of the TPEs and the increase in glycerol and olein oil content; the relationship between the former and CPO increase was the inverse.

The 30% and 50% CPO-plasticized TPE further showed high degradability, with a loss of up to 100% weight when in an environment with a temperature of 500 °C (Chi et al., 2007).

On the physical properties of linear low-density polyethylene (LLDPE) and high-density polyethylene when crude palm oil (CPO) was used as plasticizer. The study showed a general increase in material toughness which was denoted by a reduction in the density of both LLDPE and HDPE, and elongation at break and tensile strength improvement, all per increase of CPO content. Although the impact strength of HDPE increased, corroborating this, that of LLDPE reduced and this was believed to be due to its amorphous phase formation defects. Thermal stability and crystallinity of a polymer have shown to increase on a study of crude palm oil (CPO) as plasticizer on the thermodynamic properties of blown films made from polypropylene. A decrease in the transition temperature from 10 °C to 1 °C in both the storage and loss moduli with a pattern of about 3–5% was also observed (Sims, 2011).

Nasruddin & Susanto (2018) studied the use of fried palm oil and castor oil as alternative plasticizers for minarex and white oils, which are both petroleum based and used in the rubber tire composite of Natural Rubber and Styrene Butadiene Rubber (NR-SBR). The range of values obtained were 118–136 kg/cm^2 for tensile strength, 102,37–135,64 mm^3 for abrasion resistance, 1,22–1,29 g/cm^3 for density, 68–74 for Shore A hardness, 200% 19, 25–31, 16 for modulus, 65, 44–72, 35 for compression set at 25% deflection, 22 h, 70 °C, and no ozone resistance cracking at 50 pphm, 40 °C, 24 h and 20% strain. The latter had slightly better abrasion resistance, modulus, and tensile strength compared to the former. The study shown that plant oil-based plasticizers prospect for replacing petroleum-based plasticizer in Natural Rubber and Styrene Butadiene Rubber composites filled with carbon black and calcium carbonate for rubber solid tire industries.

Kamarudin et al. (2019) investigated the effects of crude palm oil (CPO) as plasticizer on the morphological, mechanical, and physical properties of blown films made from polypropylene. The CPO was added to the PP in 1%, 3%, and 5% wt, homogeneously blended, extruded, and blown into thin films. There was a direct relation between the increase in CPO content and the melt flow rate, elongation at break, and the impact and tear strengths, as opposed to the density and tensile strength. The CPO was seen to act as a lubricant, create ester linkages within the polymer matrices and reduce interchain reactions. (Raju et al., 2015).

3 Utilization of Glycerol as Plasticizers

The major source where glycerol is obtained is from plant (such as soybeans and palm oil) and animal sources where it occurs with long-chain carboxylic acid as triglycerides and esters. The primary intermediate product of biodiesel production is glycerol, which is approximately 10% of the average biodiesel production (Raju et al., 2015). The crude glycerol synthesized from the process of transesterification of triglycerides with alcohol showed low quality resulting from impurities including sodium hydroxide, free fatty acids esters proteins fatty acid salts, and sulfur compounds. Therefore, this crude glycerol undergoes refinement through multistep distillation to improve its qualities and this is highly expensive (Afolalu et al., 2019). The unique properties of glycerol (boiling point, the presence of hydroxyl group which makes it hygroscopic) have placed demand on its applications in various sectors: pharmaceutical, medicine, and petroleum industries. Glycerol is mostly burned for energy because of its low heat value. Recently, glycerol has been beneficial in its use as plasticizer to produce starch-based biodegradable films (Silva et al., 2009). The increase in plasticizer quantity was seen to vary directly with the moisture content, thickness, and solubility of the films, as well as inversely with the water absorption and density of the films. The study revealed that different plasticizers and concentrations had varying effects on the properties of the films. Another study by Darbon et al. (1999) investigated the effects of glycerol addition on the transparency, thickness, tensile strength, and percent elongation of edible film alginate. Glycerol was added in concentrations ranging from 0.3% to 1.1% of the solution volume, in three places and with 0.2% intervals. The results were compared with the Japanese Industrial Standard (JIS), and it was determined that the 0.9% glycerol-plasticized film had the best properties with values of 1.86 for transparency, 0.094 mm for average thickness, 8.25 MPa for tensile strength, and 10.83% for elongation percent. Moreover, the influence of glycerol on the microbial action enhances the biodegradation of bioplastics (Zhu & lawman 2002). One of the prospective utilizations of glycerol is the bioconversion to energy-rich compounds via microbial fermentation.

Glycerol has shown a higher level of reduction compared to sugars (Whyuningtiyas & Suryanto, 2017). The major advantage it offers is to transverse via passive diffusion across cytoplasmic membrane of microbial cells, and this has been studied in the inner membrane of *Escherichia coli* (Yusoff et al., 2016). Diffusion is also enhanced by integral membrane protein; glycerol facilitator which aids its metabolism by oxidative and reductive. These features enhance its rapid utilization by microorganisms, thereby supporting rapid biodegradation of biopolymers (Samuel et al., 2019).

The biodegradability, moisture absorption, shelf life, and morphological properties were also investigated. Samples were produced from the mixture of 5 wt% of cassava flour added to four different concentrations of distilled water (98.5, 98, 97.5, and 97 ml) mixed with glycerol as the plasticizer at various concentrations ranging from 0, 2, 2.3, and 3 wt% after which the samples were poured into an 88mm mold and heated in an oven at a temperature of 50 °C to solidify the samples (Afolalu et al., 2019).

The resulting samples produced showed significant improvement in tensile strength due to the glycerol utilized serving as an antifungal agent increasing its shelf life making them more effective as packaging materials. The biodegradation analysis of the produced samples showed that the mass of the bioplastic produced reduced to more than 50% under 6 days while it was buried (Afolalu et al., 2019). By the seventh day, the bioplastic had broken down into smaller pieces, and complete degradation was attained by the ninth day. From their research, the significant mass loss was due to the high concentration of glycerol utilized in the production of the bioplastic. The higher the level of glycerol utilized, the faster the degradation process. Another study by Whyuningtiyas & Suryanto (2017) prepared films based on starch from banana peel paste and glycerine or propane-1,2,3-triol; both were used as plasticizers. An increase in elongation with increased plasticizer content was observed. Also, the plastic was seen to be workable into different shapes when poured into different molds (Özeren et al., 2020).

The samples that were buried in soil were of the same size and shape, and were seen to degrade completely under 15 days as opposed to the initial supposed 90-day period. The samples were exhumed every 2 days and after the first 6 days, with the degradation rates of plastics increasing with time and made visible by their color darkening. The 30 ml sample was losing the most weight after 6 days, and this led to a presumption that the glycerol content was responsible for this (Sims, 2011). The control sample had the slowest degradation rate, while the synthetic plastic did not degrade (Afolalu et al., 2019).

4 Conclusion

It is convincing to a reasonable degree that renewable products from plants and animals could replace minerals oils from petroleum sources most especially in its use as plasticizers. It is imperative that more studies be conducted on its use in synthesis plastics and bio-based plastics to satisfy the environmental sustainability goals. This would also disallow the excessive purposes petroleum is serving especially as energy products while alternate products can be discovered more.

Acknowledgments We acknowledge the financial support offered by Covenant University in the actualization of this research work for publication.

References

Abioye, A. A., Fasanmi, O. O., Rotimi, D. O., Abioye, O. P., Obuekwe, C. C., Afolalu, S. A., & Okokpujie, I. P. (2019, December). Review of the development of biodegradable plastic from synthetic polymers and selected synthesized nanoparticle starches. *Journal of Physics: Conference Series, 1378*(4), 042064.

Afolalu, S. A., Oladipupo, S., Bose, M. E., Abioye, A. A., Adejuyigbe, S. B., Ajayi, O. O., & Ongbali, S. O. (2019, December). Agro waste a sustainable source for steel reinforcement-review. *Journal of Physics: Conference Series, 1378*(3), 032032.

Baltacioğlu, H., & Balköse, D. (1999). Effect of zinc stearate and/or epoxidized soybean oil on gelation. *Journal of Applied Polymer Science, 74*(10), 2488–2498.

Beninba, M., & Massardier, V. (2010). Evaluation effects of biobased plasticizer on the Thermal, mechanical, dynamical mechanical properties and permanence of plasticized. *PVC Journal of Applied Polymer Science, 118*(6), 3499–3508.

Cadogan, D., & Howick, C. (1996). Chapter: Plasticizers. In Kirk-Othmer (Ed.), *Encyclopedia of chemical technology*. Wiley.

Z. Chi, , D. Pyle, , Z. Wen, C. Frear, & S. Chen, (2007). A laboratory study of producing docosahexaenoic acid from biodiesel-waste glycerol by microalgal fermentation. Process Biochemistry, 42(11), 1537-1545

Choi, J. S., & Park, W. H. (2004). Effect of biodegradable *plasticizers* on thermal and mechanical properties of poly(3 hydroxybutyrate). *Polymer Testing, 23*(4), 455–460.

Cuq, B., Gontard, N., Cuq, J., & Guilbert, S. (1997). Selected functional properties of fish myofibrillar protein-based films as affected by hydrophilic plasticizers. *Journal of Agricultural and Food Chemistry, 45*(3), 622–626.

Darbon, E., Ito, K., Huang, H. S., Yoshimoto, T., Poncet, S., & Deutscher, J. (1999). Glycerol transport and phosphoenolpyruvate-dependent enzyme I- and Hpr-catalyzed phosphorylation of glycerol kinase in Thermus flavus. *Microbiology, 145*, 3205–3212.

Godwin, A. D. (2011). Plasticizers. In *Applied plastics engineering handbook* (pp. 487–501). Elsevier.

Hassan, M. M., Takahashi, T., & Koyama, K. (2013). Thermal and mechanical properties of novel composites of methyl silicone polymer and partially ceramized rice Bran. *Industrial & Engineering Chemistry Research, 52*, 1275–1283.

Hassan, A., Formela, K., & Wang, F. (2020). Enhanced interfacial and mechanical performance of styrene-butadiene rubber/silica composites compatibilized by soybean oil derived silanized plasticization. *Composites Science and Technology, 197*, 10827.

Johnson, K., Hathaway, R., Leung, P., & Franz, R. (1991). Effect of triacetin and polyethylene glycol on some physical properties of M.G.A. *International Journal of Pharmaceutics, 73*(3), 197–208.

Kamarudin, S. H., Jusoh, E. R., Abdullah, L. C., Ismail, M. H. S., Aung, M. M., & Ratnam, C. T. (2019). Thermal and dynamics mechanical analysis of polypropylene blown films with crude palm oil as plasticizer. *Indonesian Journal of Chemistry, 19*(3), 545.

Krauskopf, L. G. (2003). How about alternatives to phthalate plasticizers? *Journal of Vinyl and Additive Technology, 9*(4), 159–171.

Makhtar, N. S. M., Rais, M. F. M., Rodhi, M. N. M., Bujang, N., Musa, M., & Hamid, K. H. K. (2013). Tacca leontopetaloides starch: New sources starch for biodegradable plastic. *Procedia Engineering, 68*, 385–391.

Malveda, M. P. (2015, July). Chemical economics handbook report on plasticizers. Langer, E., Bortel, K., Waskiewicz, S., & Lenartowicz-Klik, M. (2020). Assessment of Traditional Plasticizers. *Plasticizers Derived from Post-Consumer PET*, 1–1.

Nasruddin, & Susanto, T. (2018). The effect of natural based oil as plasticizer towards physics-mechanical properties of NR-SBR blending for solid tyres. *Journal of Physics: Conference Series, 1095*, 012027.

Özeren, H. D., Guivier, M., Olsson, T., Richard Nilsson, F., & Hedenqvist, S. M. (2020, April 7). Ranking plasticizers for polymers with atomistic simulations; PVT, mechanical properties and the role of hydrogen bonding in thermoplastic starch. *ACS Applied Polymer Materials*.

Pedersen, G. A., Jensen, L. K., Fankhauser, A., Biedermann, S., Petersen, J. H., & Fabech, B. (2008). Migration of epoxidized soybean oil (ESBO) and phthalates from twist closures into food and enforcement of the overall migration limit. *Food Additives & Contaminants, 25*(4), 503–510.

Raju, P., Nandanan, V., & Kutty, S. K. N. (2015). Effect of selected vegetable oils on the properties of acrylonitrile-butadiene rubber vulcanizates. *Polymer, 60*, 43–56.

Ratnam, C., Min, M., Chuah, T., Suraya, A., Choong, T., & Hasamuddin, W. (2006). Physical properties of polyethene modified with crude palm oil. *Polymer Plastic Engineering and Technology, 45*, 912.

Samuel, O. D., Okwu, M. O., Amosun, S. T., Verma, T. N., & Afolalu, S. A. (2019). Production of fatty acid ethyl esters from rubber seed oil in hydrodynamic cavitation reactor: Study of reaction parameters and some fuel properties. *Industrial Crops and Products, 141*, 111658.

Silva, G., Mack, M., & Contiero, J. (2009). Glycerol: A promising and abundant carbon source for industrial microbiology. *Biotechnology Advances, 27*, 30–39.

Sims, B. (2011). Clearing the way for byproduct quality: why quality for glycerin is just as important for biodiesel. *Biodiesel Magazine*.

Vieira, M., da Silva, M., dos Santos, L., & Beppu, M. (2011). hydroxypropyl methylcellulose free films. *European Polymer Journal, 47*, 254–263.

Wei, X., Linde, E., & Hedenqvist, M. (2019). Plasticiser loss from plastic or rubber products through diffusion and evaporation. *Materials Degradation, 3*(1), 18.

Whyuningtiyas, N., & Suryanto, H. (2017). Analysis of biodegradation of bioplastics made of cassava starch. *Journal of Mechanical Engineering Science and Technology, 1*(1), 24–31.

Zhu, M. M., Lawman, P., & Cameron, D. C. (2002). Improving 1.3-propanediol production from glycerol in a metabolically engineered Escherichia coli by reducing accumulation of sn-glycerol-3-phosphate. *Biotechnology Progress, 18*, 694–699.

Manufacturing of Brake Pad Using Aluminium Silicon Carbide Reinforced with Alumina for Automobile Industry

P. O. Babalola, T. A. Okunuga, A. O. Inegbenebor, O. Kilanko, and M. O. Udo

1 Introduction

The automatic brake disc is a device used to slow or stop the wheel's motion while it rotates at a particular speed, thus making vehicles safe to drive (Maleque et al., 2010). These are components that are used to improve the car's handling, reduce the weight in an automobile and improve fuel efficiency. They are inevitable in all automobiles. Recently, the most commonly used material has been aluminium matrix composites reinforced with a ceramic particulate. They are now been applied in the industry. These materials have provided solutions to the persisting problem of weight, thermal conductivity (Radhakrishnan et al., 2015) and thermal stability of brake pad and disc. Throughout the braking stage, the heat generated from the friction at the disc interface can lead to high temperatures. The frictional heat on the surface can create high temperatures that can cause unwanted effects, leading to thermal elastic instability or even premature wear. The brake disc's performance and the mechanical component's life expectancy are vital to determine the effectiveness of the brake pad (Macke et al., 2012). The braking system of a vehicle is subjected to both mechanical and thermal stresses. Therefore, materials used in brake pads need to absorb heat and quickly dissipate it to maintain thermal stability and thermal fatigue (Agbeleye et al., 2020).

Furthermore, the performance and longevity can be improved by adding alumina to the aluminium matrix composites, thus making it more wear-resistant and also increasing the thermal stability (Wang & Hutchings, 1989; Huang & Paxton, 1998).

P. O. Babalola (✉) · T. A. Okunuga · A. O. Inegbenebor · O. Kilanko · M. O. Udo
Mechanical Engineering Department, Covenant University, Ota, Nigeria
e-mail: phillip.babalola@covenantuniversity.edu.ng; teslim.okunuga@stu.cu.ng; anthony.inegbenebor@covenantuniversity.edu.ng; oluwaseun.kilanko@covenantuniversity.edu.ng; mfon.udo@covenantuniversity.edu.ng

A. O. Ayeni et al. (eds.), *Advanced Manufacturing in Biological, Petroleum, and Nanotechnology Processing*, Green Energy and Technology,
https://doi.org/10.1007/978-3-030-95820-6_4

Many studies showed the stir casting method to be the most ideal and cheapest method of producing aluminium matrix composites at a commercial level while at the same time maintaining the uniform distribution of reinforcement (Bhandare & Sonawane, 2013; Babalola et al., 2015; Inegbenebor et al., 2015; Babalola et al., 2018). Asbestos was widely used up until the 1980s. It was later banned in most countries because the fibres from it are released into the atmosphere and were a health risk as it was traced to cancer diseases. The diseases involve uncontrolled and excessive growth of abnormal cells, which invade and destroy other tissues. From that period on, aluminium metal matrix composites (AMMC) became widely used to produce brake pads (Lemen, 2004; Yashwanth et al., 2018; Babu et al., 2020; Harshavardhan et al., 2021). Some of the reinforcement materials in the aluminium matrix are SiC, graphite, MoS_2, rice husk ash (RHA), fly ash, basalt fibre, TiB_2, B_4C, Si_3N_4, AlN, TiC, mica and Cu as reported by Yashwanth et al. (2018). Others are titanium oxide, zirconium oxide (Pinca-Bretotean et al., 2021) and titanium aluminium nitride (TiAlN). Yashwhanth et al. (2021) reported natural fibres such as jute, sisal, kenaf, hemp, bamboo, coconut, pineapple, banana and flax. These are also referred to as green additives in the composite. These are used as reinforcement or filler in the matrix. The matrix could be metal or polymer (plastic) based, though the present work is actually a metal matrix composite.

In this work, stir casting, a liquid metallurgical route, was used to reinforce commercial grade aluminium with silicon carbide, alumina, zinc and calcium. The produced composites were characterised.

2 Experimental Methods

In the course of this study, there were four distinct variations (Table 1) of the brake pad that were produced. Each sample's composition varied in the percentages of aluminium as the matrix and silicon carbide (SiC) of sizes of 1200 grits (3 μm) as reinforcement. Additional elements that were added to the production include powdered forms of zinc, alumina and calcium. The aluminium ingots were then cut using the electric power arc saw machine. A coolant was poured as the lubricant to help alleviate the heat generated and prevent metal distortion. The required silicon carbide powder for each aluminium ingot was measured with an electronic

Table 1 Composites' variation and constituent in %wt

Constituents	A	B	C	D
Aluminium	70	80	90	66
Silicon carbide	20	10	10	6
Alumina	10	10	–	8
Calcium	–	–	–	8
Chromium	–	–	–	2
Zinc	–	–	–	10

weighing scale (OHAUS Pioneer TM Model PA214). The powder was placed on the tray in a transparent glass container of the scale to eliminate the air's interference on the actual weight. This procedure was repeated for the other elements. The final product (AMMC) was produced using stir casting method. The procedure was done properly and ensured the steps were carried out in casting the composites. The aluminium ingots were put in the graphite furnace and were melted at 800 °C using the tilting furnace, fired with diesel oil though used lubricating oil is also applicable. However, the use of spent lubricating oil brought unwarranted soot in the foundry floor used for synthesising the composite. It is not advisable for clean and perfect combustion process. This tilting furnace has small hand wheel to rotate the furnace for easy and almost total evacuation of the melt after successive production of the composites. Hence, ladle for evacuating the melt is not needed. The SiC powder, Al_2O_3, Zn, Ca and Cr were all preheated before introduction into the molten aluminium and stirred for several minutes with a mechanical stirrer at about 420 revolutions per minute (RPM). The mechanical stirrer is a single-phased electric motor (Siemens) with integral three-blade impeller. The revolution could be varied from 0 to 1950 RPM and the power consumption is 1.2 kW.

Other instruments used in this work (Table 2) are thermocouple (Jenway Model 220T/C) for temperature measurement and monitoring, microprocessor tachometer (Compact Instruments Model-CT6) for revolution of the stirrer, electronic weighing scales (Ohaus and Poyear Model) and air blower for the furnace.

Thermal conductivity, compressive strength and wear resistance were the properties measured for all the samples during characterisation.

Table 2 Equipment schedule

Description	Electrical
Oil fired tilting furnace, 10 kg capacity	–
Air blower, 2.8 cbm/min at 16,000 rev/min Make: Powerdevil Model PD600SUL	0.6 kW 230 V/1PH/50 Hz
Mixer, 0–1950 min^{-1} Make: Siemens	1.2 kW 220 V/1PH/50 Hz
Fuel tank, 32 litres capacity	–
Electronic weighing scale1, 110 g maximum capacity Make: Ohaus Model PA114	4.0 VA 8–14.5 V, 50/60 Hz 9.5–20 VX4W
Electronic weighing scale2, 15 kg maximum capacity Make: Poyear	0.6 kW 230 V/1PH/50 Hz
Thermocouple Make: Jenway Model 220 T/C	9.0 V DC
Microprocessor Tachometer Make: Compact Instruments Model-CT6)	4.5 V DC

2.1 Thermal Conductivity Test

The thermal conductivity (W/mK) was obtained in the thermal laboratory at the Physics Department, Covenant University, Ota, Nigeria. The products of specific heat capacity (kJ/kg.K), thermal diffusivity (m^2/s) and density (kg/m^3) were used to get thermal conductivity (Equation 1). While Archimedes method was used to obtain the density, thermal apparatus (EQUILAB-CU/PHY/TCA 1-2) was used to obtain thermal properties.

$$\lambda = C_{\mathrm{p}} \alpha \rho \tag{1}$$

where λ is the thermal conductivity, Cp is the heat capacity, α is the coefficient of thermal diffusivity and ρ is the density.

2.2 Compressive Strength Test

The compressive strength tester (TQ SM1000) in the Department of Civil Engineering, Covenant University, Ota, Nigeria, was used to measure the samples' compressive strength. At first, the measurement of the different direction of a specimen along its length and height was determined with the vernier calliper to get the cross-sectional area. Sample test is thereafter located on the base block and moved down through the central grip to apply the load. Readings were obtained from the associated computer workstation.

2.3 Wear Resistance Test

Measuring wear resistance of the samples was done using the abrasive wear tester (Department of Metallurgical Engineering, Yabatech). The equipment has an abrasive paper on a circular wheel (ø50 mm × 12 mm thick) and a table holding the test specimen. The two slides against each other in a forward and backward reciprocating motion (1/600 (REV.)). Wearing of test material against fresh abrasive surface is ensured by turning the wheel by a fraction of one rotation at the end of each rub.

3 Results and Discussion

The samples were subjected to various tests to obtain values for comparative analysis. The parameters that were chosen for the samples were based on their physical and thermal properties, vis-à-vis, wear, compressive and thermal. The results for

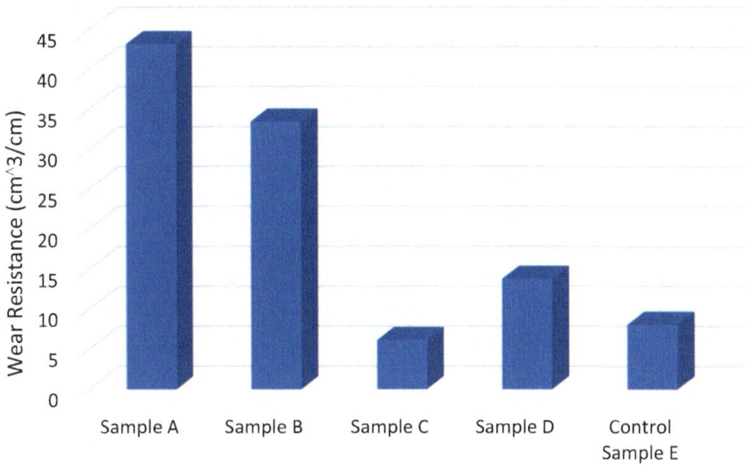

Fig. 1 Wear resistances of composites

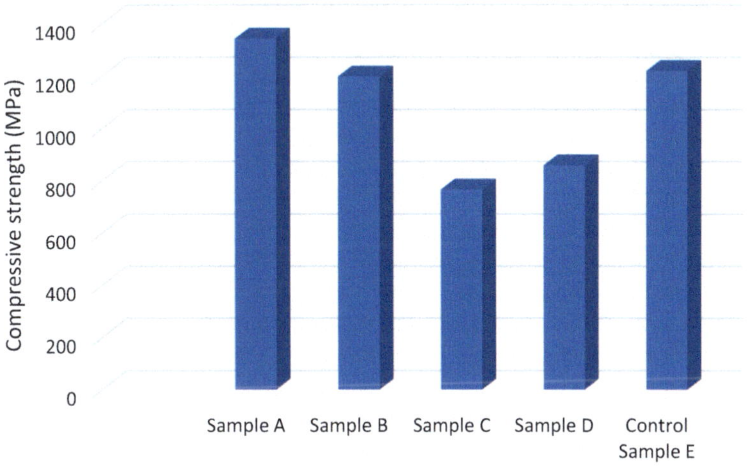

Fig. 2 Compressive strengths of composites

conventional brake pad tagged as the control sample E along with the other composite samples are shown in Figs. 1, 2 and 3, respectively.

In Fig. 1, composite sample C, which was reinforced with silicon carbide alone, has wear resistance, a critical property for brake pad, lower than that of the control sample. Also, the compressive strength (Fig. 2) is lower and hence will not be considered further.

However, composite samples A, B and D reinforced with both silicon carbide and alumina have higher wear resistance than the control sample. While sample B's compressive strength is almost at par with the control sample, sample D is lower,

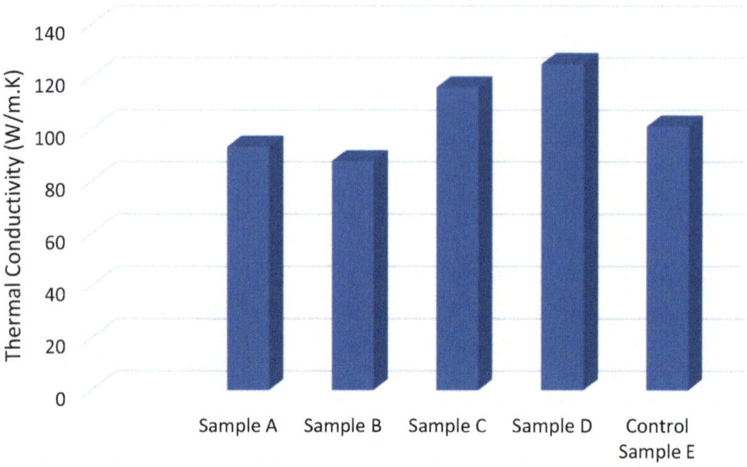

Fig. 3 Thermal conductivities of composites

whereas sample A has higher compressive strength and it is therefore preferred to the other composites (samples B and D) . The tendency of the wear resistance to increase with alumina's addition was determined by Kök & Özdin (2007) in their study and it is confirmed in this work. The inclusion of hard ceramic particles (alumina and silicon carbide) increased the surface hardness of the composites and subsequently retarded abrasive wear. Also, these hard particles produced dislocation loops (Orowan strengthening) in the soft aluminium matrix, thereby retarding the passage of dislocation stress (Inegbenebor et al., 2016). The resultant effect is the observed increase in compressive strength of the composites. Hence, they are better alternative for the control brake pad. Brake pad must resist abrasive wear so as to prolong the service life of the pad and also withstand compressive force between the disc and the drum. Figure 3 shows the thermal conductivities of all the composites with values so close to the control sample (101.2 W/m.K). This property is not as critical as the wear and compressive behaviours of the pad. The use of aluminium as the matrix material ensures a low weight brake pad. This helps in improving fuel efficiency (Macke et al., 2012).

4 Conclusion

The use of aluminium matrix reinforced with alumina and silicon carbide to produce brake pad is attempted in this work. Stir casting route was adopted to bring the constituent together to form aluminium metal composites. Sample A consisting of 70%wt of aluminium, 20%wt of SiC and 10% of Al_2O_3 is recommended as a substitute for the conventional brake pad. It has thermal conductivity of 93.3 W/m.K, wear resistance of 43.22 cm^3/cm and a compressive strength of 1343.3 MPa against

101.2 W/m.K, 8.10 cm^3/cm and 1220.4 MPa, respectively, for the control sample. In conclusion, the use of alumina as a reinforcement with aluminium silicon carbide is essential for application in vehicles.

Acknowledgement The author wishes to acknowledge the education and financial support offered by Covenant University in realisation of this research for publication.

References

Agbeleye, A. A., Esezobor, D. E., Balogun, S. A., Agunsoye, J. O., Solis, J., & Neville, A. (2020). Tribological properties of aluminium-clay composites for brake disc rotor applications. *Journal of King Saud University – Science, 32*(1), 21–28.

Babalola, P. O., Inegbenebor, A. O., Bolu, C. A., & Inegbenebor, A. I. (2015). The development of molecular-based materials for electrical and electronic applications. *The Journal of The Minerals, Metals & Materials Society (TMS), 67*, 830. https://doi.org/10.1007/s11837-015-1355-2. ISSN 1047-4838.

Babalola, P. O., Bolu, C. A., Inegbenebor, A. O., & Kilanko, O. (2018). Graphical representations of experimental and ANN predicted data for mechanical and electrical properties of AlSiC composite prepared by stir casting method. *IOP Conference Series: Materials Science and Engineering, 413*(2018), 012063. https://doi.org/10.1088/1757-899X/413/1/012063

Babu, K. V., Marichamy, S., Ganesan, P., Madan, D., Uthayakumar, M., & Rajan, T. P. D. (2020). Processing of functionally graded aluminum composite brake disc and machining parameters optimization. *Materials Today: Proceedings, 21*, 563–567.

Bhandare, R. G., & Sonawane, P. M. (2013). Preparation of aluminium matrix composite by using stir casting method. *The International Journal of Engineering and Advanced Technology (IJEAT), 3*(3), 61–65.

Harshavardhan, K., Nagendran, S., Shanmugasundaram, A., Sankar, S. R. P., & Kowshik, K. S. (2021). Investigating the effect of reinforcing SiC and graphite on aluminium alloy brake rotor using plasma spray process. *Materials Today: Proceedings, 38*, 2706–2712.

Huang, S. X., & Paxton, K. (1998). A macrocomposite Al brake rotor for reduced weight and improved performance. *JOM, 50*(8), 26.

Inegbenebor, A. O., Bolu, C. A., Babalola, P. O., Inegbenebor, A. I., & Fayomi, O. S. (2015). Influence of the silicon carbide grit size particles on the mechanical and electrical properties of stir casting aluminum matrix composite material. *Silicon*. https://doi.org/10.1007/s12633-015-9305-8

Inegbenebor, A. O., Bolu, C. A., Babalola, P. O., Inegbenebor, A. I., & Fayomi, O. S. I. (2016). Aluminum silicon carbide particulate metal matrix composite development via stir casting processing. *Silicon*. https://doi.org/10.1007/s12633-016-9451-7

Kök, M., & Özdin, K. (2007). Wear resistance of aluminium alloy and its composites reinforced by Al2O3 particles. *Journal of Materials Processing Technology, 183*(2–3), 301–309.

Lemen, R. A. (2004). Asbestos in brakes: Exposure and risk of disease. *American Journal of Industrial Medicine, 45*(3), 229–237.

Macke, A., Schultz, B. F., & Rohatgi, P. (2012). Metal matrix composites. *Advanced Materials and Processes, 170*(3), 19–23.

Maleque, M. A., Dyuti, S., & Rahman, M. M. (2010). Material selection method in design of automotive brake disc. In *Proceedings of the world congress on engineering* (Vol. 3).

Pinca-Bretotean, C., Josan, A., & Putan, V. (2021). Testing of brake pads made of non asbestos organic friction composite on specialized station. *Materials Today: Proceedings*, 1–6.

Radhakrishnan, C., Yokeswaran, K., Vengadeshprasadh, M., Vishnuhasan, A., Vimalraj, T., & Velusamy, M. (2015). Design and analysis of disc brake with titanium alloy. *International Journal of Innovative Science Engineering and Technology (IJISET), 2*(5), 1044–1050.
Wang, A. G., & Hutchings, I. M. (1989). Wear of alumina fibre–aluminium metal matrix composites by two-body abrasion. *Materials Science and Technology, 5*(1), 71–76.
Yashwanth, K. M., Dhanasekaran, R., Sreenatha, R., & Vijayakumar, B. (2018). Review on wear characteristics of aluminum silicon carbide reinforced with basalt fibre. *Materials Today: Proceedings, 5*, 26948–26954.
Yashwhanth, S., Mithun, M. M., Anandhan, R., & Senthil, K. S. (2021). Present knowledge and perspective on the role of natural fibers in the brake pad material. *Materials Today: Proceedings, 46*(2021), 7329–7337.

Overview of Nanofluid Applications and Its Sustainability

F. Iriaye, A. A. Noiki, O. O. Yusuf, S. A. Afolalu, and M. E. Egbe

1 Introduction

Nanotechnology is applied in various fields to improve the outcome of many processes. In the machining industry, nanotechnology can be applied in the form of nanofluids either via the development of nano-cutting fluids or via the use of nanoparticles to improve the performance of cutting tools via coated cutting tools (Shokoohi & Shekarian, 2015). Nanoparticles are solid particles that are nanometre-sized, meaning each particle ranges from 1 to 100 nm. Suspending nanoparticles in a carrier fluid enhances certain characteristics of the fluid based on the intended use (Williams et al., 2006). From the literature review on nanostructures by Zamani et al. (2009), nanofluids are made by combining a base/carrier fluid with nanomaterials. These nanomaterials include nanoparticles, nanotubes, nano-rods, porous materials and thin films. Nanofluid is a made up of nanometre-sized materials, referred to as nanoparticles. These are concocted colloidal suspensions of nano-sized materials in a carrier fluid. In addition, these nanoparticles contained in nanofluids are usually made up of carbides, metals, carbon nanotubes and oxides. Nanofluids are fluids created by dispersing solid nanomaterials into a base fluid. Taylor et al. (2013) found that suspending nanometre-sized particles with average diameters ranging from 1 to 100 nm in a fluid, called nanofluid, leads to a significant improvement of its thermal conductivity. Nano-coolants are made by dispersion of

F. Iriaye (✉) · A. A. Noiki · S. A. Afolalu · M. E. Egbe
Department of Mechanical Engineering, Covenant University, Ota, Nigeria
e-mail: ejeomo.iriaye@covenantuniversity.edu.ng; ayodeji.noiki@covenantuniversity.edu.ng; sunday.afolalu@covenantuniversity.edu.ng; mouyowa.egbe@covenantuniversity.edu.ng

O. O. Yusuf
Department of Microbiology, Obafemi Awolowo University, Ile-Ife, Nigeria
e-mail: omolola.yusuf@oau.edu.ng

A. O. Ayeni et al. (eds.), *Advanced Manufacturing in Biological, Petroleum, and Nanotechnology Processing*, Green Energy and Technology,
https://doi.org/10.1007/978-3-030-95820-6_5

nanoparticles of metallic or non-metallic oxide into heat dissipating conventional fluids. Simply put, any organic or inorganic fluid containing suspended nanoparticles is a nanofluid (Zamani et al. 2019). Creation of two-phase nanofluids is done to improve the characteristics of the host fluid. The fluid thermal conductivity can be increased via solid particle dispersion/suspension in fluids. Introducing even a minimal volume of nanoparticles into a base fluid leads to an enhancement in its thermophysical properties (Afolalu et al., 2019b). It has been proven severally that the introduction of nano-sized solid particles into a fluid enhances its thermal capacity and the overall performance of the fluid. Nanofluids have distinctive characteristics and properties as well as a potential for various applications involving heat transfer. Some of which include use for hybrid-powered engines, microelectronics, machining, pharmaceutical processes, vehicle thermal management, domestic refrigeration systems, fuel cells, engine cooling, chiller, boilers, heat exchangers and grinding. This enhancement is due to the properties of the nanoparticles contributing to changes in the heat, flow, transport and thermal capabilities of the fluid (Jama et al., 2016; Devendiran & Amirtham, 2016). They have improved thermal conductivities and enhanced convective heat transfer coefficients as opposed to base fluids without nanoparticles. For heat transfer applications, it is important to study the rheological behaviours of nanofluids to determine suitable applications (Chen et al., 2009). Also, they possess distinct acoustic properties. Research has shown that these fluids also exhibit additional reconversion of shear wave on an incident compressional wave in the field of ultrasonic. As the concentration of nanoparticles in the fluid increases, the reconversion effect becomes more prominent (Ajayi et al., 2018). For computational fluid dynamics (CFD) analysis, nanofluids are regarded as single-phase fluids. However, a large percentage of recent publications classify these fluids as two-phase fluids (Alizadeh & Dehghan, 2014). The properties of the nanoparticle constituents, as well as the concentration of nanoparticles contained in a single-phase fluid, are variables that determine the physical properties of the nanofluid. Kuznetsov and Nield (2010) alternatively identified an approach simulating nanofluids with the use of a two-component model. Nanofluids can be classified according to the nature of the nanoparticles used in their creation. Due to this criterion, there are two classes of nanofluids, specifically metallic-based and non-metallic-based nanofluids. Metallic nanofluids are those made from metallic nanoparticles like Al, Fe, Zn, Cu and Si. Non-metallic nanofluids contain non-metals, which consist of oxides of metals (e.g. Al_2O_3, SiO_2), semiconductors (e.g. TiO_2), carbon nanotubes (such as single-walled carbon nanotubes (SWCNT), double-walled carbon nanotubes (DWCNT), and multiple-walled carbon nanotubes (MWCNT)) and composite materials (Manikanta et al., 2018). The arrangement of nanoparticles along the contact line of a nanofluid droplet is due to diffusion; this arrangement aids in the dispersion of the nanofluid, which gives result to a structural disjoining pressure close to the contact line. However, this phenomenon does not arise for small droplets with a nanometre-scale diameter due to the greater diffusion time scale.

A new development in the field of nanofluids is concerned with the suspension of more than one nanoparticle in a base fluid giving rise to composite/hybrid nanofluids. Hybrid nanofluids combine the properties of the various nanoparticles suspended in it, giving rise to a nanofluid with greater thermal properties; however, this enhancement can only be achieved when the varying nanoparticles are synergistic (Devendiran & Amirtham, 2016).

Afolalu et al. (2019) developed a tri-nano-additive consisting of nanoparticles synthesized from agricultural wastes, which were used as carburizers and energizers to optimize the properties of mild steel. The field of nanotechnology is a growing one; hence, there are still problems associated with the preparation of nanoparticles/nanofluids consisting of the high production cost, problems associated with stability, complex production process, agglomeration of particles and thermal performance, to name a few (Devendiran & Amirtham, 2016)

2 Nanofluids: Preparation and Characteristics

Nanofluids are prepared using various techniques depending on the type of nanomaterials used. Subramani and Mohan (Kannan & Baskar, 2013) identified that the two ways of making nanofluids are single-step/one-step method and double-step/two-step method. For a unit-step method, the nano-sized particles are prepared and synthesized with the nanofluid simultaneously. It combines the synthesis and dispersion of the nanoparticles into a single stage (Uddin et al., 2016). They can be synthesized via the physical vapour deposition technique described by Uddin et al. (2016) where nanofluids are synthesized by heating the nanoparticles in a crucible and depositing the nano-sized vapour into the base fluid by condensation simultaneously. Nanofluids can also be prepared via the direct evaporation method, which is a single-step method where the nanoparticles contained in the fluid in the gaseous state are solidified, thus forming a colloidal suspension of nanoparticles in the fluid (Uddin et al., 2016). The single-step preparation method proffers less particle agglomeration and eliminates the need for ultrasonic sonication of the nanofluid as particle dispersion via this method is adequate. However, the single-step preparation method does not provide high purity nanofluids due to the presence of residue from the reactants used during its preparation. Also, nanofluids prepared via the single-step method cannot be used for high vapour pressure fluids. The double-step/two-step technique entails synthetization of the nano-sized particles in powdery form and subsequently, dispersing it in a base fluid. The dispersion can be done in a sonic homogenizer to ensure adequate dispersion of the particles and to ensure aggregation does not occur (Gupta et al., 2019). The sonication process is done to ensure the nanoparticles are adequately mixed with the base fluid. After which, adopt magnetic stirring and the use of dispersion additives/surfactant to increase the wettability and dispersion rates. These surfactants also reduce the surface tension present in the base fluid (Manikanta et al., 2018). Majority of nanofluids are prepared using the two-step method since the single-step method is not suited to large-scale applications. Here, the nanoparticles are usually purchased as a dry powder and mixed with the base fluid; however, the stability of the nanofluids developed is poor due to varying factors and specifications from different manufacturers (Jalal et al., 2010). According to Manikanta et al. (2018), the four major factors to be considered when preparing nanofluid are highlighted below:

I. Chemical compatibility between the nanofluid and the materials to be applied on
II. The thermal stability
III. The structure of the material to be applied on
IV. The textural properties of the fluid (Zamani et al., 2019)

Another factor to consider is the stability and dispersion during the preparation of nanofluids. A nanofluid is termed stable as long as the particles are small enough to overcome settling due to gravity over a suitable period.

The nanoparticles are agitated via Brownian motion during preparation to ensure the stability of the nanofluid (Devendiran & Amirtham, 2016). The stability of a nanofluid depends on the method of preparation used. It is observed that due to the van der Waals forces between the particles, nanoparticles tend to aggregate after a period (after few days) (Abioye et al., 2019a). During the preparation of nanofluids, factors such as the pH levels, presence of surfactants, presence of electrolytes as well as temperature affect the rheological properties of the nanofluid [20]. Stability of a nanofluid can be achieved using any of the following means: alteration of the pH levels of the nanofluid, use of surfactants/dispersants, or use of an ultrasonic vibrator for particle dispersion. The use of any of the processes named above will lead to fewer particle clusters in the fluid, i.e. agglomeration. Nanofluid preparation is a delicate process because many factors are considered before their preparation. Hence, suitable activators, additives and dispersants should be selected to prevent the alteration of the chemical composition and thermal properties of the nanofluid. Its zeta potential value determines the stability of a nanofluid where the higher the stability, the higher the zeta potential value indicated (Devendiran & Amirtham, 2016).

2.1 *Properties and Characterization of Nanofluids*

The physiochemical characteristics of a nanofluid are determined by the method of preparation, type of nanoparticles and its properties such as the particle size and concentration. There are many physical properties of nanofluids, including volume fraction, thermal conductivity, viscosity, size and the shape of nanoparticles. Combining nanoparticles of different metals, oxides and semiconductors with fluid using different techniques produces nanofluids. The thermal conductivity of a nanofluid is one of the most important physical properties, which evaluates the performance of a nanofluid. By dispersing nanoparticles in heat/thermal transfer fluid, the heat exchange property, i.e. warming or cooling, becomes significantly enhanced (Mohammadi et al., 2019). Jalal et al. (2010) stated that as the improved thermal property decreases, the thermal conductivity of base liquid increases. They also measured various physicochemical properties such as surface pressure, thermal conductivity and consistency of ZnO nanoparticles in glycerol and ethylene glycol carrier fluids. Nanofluids made from pure metallic nanoparticles possess a higher thermal conductivity compared to nanofluids bearing oxide nanoparticles. The thermal conductivity of metallic nanofluids increases linearly with the volume fraction (Mondragon et al., 2012). Nanoparticles such as CuO, Al_2O_3, TiO, SiO_2, carbon nanotubes, silica and alumina are dispersed into carrier fluids to improve their thermal capabilities. In addition to higher thermal conductivity and convection, research has shown that in use, nanofluids lead lesser operator and environmental hazards. In an engine, during cooling, nanofluid losses of up to 30% ensue due to cooling phenomenon. Using coolants, which have higher thermal conductivities and heat transfer coefficients, leads to a decrease in the losses. Nowadays, energy conservation is important; therefore, the use of better coolants leads to achieving a reduction in fuel consumption. Pak and Cho (1998) developed a model which determines the density of a nanofluid shown below (Eqs. 1 and 2):

$$\rho_{nf} = (1-\varphi) + \varphi\rho_{np} \tag{1}$$

The Ward model shows the viscosity of a nanofluid and can be used for nanofluids with particle volume concentration up to 35%. Equation 2 has incorporated the viscosity parameter of the base fluid and in representing the overall viscosity of the nanocluster:

$$\mu_{nf} = [1 + \eta\varphi + (\eta\varphi)^2 + (\eta\varphi)^3 L \tag{2}$$

Saxena and Raj (2017) inspected the effect of molecule size on the thermal conductivity proportion of alumina/water nanofluids. The after-effects of their work demonstrated that shrinkage of molecule size improves the thermal conductivity proportion of nanofluids (Fig. 1).

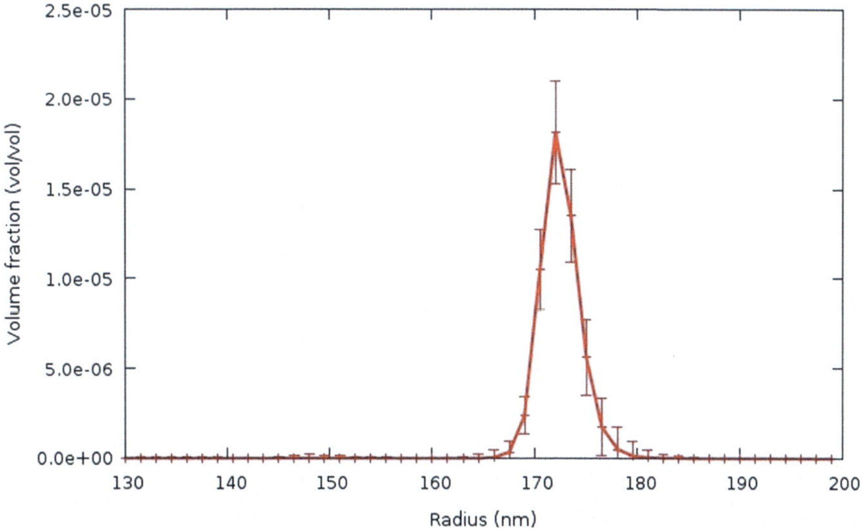

Fig. 1 Graph showing nanoparticle size distribution. (Teng et al., 2010)

Yeganeh et al. (2010) calculated the improvement in the thermal conductivity of deionized water upon the addition of nano-diamond particles with varying volumes ranging from 0.8% to 3%. At a volume concentration of 3%, the highest increase in thermal conductivity (7.2%) was attained. Nanofluids are utilized for their lubricating properties. Teng et al. (2010) stated that using advanced lubricants on machinery helps to improve its life cycle due to their friction and wear-reducing properties. Xu prepared a nanofluid using MoS_2/TiO_2 nanoparticles in a rapeseed oil base and studied its tribological properties. It was observed that the nanoparticles helped to reduce the wear rate in machine moving parts. Teng et al. (2010) reported that the tribological properties of a KH-560 machine oil with ZrO_2 nanoparticles showed enhanced anti-wear and improved friction properties. When added to a lubricating oil base, nanoparticles provide four possible effects: polishing effect (Yeganeh et al., 2010), tribofilm formation (Shen et al., 2008), rolling of nano-spheres (Gulzar et al., 2016), and self-repairing

3 Applications of Nanofluids

Nanofluids are used in a vast number of industrial applications, especially for their enhanced heat transfer and thermal properties. These areas include refrigeration and air conditioning, lubrication, car engine radiators, drilling and space applications, to mention a few (Shen et al., 2008). Nanofluids are generally used in lubrication – for refrigeration systems and automobile engine oil; cryogenic systems, such as coolant agents in heat exchangers; and heat transfer applications, such as in cutting fluids. Nanofluids are also often utilized in biomedical applications, in electronics components for cooling, in the chemical industry, for waste heat recovery as well as in environmental engineering (Ali et al., 2018) (Fig. 2).

Heat transfer has applications in aviation, waste recovery, refrigeration, chemical industries, power generation units and air conditioning, to name a few. Hence, it is essential in various engineering fields. However, low thermal conductivities of heat transfer fluids have led to a growing interest in the use of nanofluids as heat transfer fluids (Manikanta et al., 2018). Nanofluids are often used to improve the heat transfer properties of commonly used heat transfer fluids such as water and mineral oils, which have considerably lower heat transfer potentials. In many industries, the use of nanoparticle-based metalworking fluids has been adopted. In the aerospace industry, for example, nanoparticle-based cutting fluids are widely used in machining parts because they have no adverse effects on the characteristics of the workpiece or its performance. Manikanta et al. (2018) stated that cutting fluids developed with carbon nano-onion particles yielded successful results, especially with the typically "tough-to-machine materials", and are already used in some aerospace industries. The carbon nanoparticles were dispersed in oil- and water-based cutting fluids where they optimized cutting

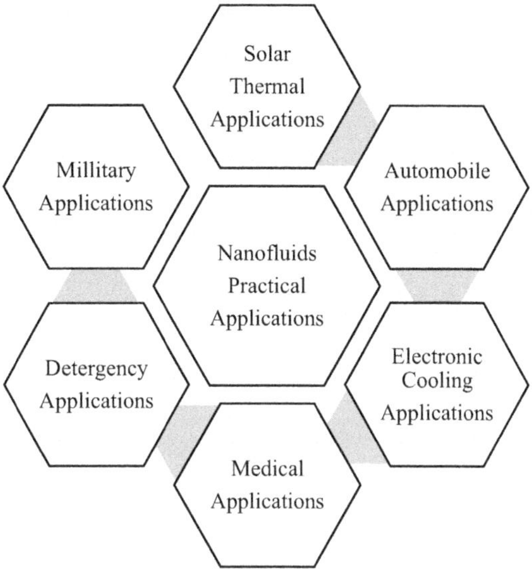

Fig. 2 Practical application of nanofluids. (Ali et al., 2018)

parameters and improved the surface finish of the materials compared to conventional cutting fluids. Ali et al. (2008) noted that nanofluids could be utilized for their lubricating properties for the following applications: biomedical applications, refrigeration and air conditioning

3.1 Advantages of Nanofluids over Solid and Liquid Suspensions in Use

As emphasized by Mondragon et al. (2012), the introduction of micro-sized solid particles into fluids to a base fluid led to problems of channel clogging and low stability. Two-phase fluid suspensions containing micrometre-sized particles are not suited to practical applications as they cause sedimentation because of their large particle size as well as erosion of the tube through which the fluid flows (Patil et al., 2016; Rashidi et al., 2018; Afolalu et al., 2018a). Patil (2016) observed that compared to solid and colloidal liquid mixtures, nanofluids are far more superior as the nanoparticles do not cause clogging when they are applied. These nanofluids also have enhanced heat transfer properties, which improve as the surface area of the particles increases. When dispersed in a base fluid, the nanoparticles have proven to offer more stability. Manikanta et al. (2018) also consigned the statement above, stating that in comparison, nanofluids offer reduced clogging due to the particle size as well as better dispersion stability with rigorous Brownian motion (Ajayi et al., 2019a). Furthermore, the nanoparticles in nanofluids cover a large area on the surface on which it is deposited, thus providing a higher rate of heat transfer (Abioye et al., 2019).

4 Conclusion

The field of nanotechnology is a growing one; however, there are still problems associated with the preparation of nanoparticles/nanofluids consisting of the high production cost, problems associated with stability, complicated production process, agglomeration of particles and thermal performance. Nanotechnology is gaining the attention of several researchers in recent times, particularly the application of nanofluids in various industrial processes due to their notable mechanical properties. However, there are still several challenges associated with its preparation techniques. In this study, we have attempted to analyse the relevance of nanofluids, its properties and characterization, as well as its numerous applications in various fields of endeavour and its comparative advantages over solid and liquid suspensions. However, there is still ongoing research aimed at addressing the factors affecting its optimization in industrial processes.

Acknowledgements We acknowledge the financial support offered by Covenant University in the actualization of this research work for publication.

References

Abioye, A. A., Fasanmi, O. O., Rotimi, D. O., Abioye, O. P., Obuekwe, C. C., Afolalu, S. A., & Okokpujie, I. P. (2019a, December). Review of the development of biodegradable plastic from synthetic polymers and selected synthesized nanoparticle starches. *Journal of Physics: Conference Series, 1378*(4), 042064.

Abioye, A. A., Rotimi, D. O., Fasanmi, O. O., Abioye, O. P., Afolalu, S. A., Owoeye, T. F., & Obuekwe, C. C. (2019b, December). Synthesis and characterization of selected starch nanoparticles as matrix reinforcements for low density polyethylene. *Journal of Physics: Conference Series, 1378*(4), 042070.

Afolalu, S. A., Asonaminasom, E. H., Ongbali, S. O., Abioye, A. A., Udo, M. O., & Salawu, E. Y. (2018a). Dataset on experimental investigation of optimum carburizing temperature and holding time of Bi-Nano additives treatment of *AISI 5130 steel. Data in Brief, 19*(1), 2279–2283.

Afolalu, S. A., Ongbali, S. O., Abiodun, A. A., Udo, M. O., & Tunde, C. A. (2018b). Experimental Investigation of the effects of Bi-Nano additives on the mechanical properties of AISI 5130 mild steel during machining. *International Journal of Mechanical Engineering and Technology, 9*(12), 264–273.

Afolalu, S. A., Efekodha, G. E., Ongbali, S. O., Abioye, A. A., Salawu, E. Y., Ajayi, O. O., & Oluwabunmi, A. P. (2019a). Experimental analysis of the effect of tri-nano additives on wear rate of mild steel during machining. *Procedia Manufacturing, 35*, 395–400.

Afolalu, S. A., Ongbali, S. O., Ohwofa, O., Ajayi, O. O., & Abioye, A. A. (2019b, November). Nanocomposite and its morphological characterization – Review. *IOP Conference Series: Materials Science and Engineering, 640*(1), 012067.

Agapiouv, J. S. (2018). Performance evaluation of cutting fluids with carbon nano-onions as lubricant additives. *Procedia Manufacturing, 26*, 1429–1440.

Ajayi, O. O., Omowa, O. F., Abioye, O. P., Omotosho, O. A., Akinlabi, E. T., & Folalu, S. A. A. (2018, March). Finite element modelling of electrokinetic deposition of zinc on mild steel with ZnO-citrus sinensis as nano-additive. In *TMS annual meeting & exhibition* (pp. 199–211). Springer.

Ajayi, O. O., Aba-Onukaogu, C. C., Salawu, E. Y., Owoeye, F. T., Akinlabu, D. K., Afolalu, S. A., & Abioye, A. A. (2019a). Effect of biomaterial (Citrullus Lanatus Peels) nanolubricant on the thermal performance and energy consumption of R600a in refrigeration system. In *Energy technology 2019* (pp. 91–102). Springer.

Ajayi, O. O., Okolo, T. I., Salawu, E. Y., Owoeye, F. T., Akinlabu, D. K., Akinlabi, E. T., & Afolalu, S. A. (2019b). Performance and energy consumption analyses of R290/bio-based nano lubricant as a replacement for R22 refrigerant in air-conditioning system. In *Energy technology 2019* (pp. 103–112). Springer.

Ali, N., Teixeira, J. A., & Addali, A. (2018). A review on nanofluids: Fabrication, stability, and thermophysical properties. *Journal of Nanomaterials, 6978130.*

Alizadeh, M. R., & Dehghan, A. A. (2014). Conjugate natural convection of nanofluids in an enclosure with a volumetric heat source. *Arabian Journal for Science and Engineering, 39*(2), 1195–1207.

Chen, H., Witharana, S., Jin, Y., Kim, C., & Ding, Y. (2009). Predicting thermal conductivity of liquid suspensions of nanoparticles (nanofluids) based on rheology. *Particuology, 7*(2), 151–157.

Devendiran, D. K., & Amirtham, V. A. (2016). A review on preparation, characterization, properties and applications of nanofluids. *Renewable and Sustainable Energy Reviews, 60*, 21–40.

Gulzar, M., Masjuki, H. H., Kalam, M. A., Varman, M., Zulkifli, N. W. M., Mufti, R. A., & Zahid, R. (2016). Tribological performance of nanoparticles as lubricating oil additives. *Journal of Nanoparticle Research, 18*(8), 223.

Gupta, M. K., Jamil, M., Wang, X., Song, Q., Liu, Z., Mia, M., & Imran, G. M. (2019). Performance evaluation of vegetable oil-based nano-cutting fluids in environmentally friendly machining of inconel-800 alloy. *Materials, 12*(17), 2792.

Jalal, R., Goharshadi, E. K., Abareshi, M., Moosavi, M., Yousefi, A., & Nancarrow, P. (2010). ZnO nanofluids: Green synthesis, characterization, and antibacterial activity. *Materials Chemistry and Physics, 121*(1-2), 198–201.

Jama, M., Singh, T., Gamaleldin, S. M., Koc, M., Samara, A., Isaifan, R. J., & Atieh, M. A. (2016). Critical review on nanofluids: Preparation, characterization, and applications. *Journal of Nanomaterials, 6717624.*

Kannan, S., & Baskar, N. (2013). Modeling and optimization of face milling operation based on response surface methodology and genetic algorithm. *International Journal of Engineering and Technology, 5*(5), 959–971.

Kuznetsov, A. V., & Nield, D. A. (2010). The onset of double-diffusive nanofluid convection in a layer of a saturated porous medium. *Transport in Porous Media, 85*(3), 941–951.

Manikanta, J. E., Gupta, B. N. V. S. G., Manikanta, K. J. T., & Pradeep, R. (2018). Integrality characterization of machining with nano cutting fluids. *International Journal of Mechanical Engineering and Technology, 10*(10), 1033–1042.

Mohammadi, M., Abadeh, A., Nemati-Farouji, R., & Passandideh-Fard, M. (2019). An optimization of heat transfer of nanofluid flow in a helically coiled pipe using Taguchi method. *Journal of Thermal Analysis and Calorimetry, 138*(2), 1779–1792.

Mondragon, R., Julia, J. E., Barba, A., & Jarque, J. C. (2012). Characterization of silica–water nanofluids dispersed with an ultrasound probe: A study of their physical properties and stability. *Powder Technology, 224*, 138–146.

Pak, C., & Cho, Y. I. (1998). Hydrodynamic and heat transfer study of dispersed fluids with submicron metallic oxide particles. *Experimental Heat Transfer, 11*, 151–170.

Patil, M. S., Seo, J. H., Kang, S. J., & Lee, M. Y. (2016). Review on synthesis, thermo-physical property, and heat transfer mechanism of nanofluids. *Energies, 9*(10), 840.

Rashidi, S., Mahian, O., & Languri, E. M. (2018). Applications of nanofluids in condensing and evaporating systems. *Journal of Thermal Analysis and Calorimetry, 131*(3), 2027–2039.

Saxena, G., & Raj, J. (2017). a critical review on applications of nano-fluid as coolant. *International Journal of Engineering and Management Research (IJEMR), 7*(1), 304–311.

Shen, B., Shih, A. J., & Tung, S. C. (2008). Application of nanofluids in minimum quantity lubrication grinding. *Tribology Transactions, 51*(6), 730–737.

Shokoohi, Y., & Shekarian, E. (2015). Application of nanofluids in machining processes – A review. *Journal of Nanoscience and Technology*, 59–63.

Taylor, R., Coulombe, S., Otanicar, T., Phelan, P., Lv, A., & W. Gunawan,...& H Tyagi. (2013). Small particles, big impacts: A review of the diverse applications of nanofluids. *Journal of Applied Physics, 113*(1), 011301–01130119.

Teng, T. P., Hung, Y. H., Teng, T. C., Mo, H. E., & Hsu, H. G. (2010). The effect of alumina/water nanofluid particle size on thermal conductivity. *Applied Thermal Engineering, 30*(14–15), 2213–2218.

Uddin, M., Kalbani, K. S. A., Rahman, M. M., Alam, M. S., Al-Salti, N., & Eltayeb, I. (2016). Fundamentals of nanofluids: Evolution, applications and new theory. *International Journal of Biomathematics and Systems Biology, 2*(1).

Williams, W. C., Bang, I. C., Forrest, E., Hu, L. W., & Buongiorno, J. (2006). Preparation and characterization of various nanofluids. In *Proceedings of the NSTI nanotechnology conference and trade show (Nanotech'06)* (Vol. 2, pp. 408–411).

Yeganeh, M., Shahtahmasebi, N., Kompany, A. H. M. A. D., Goharshadi, E. K., Youssefi, A., & Šiller, L. (2010). Volume fraction and temperature variations of the effective thermal conductivity of nanodiamond fluids in deionized water. *International Journal of Heat and Mass Transfer, 53*(15–16), 3186–3192.

Zamani, A., Marjani, A. P., & Mousavi, Z. (2019). Agricultural waste biomass-assisted nanostructures: Synthesis and application. *Green Processing and Synthesis, 8*(1), 421–429.

Influence of Cutting Fluid and Parameters on Machining and Cooling Techniques in Recent Technology

F. Ademuyiwa, S. A. Afolalu, O. O. Yusuf, and M. E. Emetere

1 Machining

Machining is a crucial aspect of metalworking since it allows for metal cutting and shaping. Because of the substantial rise in processing time and the need to fund the large capital cost, machining has become increasingly important in modern automated manufacturing systems. Many nations around the world now have the ability to manufacture products, thanks to machining. The machining process has sparked competition among manufacturing firms, resulting in a wide range of product quality. As a result, most businesses conduct research into various methods or processes that they can use to manufacture high-quality, long-lasting products at a lower cost (Okokpujie et al., 2019).

However, machining operations invariably waste material in the form of chips, output rates can be poor, and the processes may have negative effects on the surface properties and efficiency of parts if not done correctly (Estrems et al., 2008). Consequently, varieties of sectors are testing different techniques for producing high-quality, long-lasting products at a lower cost. This paper will go through a few of these techniques in more detail.

F. Ademuyiwa (✉) · S. A. Afolalu
Department of Mechanical Engineering, Covenant University, Ota, Ogun State, Nigeria
e-mail: faith.ademuyiwa@covenantuniversity.edu.ng; adeniran.afolalu@abuad.edu.ng

O. O. Yusuf
Department of Microbiology, Obafemi Awolowo University, Ile-Ife, Nigeria

M. E. Emetere
Department of Physics, Covenant University, Ota, Nigeria
e-mail: emetere.moses@covenantuniversity.edu.ng

A. O. Ayeni et al. (eds.), *Advanced Manufacturing in Biological, Petroleum, and Nanotechnology Processing*, Green Energy and Technology,
https://doi.org/10.1007/978-3-030-95820-6_6

2 Machining Operations

The application of theoretical or scientific knowledge about the actions of materials, devices, or systems in diverse working environments is central to engineering design of every endeavor. When an engineer needs to machine a component, he must first decide what kind of machining process to use. When he knows this, he uses his knowledge of the chosen technique's performance to create a design that can achieve the goal (Zurita, 2017) Turning, boring, milling, and facing are examples of traditional machining techniques, as well as abrasive methods like grinding, ultrasonic machining, and lapping. Drilling, milling, and turning are some of the most popular techniques for shaping and forming metals to satisfy industry requirements (Pirtini & Lazoglu, 2018).

2.1 Drilling Operation

Drilling is a machining process for making holes in metals, especially sheet metals. This is a crucial activity in the manufacturing industry (Onwubolu & Kumar, 2006). Drilling is a popular machining method in a variety of industries, including automobile, aircraft and aerospace, dies/molds, home appliances, medical, and electronic appliances. Because of the increased competition in the industry, drilling process cycle times must be reduced. Furthermore, strict geometric tolerance criteria in designs necessitate an improvement in drilled hole precision in output (Pirtini & Lazoglu, 2018). Drilling is a crucial component of the cutting process, accounting for more than 40% of all material removal operations. Traditionally, a drilling tool is constructed of high-speed steel (HSS). The tool dulls more rapidly and has a shorter tool life as a consequence of the high temperature generated during drilling. Additionally, the workpiece hardens during drilling, resulting in post-drilling problems. On the other side, the chips' coherence phenomenon may result in a poor hole surface and accuracy (Cantero et al., 2005a). One of the most popular machining methods is drilling. A conventional drill's tip angle, chisel edge angle, chisel edge thickness, cutting lip length, and helix angle are all design requirements. Cutting forces and drilled hole quality are affected differently by each of these variables (Pirtini & Lazoglu, 2018).

The ability to form chips that can be quickly removed from the drilled hole is critical to the performance of a drilling process. Long chips are undesirable since they tangle along the twist drill body and must be manually removed. Because of the complicated cutting-edge geometries, drilling is called a three-dimensional cutting procedure requiring complex cutting mechanisms. Drilling would be smooth if well-broken chips are formed during the process. Many ductile materials, including austenitic stainless steels, do not break during drilling and instead form a continuous chip (Sultan et al., 2015).

2.1.1 Definition of Abbreviation

Abbreviation	Definition
HSS	High-speed steel
Vc	Cutting speed
f	Cutting feed
MRR	Material removal rate
MQL	Minimum quantity lubrication
CFs	Cutting fluids
SAE	Society of Automotive Engineers
AISI 1040, E4340C	American Iron and Steel Institute
(ISO P25)	International Organization for Standardization
DNMG 150608	Deutsch Netz Marketing Gmbh
CVD	Chemical vapor deposition
HRC	High rupturing capacity
ND	Nanodiamond
CA	Compressed air
SEM	Scanning electron microscope
MoS_2 nanoparticles	Molybdenum disulfide
MWCNT	Multi-walled carbon nanotube
CNT nanofluid 1% Al_2O_3	Carbon nanotube
NPs	Nanoparticles
WC (tungsten carbide)	Chemical Formula of tungsten carbide

Sultan et al. (2015) reviewed that continuous chips are divided into spiral chips (tight helix and loose helix chips) and string chips based on their chip shaping mechanisms (short and long ribbon chips). Chips with a close helix are produced by smaller drill wear, while chips with a long ribbon are produced by larger drill wear. When comparing Figs. 1 and 2, it is clear that shifts in feed rate have the greatest impact on chip form. This can be due to higher flank wear and material removal when the process is run at a higher feed volume.

Drilling with a twist drill is the most common method for producing holes for riveting and fastening structural assemblies among all machining operations. The never-ending quest for better drill efficiency and more cost-effective hole construction demonstrates the complexity of the drilling process. Drill point geometry is well known to have a major impact on the thrust force of a twist drill (Hocheng & Tsao, 2005). The efficiency of drilling bits has a direct impact on cost, according to Peter Muchendu et al. (2014), and an improvement in the rate of penetration leads to considerable cost and time savings. Chow et al. (2007) postulated that because of its high hardness, low thermal conductivity, and high work-hardening coefficient, stainless steel has been widely adopted since the trend of technological growth. However, these properties make drilling operations extremely difficult. Dubey (2017) used TiN-coated WC (tungsten carbide) to drill stainless plates and discovered that high drilling speed and feed rate resulted in a significant surface roughness while also affecting tool life.

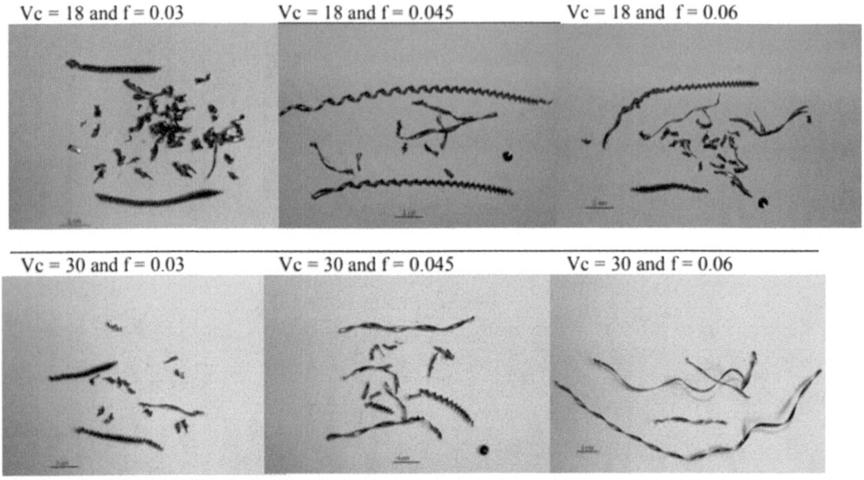

Fig. 1 Formed chip samples at different cutting speed (Vc=cutting speed in m/min) and feed rate (f, in mm/rev). (Sultan et al., 2015)

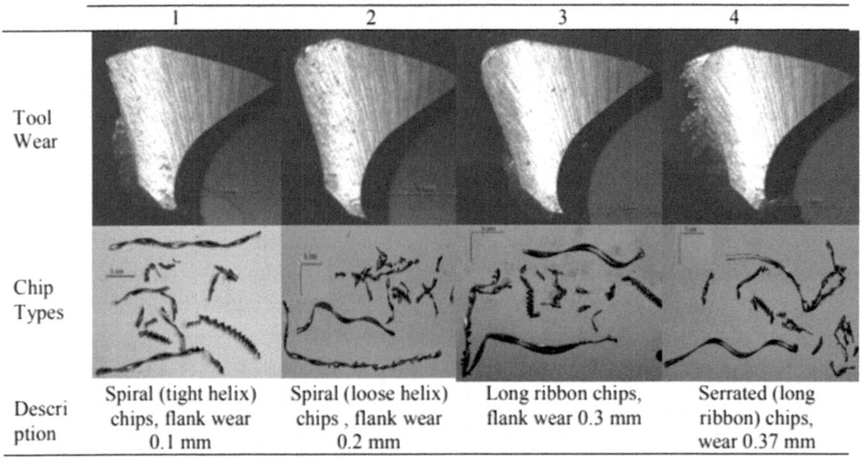

Fig. 2 Formed chip samples at different cutting speed (Vc, in m/min) and feed rate (f, in mm/rev). (Sultan et al., 2015)

2.2 *Milling Operation*

Despite the increasing adoption of contemporary industrial technology, machining is still one of the most frequently utilized technologies in the metalworking industry, and it may be employed alongside these new technologies (Sousa et al., 2021). Milling is a kind of machining that creates smooth, contoured, and helical surfaces by using a multipoint rotating cutting tool called a milling cutter. A milling cutter is a spinning body with cutting teeth placed along the surface, on the edges, or on both sides of the tool with multi-edge layering. The rotation that is delivered to the cutter is the primary cutting action. In most cases, the feed movement is sent straight to the cutter or the job . The workpiece is clamped to the worktable, and a linear feed is applied to the spinning cutter. For greater dimensional precision, most machine components are finished with a milling process (Kannan & Baskar, 2016). Milling is commonly used to manufacture parts that are not axially symmetric and have many component characteristics, such as holes, gaps, pockets, and maybe even three-dimensional surface curves, according to Jayal and Balaji (2009), and milled surfaces are mostly used to mate with other parts in die, aerospace, industrial, and machinery construction, as well as in manufacturing industries.

Peripheral or slab milling, face milling, and end milling are three distinct types of milling operations that can be used depending on the machining specifications.

2.3 *End Milling*

End milling's ability to remove metal at a faster pace while maintaining a reasonable surface roughness has led to its widespread usage for machining components. End milling procedures may be used for roughing and finishing operations on a wide range of materials that need high accuracy and surface polish. During the end milling process, the cutting tool's axis of rotation is kept perpendicular to the feed direction, i.e., parallel or perpendicular to the machined surface. Milling is a kind of intermittent cutting technique in which the teeth of the milling cutter enter and exit the workpiece at various moments throughout each rotation. As a consequence of the disturbed cutting process, the teeth are subjected to a period of impact force and thermal shock on any rotation, resulting in significant levels of noise (Shihab, 2017). According to Jayal and Balaji (2009), cutting forces have the largest impact on the output of machined components in end milling operations. As a consequence, cutting force must be addressed when developing cutting parameters that affect a surface roughness identification system in end milling operations.

2.4 *Face Milling*

Face milling is a common machining technique for creating high-quality flat surfaces. Another important feature of the process is the high rate of material removal or, in the case of one-pass milling, the high surface rate (Kundrák et al., 2018). Face milling is the process of machining surfaces parallel to the cutter axis. Face milling produces smooth surfaces, and machines may work to any length. The feed may be horizontal or vertical in face milling. Almost majority of the cutting in face milling is done by the teeth on the cutter's perimeter. When a cutter is properly ground, the face teeth merely remove a little bit of stock left over from the workpiece or cutter springing, giving in a smoother finish (Deepak, 2012).

Face milling is a well-known commercial process that is widely used in heavy industries and requires a lot of power. Face milling is a highly efficient type of machining process that is commonly used in the production of flat surfaces. Face milling also involves a large number of mill teeth at the same time, resulting in a dramatic rise in cutting power. Face milling, on the other hand, requires a considerable amount of energy, which is partly due to a greater tooth wear region (Pimenov et al., 2020). Face milling research has shown that the geometrical characteristics of the tool, the tool and workpiece material composition, and of course cutting conditions, which include the depth of cut, feed rate, and cutting speed, all have a major impact on the milling operation (Kundrak et al., 2018). With their free-cutting capabilities, the newest face milling cutters incorporate a new line of multi-sided, multi-edged insert solutions that help to boost current machine tool performance. These insert geometries extend tool life, decrease cutting forces, and provide open-form chips that reduce cutting-edge stress (Kumar, 2019).

It is also critical when planning face milling operations to ensure the design surface roughness while still minimizing power consumption and machining costs. This method allows for cost savings in the end product's manufacturing (Gupta et al., 2020).

2.5 Peripheral Milling

The fact that the chip thickness ranges from zero to its maximum value characterizes the phenomena of peripheral milling. High removal rates are also possible due to the structural stability (Estrems et al., 2008). During peripheral milling, also known as plain milling, the cutter's axis is parallel to the surface being machined, and the process is carried out by cutting edges on the cutter's outer perimeter. The rotation of the cutter is the primary movement. The workpiece receives the feed. Slab milling is the most basic type of peripheral milling, in which the cutter's breadth extends beyond the workpiece on all sides (Ratnakumar & Rao, 2013) (Fig. 3).

3 Turning Operation

Turning is an important material removal technique in today's industry. Turning is the most popular technique for metal cutting and, in particular, for completing machined components in the industrial sector. Turning is a single-point cutting technique that uses a tool to remove material from a revolving cylindrical object (Zurita, 2017). Turning is a material removal or machining technique that is used to remove undesired material from rotating components. The turning process requires the use

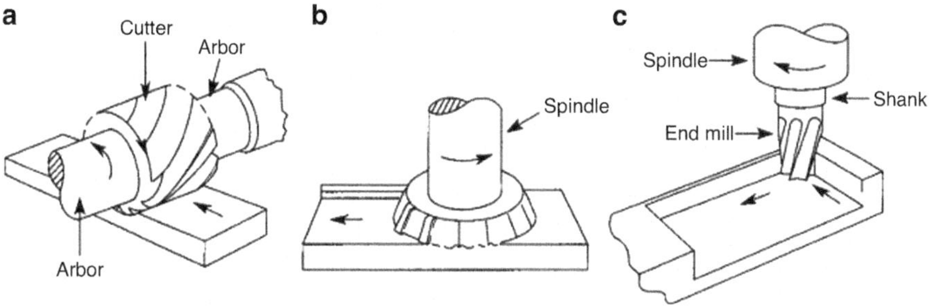

Fig. 3 Basic types of milling cutters and operations. (**a**) Peripheral milling, (**b**) face milling, (**c**) end milling. (Estrems et al., 2008)

of a turning machine or lathe, a workpiece, a fixture, and a cutting tool. The workpiece is a reshaped piece of material that is attached to the fixture and turned at high speeds on the turning machine (Sahu et al., 2017). The cutter is usually a single-point cutting instrument that is fastened in the machine. The cutting tool feeds into the spinning workpiece and cuts away material in the form of tiny chips to produce the desired shape. When turning, a number of factors govern the speed and motion of the cutting tool. These parameters are selected for each operation depending on the workpiece material, tool material, tool size, and other variables (Dubey, 2017).

Reducing power consumption and increasing the material removal rate (MRR) in turning operations may improve mechanical component performance and lower manufacturing costs (Kumar, 2019). To enhance machining conditions, the optimum turning process parameters, such as cutting speed, depth of cut, and feed rate, must be selected in order to minimize power consumption while improving MRR.

Nur et al. (2015) reviewed that in order to reduce energy consumption, it is necessary to be able to monitor the energy consumption of the machining process. We need to include the machining conditions in the energy consumption calculation to get a more accurate result. Kara and Li (2011) presented an empirical approach to developing models for machining processes and predicting their energy consumption for each unit process. They demonstrated that the energy consumption of the machining process could be predicted using empirical models within the established cutting parameters for the chosen machine tools. Their methodology can estimate the amount of energy required to machine a product using turning operation techniques.

Dahlman et al. (2009) investigated the effects of rake angle, cutting speed, and cutting depth on residual stresses in hard turning. Their findings revealed that a greater negative rake angle results in higher compressive stresses, which increase with increased feed rate. With ceramic tools, Chou and Song (2004) investigated the effects of tool nose radius on finish hard turning. Surface finish, tool wear, cutting forces, and, in particular, white layers were all evaluated at various machining conditions in this study. Their findings revealed that large tool nose radii not only produce a finer surface finish but also cause significant tool wear when compared to tools with small nose radii.

4 Cutting Fluids

Since the early twentieth century, when F.W. Taylor utilized water to cool the machining process for the first time and determined that it enhanced tool life, a wide range of cutting fluids have been employed for this and other purposes. However, due to the expense of cutting fluids, environmental concerns, human health, and other factors, much has been done in the previous decade to limit their usage in manufacturing (Priarone et al., 2015). Machining is a widespread industrial procedure in which cutting fluids often take a significant role for quality and efficient machining because of their lubrication, cooling, and chip removal capabilities (Jiang et al., 2015). Cutting fluids greatly increase the productivity of work processes, tool life, and the quality of the workpiece. They also minimize overheating of cutters and machine tools.

Cutting fluids are commonly utilized in operations of machining. Cooling, lubrication, removal of the chips, and protection of the workpiece are the key purposes of the cutting fluid. In a machining operation, the cutting fluid is essential to enhance performance productivity (Mane & Kumar, 2018). A major concern in sustainable manufacturing is the environmental impact of lubricants. Although dry cutting may be regarded as the ultimate objective for achieving it, lubrication remains an industry standard that is hard to overcome when tough to cut alloys are being machined. For oils used as cutting fluids, biodegradability, renewability, and lubrication are all key elements to consider (Raju et al., 2020).

Cutting fluids are used in machining for cooling and lubrication, but while the decrease in cutting temperature enhances tool life, cutting fluids have a detrimental influence on the environment, according to Priarone et al. (2015). This is because numerous chemical additives, such as calcium sulfonate and alkanoamides, pollute the environment and pose a health risk to workers. Furthermore, when compared to overall production expenses, the expenses of cutting fluid supply and disposal are not insignificant. Cutting fluids are used in machining operations, and several novel techniques have been tried to reduce or even eliminate their use in order to reduce the high cost and reduce the environmental burden associated with their use, treatment, and disposal. Two of these techniques are dry machining and machining with minimum quantity lubrication (MQL) (Z. Jiang et al., 2015).

4.1 Machining with Minimum Quantity Lubrication Technique

MQL (minimum quantity lubrication) refers to a machining system that uses a small amount of cutting liquid. Cutting fluid flow rates in MQL applications are usually between 50 and 500 ml/h, which is 10,000 times lower than flood cooling (Okokpujie et al., 2019). Standard machining techniques are heavily reliant on the availability of cutting fluid in order to reduce cutting temperatures and cutting forces. These cutting fluids, unfortunately, have a long history of causing environmental and health problems (Boswell et al., 2017). Dry or near-dry machining is often considered as a good way to reduce the environmental effects of cutting operations. However, due to the shortcomings of dry or near-dry machining in terms of application, reducing cutting fluid supply by machining parameter optimization is a cost-effective option (Jiang et al., 2015). Simultaneously, minimum quantity lubrication (MQL) use, also known as near-dry machining, has been explored as environmentally safe and economically beneficial, thereby sustainable and alternative to flooding of CFs in situations where dry machining is not possible (Jayal & Balaji, 2009). Since just a limited amount of lubricant is required, minimum quantity lubrication (MQL) could be used as a replacement for both dry and flood cooling conditions (Agarwal & Agarwal, 2021).

Sharma and Mital (2016) discussed the MQL results in the machining process (2016). MQL mist is effective at penetrating between tool and chip interfaces, resulting in a lower friction coefficient, according to the researchers. Researchers recently investigated the most recent state of the art in MQL, with a focus on machining operations like drilling, turning, milling, and grinding. Overall, MQL appears to have a lot of potential as a flood-cooling solution (Boswell et al., 2017).

4.2 Machining with Solid Lubricant

The extreme heat produced during machining is crucial to the efficiency of the workpiece. As a result, coolant and lubrication are critical in machining. Cutting fluids used in traditional machining have some drawbacks in terms of their application, both environmentally and economically (V. Agarwal & Agarwal, 2021). As a result, eco-friendly and user-friendly alternatives to traditional cutting fluids are needed. Solid lubricants have become a viable alternative to cutting fluids, thanks to advances in tribology. The most common solid lubricants used in machining are graphite, calcium fluoride, molybdenum disulfide, and boric acid (Krishna & Rao 2008). Solid lubricants are usually added to oils and greases as additives. Because of their distinct properties and features, solid lubricants are often used as a dry powder or as a component of coatings (Zailani et al., 2011).

Instead of flooding coolant, this is achieved mostly by clean machining methods that use a small amount of lubricant, such as MoS_2 powder, grease-based graphite combined with water, and SAE 20 oil in various proportions. The solid lubrication technique has resulted in a substantial reduction in tool wear rate, dimensional inaccuracy, and surface roughness, owing to a reduction in cutting zone temperature and a beneficial improvement in chip-tool and work-tool contact (Krishna & Rao, 2008).

4.3 Machining with Cryogenic Cooling System

Metal cutting generates heat, which has an impact on the quality of the final product, the cutting force, and the life of the cutting tool. To achieve the greatest machining performance and workpiece quality, researchers from all around the globe are attempting to find out the method and principle underlying temperature buildup during milling (Ahmad-yazid et al., 2010). The method of implementing and supplying non-oil-based cooling media such as liquid nitrogen (LN_2) and carbon dioxide (CO_2) to the cutting zone is known as cryogenic machining. It is gaining prominence as a result of its ability to reduce temperature while also being a clean operation (Aramcharoen 2016).

The word "Kryos" comes from the Greek language and means "cold." As a result, cryogenics entails operating in a cold climate. Cryogenic-assisted machining, or cryogenic machining, is an emerging renewable method of machining superior quality materials (Zindani & Kumar, 2020). Cryogenics, in the modern sense, refers to phenomena that occur at temperatures of −153 °C or below. Chemicals such as liquid nitrogen, oxygen, hydrogen, methane, carbon dioxide, and even argon are used in the definition of cryogenic liquids (Kaushal et al., 2016).

Because of the lower temperature, tool hardness, and tool life maintained, cryogenic gases such as nitrogen or carbon dioxide are a more recent trend in machining. This allows for higher cutting rates. Furthermore, the nitrogen evaporates, leaving no negative environmental impact, while the concept of dry cutting is to apply a tiny mist of air-fluid combination containing a very little amount of cutting fluid, such as vegetable oil (Zailani et al., 2011). Cryogenic cooling using liquid nitrogen jets resulted in a lower average chip-tool interface temperature, a high reduction in flank wear, better surface quality, and dimensional accuracy when compared to dry machining of AISI 1060, AISI 1040, E4340C, AISI 4140, and AISI 4037 steels. Cryogenic machining using liquid nitrogen and a customized tool holder outperformed dry machining in terms of tool life and wear resistance (Krishna & Rao, 2008).

5 Reported Effect of Cutting Fluid on Machining Parameters and Tool Life

Vegetable oils as suitable cutting fluids were investigated by Shashidhara and Jayaram (2010). They tested three vegetable-based cutting fluids, including soya bean, sunflower, and rapeseed oil. They concluded that, due to their environmentally beneficial properties, vegetable oils are a viable alternative to mineral-based oils.

Jiang et al. (2015) examined the impact of cutting fluids on tool wear and. For the turning process, they utilized AISI 304 workpiece material. They used three vegetable oil-based cutting fluids: coconut oil, soluble oil, and straight cutting oil. They got to the conclusion that feed rate affects surface roughness, while cutting speed affects tool wear. When compared to conventional mineral oils, coconut oil is a better cutting fluid for reducing tool wear and surface roughness.

Nur et al. (2015) assess the efficacy of MQL when castor oil is used as a cutting fluid. The outcomes are comparable to those of dry cutting. They found that applying a little amount of lubricant (50 ml/h) during a specific turning operation improves tool life compared to dry cutting.

Because the effect of oil mist dissipates at high speeds, they find that machining under MQL seems to be limited by cutting temperature.

An experimental investigation on the effect of cutting fluids in turning with coated carbide tool was undertaken by Yahyalsik (2010). In this study, a CVD-coated carbide TiC+AI_2O_3+TiN insert was employed as the cutting tool (ISO P25). DNMG 150608 was the type of insert. Dry and wet-cooled turning were used to compare the findings. During wet machining, he found that the CVD-coated carbide TiC+AI_2O_3+TiN cutting tool performed better.

Johnson et al. (2014) optimized cutting parameters and fluid application parameters during the turning of OHNS steel. The optimized results were compared to dry turning and traditional wet turning under comparable cutting circumstances. On a lathe machine, they utilized OHNS steel with a hardness of 34 HRC to complete turning operations. They got to the conclusion that turning with the least amount of cutting fluid improves cutting performance and improves surface quality. Ozcelik et al. (2013) investigated the effects of vegetable-based cutting fluids on drilling wear. They worked with sunflower and canola oils, as well as semisynthetic commercial cutting fluid. Due to its superior lubricating qualities, their experimental results suggest that canola-based cutting fluid provides the highest performance.

5.1 Effect on Cutting Forces

Sarhan et al. (2012) used a lubricant made up of SiO_2 nanoparticles as well as ordinary mineral oil. On the vertical milling machine, machining proceeded. They used two types of lubricants: ordinary mineral oil and SiO_2 nanoparticles (0.2 vol%) in the same mineral oil. In comparison to ordinary mineral oil, the inclusion of nanoparticles lubrication results in a significant reduction in power, cutting force, as well as specific energy requirements.

Setti et al. (2012) examined the influence of Al_2O_3 nanofluid in combination with MQL in a grinding operation on a surface grinder with only a SiC grinding wheel. The researchers used the Taguchi experimental design method to create a model for predicting the quality of surface finish and also the grinding forces involved. The findings obtained with nanofluid were compared to those obtained with pure water and traditional coolant in this research. With the MQL system, it was discovered that nano-Al_2O_3 particles reduce grinding forces far beyond traditional fluid and pure water. It was also discovered that Al_2O_3 nanoparticles with a 4% vol. concentration was shown to be more successful than other methods.

Vasu and Reddy (2011) investigated the machinability of Inconel alloy in various cutting conditions, including dry, MQL, and Al_2O_3 nanoparticles in vegetable oil. Cutting forces, surface roughness, tool life, and temperature dissipation were used to classify the materials. The results of the experiment aid in determining the significant parameters that have a greater impact on tool wear, feed rate, and cut depth. When compared to other cutting conditions, Al_2O_3 in a 6% volume concentration produced the best desired results.

Sayuti et al. (2013) investigated the milling outcomes of an aerospace duralumin workpiece. The procedure was carried out using a vertical milling machine. The study used a cutting fluid including onion-enriched nanofluid, which resulted in decreased surface roughness as well as cutting force. The maximum amount of carbon onion (1.5 wt percent) lubricant led to decreased cutting force and surface roughness. When compared to conventional lubricating oil, they were reported to have a 21.99 decrease in cutting forces and a 46.32% reduction in surface roughness.

Nam et al. (2011a) investigated the effect of nanofluids in a micro-drilling process using aluminum as the workpiece material using the MQL technique. Nanodiamond (ND) has been added to vegetable and paraffin oils to create nanofluids. Various types of cooling environments were used during the machining process, including compressed air (CA) lubrication, base oil with MQL method, and nanofluid-based MQL. When CA lubrication was used, the minimum number of holes through micro-drilling was 87 before tool failure. However, when using normal base oil lubrication and the nanofluid-based MQL technique, it was able to drill 150 holes without failure. To examine the effect of ND particles, observations were conducted and comparisons were done based on thrust force and drilling torques up to 86 holes. To inspect the quality of drilled holes, SEM pictures were obtained. Burrs were found around the perimeter of drilled holes during the execution of the CA lubrication technique and remained behind the chip in the drilled holes, resulting in poor hole quality. Because of the desired ball bearing effect, this issue has not been observed with nanofluid using MQL method. According to research, 1 vol percent of ND particles in paraffin oil and 2 vol percent of ND particles in vegetable oil were shown to be more effective in reducing torque and thrust force. The minimum force and drilling torque were obtained when ND particles were included in paraffin oil at a concentration of 1 vol percent.

5.2 *Effect on Surface Roughness*

Surface roughness is a common quality indicator and, in most instances, a technical necessity for mechanical products. The operational behavior of a component depends heavily on achieving the required surface quality (Benardos & Vosniakos, 2003). In order to grind silicon carbide (SiC), Gopal and Venkateswara Rao (2004) used graphite as a solid lubricant . When compared to a dry cutting environment, the findings indicate a decrease in cutting force, temperature, surface finish, as well as specific energy. Under graphite solid lubrication, a higher rate of material removal and less wheel wear have also been observed. As a result, grinding in a graphite environment was more productive and produced higher-quality products.

Sayuti et al. (2014) performed an experiment on tool life and surface roughness during turning operation of AISI 4140 steel in order to determine the impact of SiO_2 nanoparticles. Under the inclusion of nanoparticles, it was discovered that 0.5% inclusion, 60-degree nozzle angles, and 2 bar air pressure were beneficial in terms of reducing tool wear. The optimum surface polish is obtained with a 0.5% nanoparticle concentration and a 30-degree nozzle angle orientation.

Rahmati et al. (2013) to study surface morphology used end milling with MoS_2 nanoparticles. In the first mode, conventional cutting fluid was used, and in the second mode, cutting fluid with MoS_2 nanoparticles was used. By polishing and filling the machined surface, MoS_2 nanoparticles have been shown to decrease surface roughness. The findings clearly show that including 0.5 wt% MoS_2 nanoparticles in the surface yields the highest surface quality. Further nanoparticle inclusions, on the other hand, degrade surface quality. Saravanakumar et al. (2014) examined the impact of a cutting fluid based on silver nanoparticles in the turning process. During the experiment, the lubricating and cooling characteristics of silver nanoparticles were examined, and a comparison was performed using various cutting environments. Reduced surface roughness, cutting pressures, and cutting temperatures were discovered to be positive effects of nanoparticles. Lee et al. (2012) conducted a study to see how Al_2O_3 nanoparticles and nanodiamond (ND) might be used in a variety of applications utilizing micro grinding. ND particles were shown to be more effective than Al_2O_3 nanoparticles in terms of decreasing grinding forces and giving excellent surface quality. Al2O3 nanoparticles, on the other hand, were shown to be more successful in reducing surface roughness.

5.3 *Effect on Cutting Temperature*

Maoj et al. (2014) conducted an experiment in order to study the colloidal stability of Al_2O_3 nanoparticles using the MQL method in grinding, which they found to be successful. The obtained findings suggest that Al_2O_3 nanoparticles have poor suspension stability when subjected to short-duration ultrasonic vibration, as demonstrated. It has been claimed that using a 0.5% concentration of Al_2O_3 nanoparticles and running the ultrasonic vibrator for at least an hour can enhance the suspension stability of Al_2O_3 nanoparticles. A study carried out by Sayuti et al. (2014) examined the impact of SiO_2 nanoparticles in combination with coconut oil-based cutting fluid in end milling operations on the aluminum alloy AISI 6061-T-6. To obtain a difference in observation, the concentration of nanoparticles was varied between 0% and 1%. It has been observed that when the quantity of nanoparticles is increased, a thin film of SiO_2 develops between the contact surfaces of the particles. It has been observed that this kind of film usually produces a ball bearing effect, which helps to decrease the values of cutting forces, reduces the temperature of the cutting zone, and improves the quality of the machined surface.

They conducted a study on AISI 1040 steel during high-speed turning with multilayers of cemented carbide inserts. The experiment was carried out using nanofluid and included the use of MWCNT. It was noted that the tests were carried out in three distinct cutting environments: a dry environment, a wet environment, and the MQL method of lubricating. It was discovered that the most significant decrease in cutting zone temperature was observed with nanofluid inclusions whose concentrations ranged between 10% and 30%. In addition, the MQL environment was shown to be much superior than the wet cutting environment in terms of performance.

A turning experiment using a MQL method was carried out by Rao et al. (2011) in order to During the machining process, different concentrations of CNT inclusions were utilized in conjunction with regular water as the cutting base fluid. The obtained data clearly demonstrated that tool wear occurs quickly during the first few minutes of machining, but that it becomes steady as the machining process continues. It was discovered that high cutting zone temperatures were linked with carbide tools when compared to HSS tools while cutting in a dry cutting environment, and this was confirmed. The same trend in nodal temperature has been seen as in the case of tool wear when the cutting duration has been increased. Additionally, it has been shown that the temperature decreases with an increase in nanoparticle inclusion. The best results were achieved with a CNT incorporation of 2%. When CNT is included in the tool, the rate of tool wear decreases rapidly up to a 2% concentration.

Saravanakumar et al. (2014) examined a turning operation that was carried out using silver nanoparticles and conventional cutting fluid in a laboratory setting. For the purpose of establishing a comparison between the two cutting fluids, the cooling and lubricating effects of nanofluid and conventional cutting fluid were examined. After examining the results, it was discovered that nanofluid inclusion demonstrates its beneficial aspect by reducing the values of cutting forces and cutting temperatures, as well as producing results with improved surface quality.

Shen et al. (2008) carried out an experiment with Al_2O_3 nanoparticles and nanodiamond-based cutting fluid in a MQL environment to compare the results obtained with dry and wet lubrication, and they found that the results were similar. The experiment was carried out utilizing a cast-iron grinding technique to get the desired results. It was discovered that, when compared to dry and regular cooling machining conditions, nanofluids' environments performed better in terms of reducing grinding forces and improving surface quality. Additionally, the efficient cooling effect of nanofluids' environments inhibits the combustion phenomenon of the workpiece.

5.4 *Effect on Tool Wear*

According to Kalita (2012), the impact of MoS_2 nanoparticles based on two distinct cutting fluids, namely, soybean oil and paraffin oil, was studied, followed by a grinding process utilizing steel and cast iron as the machining materials. According to the findings, the introduction of nanoparticles decreases frictional losses, resulting in a decrease in tool wear as well as a reduction in the specific energy required. Amrita et al. (2014) utilized a soluble oil cutting fluid based on nanographite powder to study the impact of inclusion during the machining of steel using a tool insert made of cemented carbide, according to their findings. In order to evaluate the effectiveness of the machine, cutting forces, flank wear, chip formation, cutting zone temperature, and machined surface quality were measured. It has been discovered that the application of nanofluid produces the highest quality results when compared to dry, regular flood cooling, and the usual MQL cutting environment.

Khandekar et al. (2012) conducted an experiment to assess the cutting performance and wettability characteristics of cutting fluids containing 1% Al_2O_3 NPs, water, and conventional cutting fluid. The experiment was performed on AISI 4340 steel, using an untreated cemented carbide tool, as the work material. It has been determined that adding Al_2O_3 nanoparticles at a concentration of 1% is more helpful in terms of increasing the base fluid's wettability characteristics. Additionally to this benefit, Al_2O_3 nanoparticles-based cutting fluids have shown significant decrease in cutting forces, surface roughness, and tool wear when compared to other machining environments. The experiment by Prasad and Srikant (2013) carried out turning operation on AISI 1040 steel on a lathe machine utilizing HSS and cemented carbide cutting tools. The purpose of this study was to experiment on nano graphite immersion with MQL. Increasing nanoparticle concentrations lead to decreasing cutting forces, tool wear, and surface roughness values, as previously documented. A cemented carbide tool outperforms an HSS tool when it comes to surface quality, tool life, cutting forces, and cutting zone temperature.

6 Conclusion

Despite the increasing adoption of contemporary industrial technology, machining is still one of the most frequently utilized technologies in the metalworking industry, and it may be employed alongside these new technologies. They are traditionally employed for shaping and forming metals to satisfy industry requirements. To enhance machining conditions, the optimum parameters, such as cutting speed, depth of cut, and feed rate, must be selected in order to minimize power consumption while improving the machining process. In a machining operation, the cutting fluid is essential to enhance performance productivity. Cutting fluids often take a significant role for quality and efficient machining because of their lubrication, cooling, and chip removal capabilities. A major concern in sustainable manufacturing is the environmental impact of lubricants. Although dry cutting may be regarded as the ultimate objective for achieving quality machining without detrimental influence on the environment, lubrication remains an industry standard that is hard to overcome when tough alloys are machined.

Acknowledgments We acknowledge the financial support offered by Covenant University in the actualization of this research work for publication.

References

Agarwal, V., & Agarwal, S. (2021). Performance profiling of solid lubricant for eco-friendly sustainable manufacturing, *7*, 22–45.

Ahmad-Yazid, A., Taha, Z., & Almanar, I. P. (2010). A review of cryogenic cooling in high speed machining (HSM) of mold and die steels. *Scientific Research and Essays, 5*(5), 412–427. academicjournals.org

Amrita, M., Srikant, R. R., & Sitaramaraju, A. V. (2014). Performance evaluation of nanographite-based cutting fluid in machining process. *Materials and Manufacturing Processes, 29*(5), 600.

Aramcharoen, P. (2016). Influence of cryogenic cooling on tool wear and chip formation in turning of titanium alloy. *Procedia CIRP, 46*, 83.

Benardos, P. G., & Vosniakos, G. C. (2003). Predicting surface roughness in machining: A review. *The International Journal of Machine Tools and Manufacture, 43*(8), 833.

Boswell, B., Islam, N., & Davies, I. J. (2017). A review identifying the effectiveness of minimum quantity lubrication (MQL) during conventional machining. *The International Journal of Advanced Manufacturing Technology, 3*, 20–144.

Cantero, J. L., Tardío, M. M., Canteli, J. A., Marco, M., & Miguélez, M. H. (2005a). Dry drilling of alloy Ti-6Al–4V. *International Journal of Machine Tools and Manufacture, 45*(11), 1246–1255.

Cantero, J. L., Tardío, M. M., Canteli, J. A., Marco, M., & Miguélez, M. H. (2005b). Dry drilling of alloy Ti-6Al–4V. *International Journal of Machine Tools and Manufacture, 45*(11), 1246–1255.

Chou, Y. K., & Song, H. (2004). Tool nose radius effects on finish turning of hardened AISI 52100 steels. *Journal of Materials Processing Technology, 148*(2), 259.

Chow, H., Lee, S., & Yang, L. (2007). Machining characteristic study of friction drilling on AISI 304 stainless steel. *Journal of Materials Processing Technology, 207*, 180–186.

Dahlman et al. (2009). Effects of rake angle, cutting speed, and cutting depth on residual stresses in hard turning. *Journal of Materials Engineering and Performance, 19(6), 3*, 1–27.

Deepak, S. S. K. (2012). Applications of different optimization methods for metal cutting operation – A review. *Research Journal of Engineering Sciences, 1*(3), 52–58. ISBN 2278 – 9472.

Dhawan, A., Shanker, R., Das, M., & Gupta, K. C. (2011). Guidance for Safe Handling of Nanomaterials. *40*, 22–34.

Dubey, A. (2017). Optimization of cutting parameters by turning operation in lathe machine. *International Journal of Mechanical and Production Engineering, 5*(11), 46–51.

Elmunafi, M. H. S., Kurniawan, D., & Noordin, M. Y. (2015). Use of castor oil as cutting fluid in machining of hardened stainless steel with minimum quantity of lubricant. *Procedia CIRP, 26*, 408–411.

Estrems, M., Germa, S., Sánchez, H. T., Franco, P. (2008). Cutting forces and vibration interactions in slab milling. *Journal of Manufacturing Technology, 2–6*, 100–244.

Gopal, A. V., & Venkateswara Rao, P. (2004). Performance improvement of grinding of SiC using graphite as a solid lubricant. *Materials and Manufacturing Processes, 19*(2), 177.

Gupta, S. M., Güngör, A., Govindan, K., & Özceylan, E. (2020). Responsible & sustainable manufacturing. *International Journal of Production Research, 58*, 2020–2023.

Hocheng, H., & Tsao, C. C. (2005). The path towards delamination-free drilling of composite materials. *Journal of Materials Processing Technology, 167*, 251–264.

Jayal, A. D., & Balaji, A. K . (2009). Effects of cutting fluid application on tool wear in machining : Interactions with tool-coatings and tool surface features. *Wear, 267*(9–10), 1723–1730.

Jiang, Z., Zhou, F., Zhang, H., Wang, Y., & Sutherland, J. W. (2015). Optimization of machining parameters considering minimum cutting fluid consumption. *Journal of Cleaner Production, 23*, 11–22.

Johnson, R. D. J., Wins, K. L. D., Raj, A., & Beatrice, B. A. (2014). Optimization of cutting parameters and fluid application parameters during turning of OHNS steel. *Procedia Engineering, 97*, 172–177.

Kalita. (2012). Application of 2% and 8% mass fractions by adding MoS_2 nanoparticles into paraffin oil and soybean oil in grinding process utilizing steel and cast iron as the machining materials. manufacturing. In Nanomaterials and environmental biotechnology (pp. 333–380). Springer, Cham.

Kannan, S., & Baskar, N. (2016). Modeling and optimization of face milling operation based on response surface methodology and genetic algorithm. *International J Eng. and Tech, 20*, 12–44.

Kara, S., & Li, W. (2011). Unit process energy consumption models for material removal processes. *CIRP Annals – Manufacturing Technology, 60*, 37–40.

Kaushal, A., Vardhan, A., Tiwari, A. C., & Saluja, S. K. (2016). Study the effect of cryogenic cooling on orthogonal machining process. *American Journal of Engineering Research (AJER), 5*(3), 96–104.

Khandekar, S., Sankar, M. R., Agnihotri, V., & Ramkumar, J. (2012). Nano-cutting fluid for enhancement of metal cutting performance. *Materials and Manufacturing Processes, 27*(9), 963–967.

Krishna, P. V., & Nagesawa Rao, D. (2008). Performance evaluation of solid lubricants in terms of machining parameters in turning. *The International Journal of Machine Tools and Manufacture, 48*, 1131.

Kumar, M. (2019). *A study paper of the new trend of the face milling method, IEEE Access 22*, 123–145.

Kundrák, J., Markopoulos, A. P., Makkai, T., Deszpoth, I., & Nagy, A. (2018). Analysis of the effect of feed on chip size ratio and cutting forces in face milling for various cutting speeds. *Manufacturing Technology, 40*, 34–50.

Mane, S., & Kumar, S. (2018). Cutting fluids and cutting fluid application techniques in machining: A review. *Journal of Advanced Manufacturing Technology, 9*(3), 60–63.

Muchendu, P., Githiri, J. G., & Njeri, E. (2014). Determination of optimum drilling parameters using 8.5 inch tricone bits in olkaria geothermal steamfield, Kenya. *IOSR Journal of Applied Geology and Geophysics (IOSR-JAGG), 2*, 11.

Nam, J. S., Lee, P.-H., & Lee, S. W. (2011a). Experimental characterization of micro-drilling process using nanofluid minimum quantity lubrication. *International Journal of Machine Tools and Manufacture., 51*(7–8), 649–652.

Nam, J. S., Lee, P.-H., & Lee, S. W. (2011b). Experimental characterization of micro-drilling process using nanofluid minimum quantity lubrication. *International Journal of Machine Tools and Manufacture., 51*(7–8), 649–652.

Nur, R., Kurniawan, D., Noordin, M. Y., & Izman, S. (2015). Optimizing power consumption for sustainable dry turning of treated aluminum alloy. *Procedia Manufacturing, 23*, 12–23.

Okokpujie, I. P., Bolu, C. A., Ohunakin, O. S., Akinlabi, E. T., & Adelekan, D. S. (2019). A review of recent application of machining techniques, based on the phenomena of CNC machining operations. *Procedia Manufacturing, 34*, 12–18.

Onwubolu, G. C., & Kumar, S. (2006). Response surface methodology-based approach to CNC drilling operations. *Journal of Materials Processing Technology, 22*, 23–39.

Ozcelik, B., Kuram, E., Demirbas, E., & Şik, E. (2013). Effects of vegetable-based cutting fluids on the wear in drilling. *Sadhana, 38*, 687–706.

D. Y. Pimenov, A. T. Abbas, M. K Gupta, I. N. Erdakov, M. S. Soliman, , & M. M. El Rayes, (2020). Investigations of surface quality and energy consumption associated with costs and material removal rate during face milling of AISI 1045 steel The International Journal of Advanced Manufacturing Technology, 107 7–8 3511-3525

Pirtini, M., & Lazoglu, I. (2018). Forces and hole quality in drilling. *International Journal of Machine Tools and Manufacture, 4*(23), 40.

Prasad, M. M. S., & Srikant, R. R. (2013). Performance evaluation of nano graphite inclusions in cutting fluids with mql technique in turning of aisi 1040 steel. *International Journal of Research in Engineering and Technology*. eISSN 2319-1163 I pISSN: 2321-7308.

Priarone, P. C., Robiglio, M., Settineri, L., & Tebaldo, V. (2015). Effectiveness of minimizing cutting fluid use when turning difficult-to-cut alloys. *Procedia CIRP, 29*, 341–346.

Rahmati, B., Sarhan, A. A. D., & Sayuti, M. (2013). Morphology of surface generated by end milling AL6061-T6 using molybdenum disulfide (MoS_2) nanolubrication in end milling machining. *Journal of Cleaner Production, 66*(1), 685–691.

Raju, R. S. U., Satyanarayana, G., Prakash, M. A., & Subba, C. (2020). Materials today: Proceedings wear characteristics of alternative, bio-degradable cutting fluids. *Materials Today: Proceedings, 3*, 23–45.

Ramana, M. V., Rao, G. K. M., & Rao, D. H. (2011). Experimental investigations and selection of optimal cutting conditions in turning of Ti-6Al-4V alloy with different cutting fluids by minimum quantity lubrication (MQL) methodology. *Journal on Mechanical Engineering, 2*(1), 45–52.

Ratnakumar, G., & Rao, T. B. (2013). Modeling and analysis of a plain milling cutter using finite element analysis. *International Journal of Engineering Research and Development, 6*, 80–87.

Sahu, P. K., Sahu, N. K., & Dubey, A. (2017). Optimization of cutting parameters by turning. *International Journal of Mechanical and Production Engineering, 30*, 35–50.

Saravanakumar, N., Prabu, L., Karthik, M., & Rajamanickam, A. (2014). Experimental analysis on cutting fluid dispersed with silver nano particles. *Journal of Mechanical Science and Technology, 28*, 645–651.

Sarhan, A. A. D., Sayut, M., & Hamdi, M. (2012). Reduction of power and lubricant oil consumption in milling process using a new SiO_2 nanolubrication system. *The International Journal of Advanced Manufacturing Technology, 63*, 505–512.

Sayuti, M., Sarhan, A. A. D., Tomohisa Tanaka, M., & Hamdi, Y. S. (2013). Cutting force reduction and surface quality improvement in machining of aerospace duralumin AL-2017-T4 using carbon onion nanolubrication system. *The International Journal of Advanced Manufacturing Technology, **65***, 1493–1500.

Setti, D., Ghosh, S., & Rao, P. V. (2012). Application of nano cutting fluid under minimum quantity lubrication (MQL) technique to improve grinding of Ti – 6Al – 4V alloy. *International Journal of Mechanical and Mechatronics Engineering, 6*(10), 444–666.

Setti, D., Sinha, M. K., Ghosh, S., & Venkateswara Rao, P. (2014). An investigation into the application of Al_2O_3nanofluid–based minimum quantity lubrication technique for grinding of Ti–6Al–4V. *International Journal of Precision Technology, 4*(3-4), 268–279.

Sharma, S. K., & Mital, S. (2016). Preparation and evaluation of stable nanofluids for heat transfer application : A review. *Experimental Thermal and Fluid Science, 5*, 30–50.

Shashidhara, Y. M., & Jayaram, S. R. (2010). Vegetable oils as a potential cutting fluid—An evolution. *Tribology International, 43*, 5–6.

Shen, B., Kalita, P., Malshe, A. P., et al. (2008). Performance of novel MoS_2 nanoparticles based grinding fluids in minimum quantity lubrication grinding. *Transactions of NAMRI/SME, 36*, 357–364.

Shihab, S. K. (2017). An investigation on effects of end milling process on noise during. *Machining of Copper., 5*(22), 30.

Sousa, V. F. C., Silva, F. J. G., Alexandre, R., Fecheira, J. S., & Silva, F. P. N. (2021). Study of the wear behaviour of TiAlSiN and TiAlN PVD coated tools on milling operations of pre-hardened tool steel. *Wear, 23*, 34–60.

Strenkowski, J. S., Shih, A. J., & Lin, J. C. (2002). An analytical finite element model for predicting three-dimensional tool forces and chip flow. *International Journal of Machine Tools and Manufacture, 42*(6), 723–731.

Sultan, A. Z., Sharif, S., & Kurniawan, D. (2015). Chip formation when drilling AISI 316L stainless steel using carbide twist drill. *Procedia Manufacturing, 3*, 33–50.

Talabi, S. I., Owolabi, O. B., Adebisi, J. A., & Yahaya, T. (2016). Effect of cutting parameter on surface roughness carbon steel S45C. *Journal of Material Processing Technology, 84*, 122–129.

Taylor, F. W., Processes, M. T., Press, M., & Engineers, M. (2005). Ch 13.4 machining processes and machine tools. In *Handbook, machining technology processes*. Machining Press. 22:12–34.

Vasu, V., & Reddy, P. K. G. (2011). Effect of minimum quantity lubrication with Al_2O_3 nanoparticles on surface roughness, tool wear and temperature dissipation in machining Inconel 600 alloy. *The Journal of Nanomaterials, Nanoengineering and Nanosystems, 225*(1), 3.

Zailani, Z. A., Hamidon, R., Hussin, M. S., Hamzas, M. F. M. A., Hadi, H., & Sulaiman, S. (2011). The influence of solid lubricant in machining parameter of milling operation. *International Journal of Engineering Science and Technology, 3*(6), 66–70.

Zindani, D., & Kumar, K. (2020). A brief review on cryogenics in machining process products. *SN Applied Sciences, 2*, 1107. Springer 42452-020-2899-5.

Zurita, O. (2017). Effect of cutting parameters on surface roughness in turning of annealed AISI-1020 steel de aço AISI-1020 recozido. *60*, 100.

Nanofluid: A Sustainable Alternative Coolant for Metalworking and Machining Operations

M. Udo, A. A. Noiki, O. O. Yusuf, M. E. Emetere, S. A. Afolalu, and S. O. Ongbali

1 Introduction

Machining plays a substantial role in the manufacturing world today. Irrespective of the fabrication method of an object, machining is necessary for the surface finishing of a product (Zia et al., 2019; Kumar et al., 2021). Machining operations involve removing unwanted or excess material from a workpiece by chipping away until the desired shape and geometry are attained (Salem et al., 2021). Machine operation settings such as depth of cut, feed rate and cutting speed, as well as the cutting tool's geometry and material, and the cutting fluid used determine the overall surface finish of the workpiece (Chou et al., 2002; Kannan and Baskar, 2013). The most common machining operations are turning, milling, drilling and grinding. Kannan and Baskar (2013) defined milling as a machining process involving the use of a milling cutter, a rotating multipoint cutting tool, to create flat, helical and contoured surfaces on a workpiece. In the process of machining, films are created on contact surfaces via the introduction of metalworking fluid, thus providing lubrication and also serving as a cooling medium for dissipation of heat generated (Wickramasinghe et al., 2016; Sadeghi et al., 2008). Consequently, the chips formed during the

M. Udo (✉) · A. A. Noiki · S. A. Afolalu · S. O. Ongbali
Department of Mechanical Engineering, Covenant University, Ota, Nigeria
e-mail: mfon.udo@covenantuniversity.edu.ng; ayodeji.noiki@covenantuniversity.edu.ng; sunday.afolalu@covenantuniversity.edu.ng; samaon.ongbali@covenantuniversity.edu.ng

O. O. Yusuf
Department of Microbiology, Obafemi Awolowo University, Ile-Ife, Nigeria
e-mail: olabisi.yusuf@covenantuniversity.edu.ng

M. E. Emetere
Department of Physics, Covenant University, Ota, Nigeria
e-mail: moses.emetere@covenantuniversity.edu.ng

A. O. Ayeni et al. (eds.), *Advanced Manufacturing in Biological, Petroleum, and Nanotechnology Processing*, Green Energy and Technology,
https://doi.org/10.1007/978-3-030-95820-6_7

machining process are flushed away (Mia et al., 2016). Shokoohi (2015) depicted cutting fluids as a channel for reducing the surface contact between the workpiece and the tool, thereby inhibiting internal friction throughout the machining operations. Heat generated from the chips, workpiece and tools is being removed via the cutting fluid, thereby preventing temperature build-up. Manikanta et al. (2018) classified cutting fluids into two, namely, metallic nanofluids and non-metallic cutting fluids. The metallic nanofluids contain metallic nanoparticles like aluminium, copper, silver, gold, tin, silicon, iron, zinc and nickel. The non-metallic nanofluids are those containing non-metallic nanoparticles, some of which include SiC, CuO, ZnO, Al_2O_3 and TiO_2. According to Chou et al., cutting fluids possess the ability to prevent cutting tool from being heated to a certain temperature whereby its hardness is reduced. They should also serve as corrosion inhibitors, thereby protecting the piece from rusting and improving the surface finish of the workpiece. Cutting fluids significantly reduce tool wear rate by counteracting the heat dissipation challenges that arise during machining (Afolalu et al., 2018a). They prevent the formation of built-up edges on the cutting tool and aid satisfactory chip formation. Cutting fluids remove chips from the cutting zone, especially during deep hole drilling, milling and grinding operations (Chou et al., 2002). The power used up in cutting a piece is converted mostly into heat which is carried away by the chip. The rest of the heat is shared by the cutting tool and the workpiece. Therefore, an increment in either the cutting speed or feed will lead to a rise in the heat generated by the machining operation. On average, the temperature generated from the interface of the tool during metal cutting ranges from 800 °C to 1100 °C (Chou et al). During machining processes, cutting fluids with nanoparticles tend to provide more cooling and better lubrication, thereby improving the cutting operation and overall workflow (Chou et al., 2002). The microstructure and chemical composition of the material and the cutting fluid, respectively, can influence the overall surface finish of a workpiece. Introducing millimetre or micrometre-sized solid particles to a base fluid enhances its thermo-physical properties. Nevertheless, owing to the large molecular size of these particles which often result into surface clogging, thereby causing abrasiveness. This limitation makes them unfit for practical use. In recent times, development of nanofluids has been made possible through numerous research efforts and technological advancement (Wickramasinghe et al., 2016).

2 Roles of Cutting Fluids in Machining

The manufacturing sector is one of the major energy-consuming sectors on earth. Machining operations need to be optimized to minimize the energy consumption rate for manufacturing processes. Improved operational efficiency of machining process tools will guarantee a reduction in the energy consumption rate (Hanif et al., 2017). There are several ways through which materials are manufactured such as forging, casting and extrusion. However, these manufacturing methods create workpieces without precise dimensional tolerances. Machining produces workpieces with higher accuracy and increases overall productivity (Afolalu et al., 2018b). The utilization of cooling and lubricating properties of cutting fluids is of immense benefit to metalworking operations, for example, turning, drilling, grinding and milling (Moradnazhad & Unver, 2016). Furthermore, cutting fluids enhance the washing away of chips generated due to material removal. These fluids also play a crucial role in determining the effectiveness of the overall machining processes and consequently, improving the efficiency of the systems in mention. The use of cutting fluids contributes about 12–17% of the total machining costs (Afolalu et al., 2018b). Although metalworking fluids with nanoparticles are generally more expensive than conventional cutting fluids, the benefits, in the long run, outweigh the initial costs. The lubrication and cooling effects they provide can lead to better returns compared to using conventional cutting fluids. Cutting fluids are mainly classified into oil-based and water-based cutting fluids. Water-based cutting fluids have comparative advantage over the oil-based cutting fluids for cooling operations. Contrarily, oil-based cutting fluids are more appropriate in areas of applications where lubrication is vital consideration. Water-based metalworking fluids are categorized into the following: synthetic fluids, emulsifiable oils or semi-synthetic fluids. Oil-based fluids encompass mineral oils, synthetic oils and vegetable oils. For operations where high cooling abilities for high-speed machining are required, for instance, grinding and honing, water-based cutting fluids are highly recommended. Machining operations require low speeds such as broaching, reaming, gun drilling and tapping due to their low cooling abilities (Afolalu et al., 2018b). The volume of cutting fluid used determines its overall cooling effect as well as the base fluid characteristics. The method used in applying the cutting fluid as well as its chemical composition lubrication effects the cutting fluid will provide. The technique used in applying the cutting fluid also determines the amount of chip removed during a machining operation. Subramani et al. (2013) noted that several sustainable measures which use lubri-cooling techniques had been put in place to serve as alternatives to the use of conventional cutting fluids, some of which include minimum quantity lubrication (MQL), nanofluid-assisted machining and cryogenic machining. These techniques are adopted because they are controlled means that reduce the risks on the operator's health and the environment. The Environmental Protection Agency highly recommends them. MQL machining operates on the principle of micro lubrication, where a mist of lubricant mixed with compressed air targets the specific points that the tool interfaces with the chips formed for efficient cooling and lubrication. This technique helps to save costs and improve machinability as the cutting fluid is only applied where it is needed as opposed to flood cooling. Ramakrishnan et al. investigated the effects MQL during a turning operation compared to dry machining and flood cooling considering the quality of the surface finish as well as wear rate of the tool flank. The piece turned via the MQL machining technique yielded better surface finishing and less wear. To further improve the performance of MQL machining, for hard to machine materials, methods such as nanofluid-assisted MQL machining can be adopted. The effect of nano-additives in machining manifests in a reduction of energy consumption, temperatures and surface roughness, amplified tribology, surface finish, polishing, heat transfer rate and ball-bearing effect. Metalworking fluids enhanced with nanoparticles have superior heat transfer and friction reduction characteristics leading to lower overall heat generation for the processes involved (Hanif et al., 2017). Agapiou (2018) stated that cutting fluids created using nanoparticles are said to degrade easier than the conventional counterparts (biodegradable) and are generally safer and pose fewer health challenges to the machine operators

2.1 *Impact of Cutting Fluids on Cutting Tool and Workpiece*

During metalworking operations such as milling, grinding and especially deep hole drilling, cutting fluids are used to remove chips formed from the metal away from the cutting zone (Chou et al., 2002). The utilization of nano-metalworking fluids leads to a depletion of the cutting forces for machining processes such as drilling, turning, milling and grinding due to the effect of the characteristics of the nanoparticles used (Moradnazhad & Unver, 2016). Cutting fluids prevent and reduce tool wear, corrosion and built-up edges on cutting tools reducing the time spent on machining a piece and therefore saving costs. They are used during machining operations in large amounts to ensure proper lubrication, cooling and chip removal, and this plays a role in determining the finishing of the workpiece (Afolalu et al., 2018b). The main parameters that influence cutting force are feed rate, depth of cut and cutting speed in machine operations. These parameters determine the tool positioning and dynamics. The cutting forces can be reduced by the presence of cutting fluids and other cooling conditions. Subramani et al. (2013) observed that an increase in cutting speed and feed rate increases the cutting forces since more heat is generated as an outcome of the increments, thereby causing hardening of the workpiece. Tool wear occurs after subsequent machining operations due to the material of the workpiece as well as its surface. This phenomenon eventually leads to tool failure (Agapiou, 2018) (Fig. 1).

Tool failure arises because of friction between surfaces of the workpiece and the tool. Cutting fluids are introduced to dissipate the heat that arises due to the said

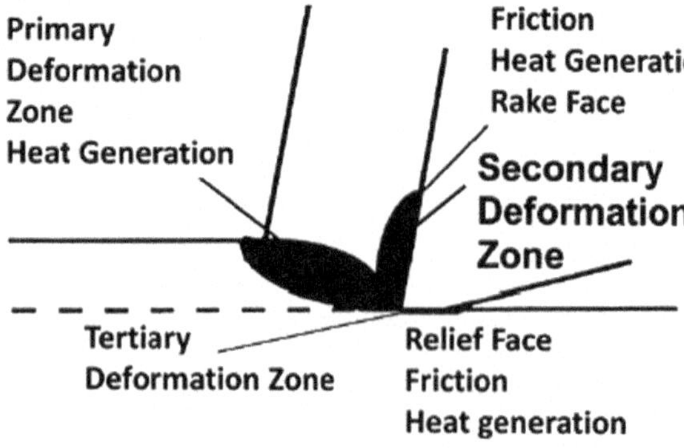

Fig. 1 Heat generation at the cutting tool and workpiece interface

friction. Chou et al. (2002) stated that cutting fluids should be capable of preventing the tool from being heated to a temperature where its hardness will reduce. In machining, the cutting tool is to be harder than the workpiece in mention. Controlling the temperatures of the chips, tool and workpiece generated during machining would help achieve that. There are a variety of parameters that affects the tool life which are subjected to the cutting fluid. These include the flow rate of the cutting fluid, application technique and the cutting fluid pressure. In order to minimize operational cost and save energy, the tool wear rate should be minimal. More so, the tool life has direct effects on cost of machining a workpiece; therefore, proper setting of the machining parameters will help in maximizing tool life. The surface quality of the material produced by the nano-cutting fluid is superior compared to that of conventional cutting fluids.

3 Nanotechnology and Its Applications in Machining Operations

Nanotechnology is a technology that operates on a nanoscale (spanning individual atoms/molecules to submicron) and has several applications in the real world. Nanotechnology integrates the resulting nanostructures into larger systems (Kadirgama, 2021). Nanofluids are created by dispersing nanomaterials in solid form into a carrier fluid, usually oil, ethylene glycol or water. For this application, the main reason for doing so is to improve the heat transfer capabilities of the fluid by increasing the thermal conductivity (Ramakrishnan, et al., 2019). A nanofluid is one that contains nanometre-sized components (less than 100 nm) called nanoparticles that are dispersed/suspended in a carrier fluid to enhance select characteristics of the fluid based on the intended use (Ajayi et al., 2019b). The nanoparticles mentioned above can be either metals, oxides of metals or non-metals, nanotubes and nanofibres. According to Kadirgama (2021), nanofluids have far more excellent thermal conductivities than ordinary fluids. For instance, the thermal conductivity of fluid increases by adding nanoparticles like copper, whose thermal conductivity is about 700 times greater than water, as opposed to using the fluid without nanoparticles. Nano-cutting fluids are made by adding select nanoparticles to conventional cutting fluids to alter its thermo-physical properties as well as improve its wettability and lubricating properties (Manikanta et al., 2018). According to Hanif et al. (2017), there are four types of nanofluids based on the composition of their nanoparticles and their source. These are carbon-based, pure metallic, alloy and ceramic nanofluids. Ceramic nanofluids consist of non-oxides such as silicide, oxides such as alumina and composites made up of both. In the early stages of nanotechnology, ceramic nanoparticles were used widely in the manufacture of nanofluids. Ceramic nanofluids enhance the thermal conductivity of their base fluids. Upon the addition of Al_2O_3 nanoparticle to water, the thermal conductivity increased by up to 30% (Venkatesh et al., 2018; Shekarian et al., 2014; Bhushan, 2010). This study focuses on utilizing nanofluids as machine cutting fluids and their influence on machining operations (cutting, milling, turning, drilling and grinding) as well as a measure of their cooling and lubricating effects. For the sake of this project, only water-based cutting fluids are considered because the nanofluid to be developed would serve as a coolant. Cutting fluids are used for metalworking and machining operations to improve the quality of the produced materials, enhance tool life and improve the overall machining process (Shekarian et al., 2014).

3.1 *Nano-cutting Fluids in Machining Operations*

Nano-cutting fluids are a new class of cutting fluids created by combining nanomaterials such as nanoparticles, nanorods and nanotubes with a base fluid (Brinksmeier et al., 2015; Zamani et al., 2019; Ajayi et al., 2019). These nanomaterials can be synthesized from agricultural waste as they have proven to be cheaper and environmentally sustainable as opposed to synthesis using expensive and toxic substances. Nano-cutting fluids (or metalworking fluids) are cutting fluids with enhanced characteristics due to the addition of nanoparticles. The main reason for utilizing them is to improve the heat transfer capabilities of the fluid by increasing the thermal conductivity. During machining, nanoparticles from the cutting fluid deposit on the

Table 1 Commonly used nanoparticles for machining operations (Shekarian et al., 2014; Venkatesh et al., 2018)

S/N	Nanoparticle	Machining operation	References
1	Molybdenum disulphide (MoS_2)	Milling	Shekarian et al. (2014)
2	Multi-walled carbon nanotubes (MWCNT)	Grinding, turning	Ajayi et al. (2019)
3	Single-walled carbon nanotubes (SWCNT)	Grinding, turning	Ramakrishnan et al. (2019)
4	Aluminium oxide (Al_2O_3)	Grinding, turning, hobbing	Subramani et al. (2013)
5	Aluminium nitride (AlN)	Turning	Subramani et al. (2013)
6	Titanium dioxide (TiO_2)	Surface	Ajayi et al. (2019)
7	Copper oxide (CuO)	Facing operation, turning	Venkatesh et al. (2018)
9	Silicon dioxide (SiO_2)	Facing, grinding	Ramakrishnan et al. (2019)
10	Zirconium dioxide (ZrO_2)	Turning	Ajayi et al. (2019)
11	Hexagonal boron nitride (hBN)	Grinding	Venkatesh et al. (2018)
12	Graphite nano-platelets (GnP)	Micro milling	Bhushan (2010)
13	Nano-boric acid (NBA)	Turning	Shekarian et al. (2014)
14	Nano-diamond (ND)	Grinding	Agapiou (2018)

tool, workpiece and chip surfaces. These nanoparticles serve as bearings as they lubricate and dissipate the heat generated during machining. Although conventional nano-cutting fluids have been worked upon to reduce the environmental hazards, they still can be improved upon. In line with enhancing the eco-friendliness of nano-cutting fluids, their nanoparticles can be synthesized from agricultural waste. The table below shows the nanoparticles commonly used for the synthesis of nano-cutting fluids and the respective machining operations (Padmini et al., 2016) (Table 1).

3.2 *Development of Silica Nanofluids for Cooling Applications*

Silica is a material utilized in various industries including but not limited to the manufacturing industry, the cement industry and the ceramics industry, to name a few. Due to its wide range of applications, researchers are seeking ways to extract high-quality silica from various sources using cost-effective and eco-friendly measures (Hossain et al., 2018). It was noticed that silica nanoparticles are favoured for their high surface area to volume ratio, excellent chemical stability and also because their toxicity is low and they are flexible for a range of applications (Brinksmeier et al., 2015). Traditionally, silica nanoparticles manufactured for industrial use are made from alkaline silicates. These silicates are gotten from a series of reactions involving carbonates and silica sand; the silicates are precipitated to produce porous silica (Ghorbani et al., 2015). Materials such as tetraethylorthosilicate are useful for nanotechnological applications such as for the development of silica-based nanofluids. Unfortunately, the cost of manufacturing tetraethylorthosilicate is expensive compared to the cost of development while making use of agro-waste (Zia et al., 2019). Silica nanoparticles can be prepared by various means, one of which involves smelting sodium carbonate and sand quartz to produce sodium silicate. However, the majority of the processes undergone in preparing SNP from methods like this consume a significant amount of energy and contribute significantly to environmental pollution. Preparation of silica particles from rice husk ash uses an environmentally friendly technique. The silica nanoparticles produced gave off vibration signals around 1075, 780 and 665cm^{-1} from the FTIR analysis, which is an occurrence with Si-O-Si bands which signifies the successful development of silica nanoparticles (Vaibhav et al., 2015; Abioye et al., 2019a). From the FTIR analysis, the nanoparticles produced had an average pore size of about 10.98 nm. The SEM analysis of the rice husk ash samples, as well as the silica nanoparticles, developed showed that the silica particles formed were spherical with irregular fibre-like structures and had an average particle size ranging from 200 nm to 20 nm. It was noted that nanofluids could be utilized for their cooling properties in the following applications: vehicles, electronics, transformers, nuclear systems and cutting fluids (Abioye et al., 2019b).

4 Conclusion

Cutting fluids play a vital role in determining the smooth flow of the overall machining operation and increasing the efficiency of the cutting process. Cutting fluids are utilized for their cooling and lubricating properties which are beneficial for metal-working operations such as drilling, milling, grinding and turning. Nanofluids are gaining the attention of numerous researchers due to its various applications in several industries, most notably in the production industry due to their excellent cooling properties and superior heat transfer performance. Although conventional nano-cutting fluids have been worked upon to reduce their environmental hazards, they can be further improved. In line with improving the eco-friendliness of nano-cutting fluids, synthesizing their nanoparticles from agricultural waste is an available option. The main reason for utilizing them is to improve the heat transfer capabilities of the fluid by increasing the thermal conductivity. However, nano-cutting fluid is an area yet to be explored, despite its role in optimizing machining operations.

Acknowledgements We acknowledge the financial support offered by Covenant University in the actualization of this research work for publication.

References

Abioye, A. A., Rotimi, D. O., Fasanmi, O. O., Abioye, O. P., Afolalu, S. A., Owoeye, T. F., & Obuekwe, C. C. (2019a, December). Synthesis and characterization of selected starch nanoparticles as matrix reinforcements for low density polyethylene. *Journal of Physics: Conference Series, 1378*(4), 042070.

Abioye, A. A., Fasanmi, O. O., Rotimi, D. O., Abioye, O. P., Obuekwe, C. C., Afolalu, S. A., & Okokpujie, I. P. (2019b). Review of the development of biodegradable plastic from synthetic polymers and selected synthesized nanoparticle starches. *Journal of Physics: Conference Series, 1378*(4), 042064.

Afolalu, S. A., Abioye, A. A., Udo, M. O., Adetunji, O. R., Ikumapayi, O. M., & Adejuyigbe, S. B. (2018a). Data showing the effects of temperature and time variances on nano-additives treatment of mild steel during machining. *Data in Brief, 19*(2018), 456–461.

Afolalu, S. A., Asonaminasom, E. H., Ongbali, S. O., Abioye, A. A., Udo, M. O., & Salawu, E. Y. (2018b). Dataset on experimental investigation of optimum carburizing temperature and holding time of Bi-Nano additives treatment of AISI 5130 steel. *Data in Brief, 19*(1), 2279–2283.

Agapiou, J. (2018). Performance evaluation of cutting fluids with carbon nano-onions as lubricant additives. *Procedia Manufacturing, 26*, 1429–1440.

Ajayi, O. O., Okolo, T. I., Salawu, E. Y., Owoeye, F. T., Akinlabu, D. K., Akinlabi, E. T., & Afolalu, S. A. (2019a). Performance and energy consumption analyses of R290/bio-based nano lubricant as a replacement for R22 refrigerant in air-conditioning system. In *Energy technology 2019* (pp. 103–112). Springer.

Ajayi, O. O., Aba-Onukaogu, C. C., Salawu, E. Y., Owoeye, F. T., Akinlabu, D. K., Afolalu, S. A., & Abioye, A. A. (2019b). Effect of biomaterial (Citrullus Lanatus Peels) nanolubricant on the thermal performance and energy consumption of R600a in refrigeration system. In *Energy technology 2019* (pp. 91–102). Springer.

Bhushan, B. (2010). *Introduction to nanotechnology in Lecture Note in Mechanical- Springer handbook of nanotechnology, 33*(6), 1100–1177.

Brinksmeier, E., Meyer, D., Huesmann-Cordes, A. G., & Herrmann, C. (2015). Metalworking fluids—Mechanisms and performance. *CIRP Annals, 64*(2), 605.

Chou, Y. K., Evans, C. J., & Barash, M. M. (2002). Experimental investigation on CBN turning of hardened AISI 52100 steel. *Journal of Materials Processing Technology, 124*(3), 274–283.

Dizaji, H. B., Zeng, T., Hartmann, I., Enke, D., Schliermann, T., Lenz, V., & Bidabadi, M. (2019). Generation of high-quality biogenic silica by combustion of rice husk and rice straw combined with pre-and post-treatment strategies—A review. *Applied Sciences, 9*(6), 1083.

Ghorbani, F., Sanati, A. M., & Maleki, M. (2015). Production of silica nanoparticles from rice husk as agricultural waste by environmental friendly technique. *Environmental Studies of Persian Gulf, 2*(1), 56–65.

Hanif, G., Saxena, G., & Raj, J. (2017). A critical review on applications of nano-fluid as coolant. *International Journal of Engineering and Management Research, 7*(1), 304–311.

Hossain, S. S., Mathur, L., & Roy, P. K. (2018). Rice husk/rice husk ash as an alternative source of silica in ceramics: A review. *Journal of Asian Ceramic Societies, 6*(4), 299–313.

Kadirgama, K. (2021). A comprehensive review on the application of nanofluids in the machining process. *The International Journal of Advanced Manufacturing Technology*, 1–13.

Kannan, S., & Baskar, N. (2013). Modelling and optimization of face milling operation based on response surface methodology and genetic algorithm. *International Journal of Engineering and Technology, 5*, 4164–4175.

Kumar, R., Singh, S., Bilga, P. S., Jatin, K., Singh, J., Singh, S., & Pruncu, C. I. (2021). Revealing the benefits of entropy weights method for multi-objective optimization in machining operations: A critical review. *Journal of Materials Research and Technology.*

Manikanta, J. E., Gupta, B. N., Manikanta, J. T., & Pradeep, R. (2018). Integrality characterization of machining with nano cutting fluids. *International Journal of Mechanical Engineering and Technology, 9*, 1033–1042.

Mia, M., Khan, M., Rahman, S., & Dhar, N. (2016). Mono-objective and multi-objective optimization of performance parameters in high pressure coolant assisted turning of Ti-6Al-4V. *The International Journal of Advanced Manufacturing Technology, 90*(1-4), 109–118.

Moradnazhad, M., & Unver, H. (2016). Energy efficiency of machining operations: A review. *Proceedings of the Institution of Mechanical Engineers, Part B: Journal of Engineering Manufacture, 231*(11), 1871–1889.

Padmini, R., Krishna, P. V., & Rao, G. K. M. (2016). Effectiveness of vegetable oil based nanofluids as potential cutting fluids in turning AISI 1040 steel. *Tribology International, 94*, 490–501.

Ramakrishnan, H., Balasundaram, R., Selvaganapathy, P., Santhakumari, M., Sivasankaran, P., & Vignesh, P. (2019). Experimental investigation of turning Al 7075 using Al2O3 nano-cutting fluid: ANOVA and TOPSIS approach. *SN Applied Sciences, 1*(12), 112–120.

Sadeghi, M., Haddad, M., Tawakoli, T., & Emami, M. (2008). Minimal quantity lubrication-MQL in grinding of Ti–6Al–4V titanium alloy. *The International Journal of Advanced Manufacturing Technology, 44*(5-6), 487–500.

Salem, A., Hegab, H., & Kishawy, H. A. (2021). An integrated approach for sustainable machining processes: Assessment, performance analysis, and optimization. *Sustainable Production and Consumption, 25*, 450–470.

Shekarian, E., Tarighaleslami, A., & Khodaverdi, F. (2014). Review of effective parameters on the nanofluid thermal conductivity. *Journal of Middle East Applied Science and Technology, 15*(4), 776–780.

Shokoohi, Y., Khosrojerdi, E., & Shiadhi, B. R. (2015). Machining and ecological effects of a new developed cutting fluid in combination with different cooling techniques on turning operation. *Journal of Cleaner Production, 94*, 330–339.

Subramani, N., Mohan, A., & Prakash, M. (2013). Performance studies on a vapour compression refrigeration system using nano-lubricant. *International Journal of Innovation, Research, Science, Engineering and Technology, 2*(1), 522–530.

Vaibhav, V., Vijayalakshmi, U., & Roopan, S. M. (2015). Agricultural waste as a source for the production of silica nanoparticles. *Spectrochimica acta part A: Molecular and Biomolecular Spectroscopy, 139*, 515–520.

Venkatesh, K., Sriram, G., Reddy, A. U. K., & Nikhil, S. (2018). Influence of nano based metal working fluids on machining process – A review. *IOP Conference Series: Materials Science and Engineering, 390*, 012094.

Wickramasinghe, K., Perera, G., & Herath, H. (2016). Formulation and performance evaluation of a novel coconut oil–based metalworking fluid. *Materials and Manufacturing Processes, 32*(9), 1026–1033.

Zamani, A., Marjani, A. P., & Mousavi, Z. (2019). Agricultural waste biomass-assisted nanostructures: Synthesis and application. *Green Processing and Synthesis, 8*(1), 421–429.

Zia, M., Pervaiz, S., Anwar, S., & Samad, W. (2019). Reviewing sustainability interpretation of electrical discharge machining process using triple bottom line approach. *International Journal of Precision Engineering and Manufacturing-Green Technology, 6*(5), 931–945.

The Essence of Intermetallic Phases in AA6061/Clay Composites

N. E. Udoye, O. S. I. Fayomi, and A. O. Inegbenebor

1 Introduction

Metal matrix composites (MMCs) are utilized immensely in the industry that requires lightweight with stiffness, super-specific strength, ductility and resistance due to heat (Basil Quent et al., 2018). For decades, SiO_2, MgO, SnO_2, CuO, TiO_2, etc. are some of the ceramic oxides commonly used as reinforcements. The fly ash, fibres and whiskers are the limited particulates of agro-industrial waste used recently for support (Joseph & Babaremu, 2019). These reinforcing particles improve the strength and corrosion properties of aluminium alloy. The requirement to produce an economically gorgeous composite has triggered the numerous inventions in the production technique used in the composite industry. AA6061 has rough particles and large needle-like eutectic silicon that are responsible for low mechanical properties (Baskaran et al., 2019). The interest in the upgrading of MMCs is motivated by its mechanical applications for light materials with an exceptional quality, malleability and thermal resistance (Kumar et al., 2018). The irresistible search for cheaper reinforcement triggered the urge towards building

N. E. Udoye (✉) · A. O. Inegbenebor
Department of Mechanical Engineering, College of Engineering, Covenant University, Ota, Lagos state, Nigeria
e-mail: nduka.udoye@covenantuniversity.edu.ng; anthony.inegbenebor@covenantuniversity.edu.ng

O. S. I. Fayomi
Department of Mechanical and Biomedical Engineering, Bells University of Technology, Ota, Ogun State, Nigeria

Department of Chemical, Metallurgical and Materials Engineering, Tshwane University of Technology, Pretoria, South Africa
e-mail: osfayomi@bellsuniversity.edu.ng

A. O. Ayeni et al. (eds.), *Advanced Manufacturing in Biological, Petroleum, and Nanotechnology Processing*, Green Energy and Technology,
https://doi.org/10.1007/978-3-030-95820-6_8

and utilizing the agro-industrial waste product as reinforcement since they are readily accessible, sustainable and cheaper to acquire (Fayomi et al., 2016). The physical, mechanical and chemical characteristics of the material are enhanced as the aluminium alloy is strengthened with particulate of a second stage or it may save a lot of elements with minimal harmful effect to the properties desired. Clay is robust and low cost, readily available and environmentally friendly. Clay is a good substitute in replacing the high cost of reinforcement, because of its accessibility and its major elements in the form of Al_2O_3, SiO_2, Fe_2O_3, TiO_2 and Na_2O. Clay could also be a substitute to SIC and Al_2O_3 as a result of their cheaper rate and hardness (Rao et al., 2009). Clay is a prospective material as reinforcement for the fabrication of MMC for wear usage. In the present study, AA6061/clay composites were used to improve the material's properties. The intermetallic phases obtained from the process have nanocrystalline that has the aluminate made in the outermost layer of the composite emanating from the oxidized progression (Ossowski et al., 2008). The AA6061/clay composites equipped with the aluminium silicate, silicon IV oxide and aluminium oxide intermetallic phases have a greater wear susceptibility than the aluminium alloy (Gupta et al., 2018). The intermetallic particles with liquefied aluminium are necessary for obtaining perfect interfacial bonding. Aluminium silicate, silicon IV oxide and iron III oxide are intermetallic phases appropriate for reinforcement base for AMCs. AMCs fortified with clay particulates have been manufactured through liquid metallurgy routes, and characterizations were carried out as noticed in literature (Balakrishnan et al., 2020).

The aim of the study was to determine the effective performance of intermetallic phases in the reinforcement of AA6061 aluminium alloy.

2 Materials and Methods

2.1 Material Selection

In this work, aluminium billet, which is the base material, was sourced from aluminium rolling mill (ARM), Ogun Housing Estate, Ota, Ogun State. Ota is located within longitudes 2°53 E and 3°14 E and latitudes 6°39 N Southwestern part of Nigeria (Ogunyemi et al., 2017). The as-received samples are high-quality AA6061 aluminium alloy because of their structural responses in service. AA6061

Table 1 Chemical configuration of AA6061 (wt%)

Element	Al	Mg	Si	Fe	Cu	Zn	Mn	O
Composition	85.0	3.3	2.25	2.13	1.5	0.25	0.12	5.0

was selected because of the inherent properties it displayed. It comprises magnesium and silicon as its foremost alloying elements, which determine their mechanical properties (Ahamed et al., 2016). The chemical configuration is displayed in Table 1

2.1.1 Clay Kaolinite

The clay kaolinite ($Al_2Si_2O_5(OH)_4$) used was obtained from Abule Onikosin Road, Abeokuta area of Ogun State. Abeokuta is located between latitude 7°30′N and longitude 3°54′E Southwestern part of Nigeria (Ufoegbune & Fabiyi, 2016). The clay kaolinite is whitish, having slight red colour impurity due to iron III oxide and blue/brown colour due to presence of other minerals. Figure 1 shows the as-received clay.

2.2 Sample Treatment

The tiny stones were separated from the broken dried kaolinite clay. Then, the clay was passed through wet ball milling in a high-energy ball milling machine for 5 h and transformed into a slurry. The obtained mixture of water and solid particles was passed through a 75 μm and 150 μm mesh sieve. The sieved blend of water and solid particles was oven-dried until negligible moisture content was realized. Thereafter, the dried blend was grounded into powder. The flow chart of the processing of clay is shown in Fig. 2.

Fig. 1 As-received clay

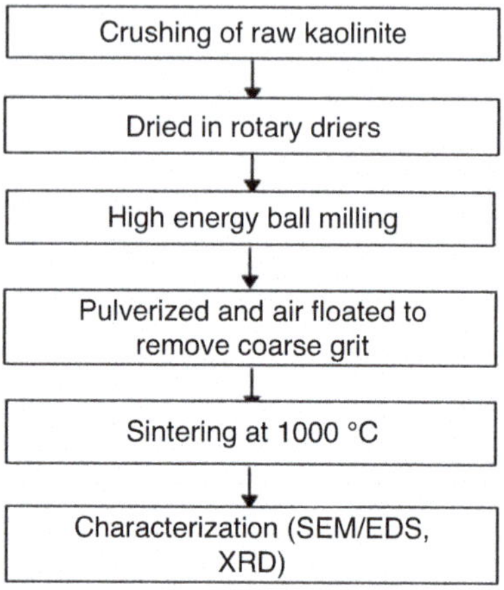

Fig. 2 The flow chart for the processing of clay

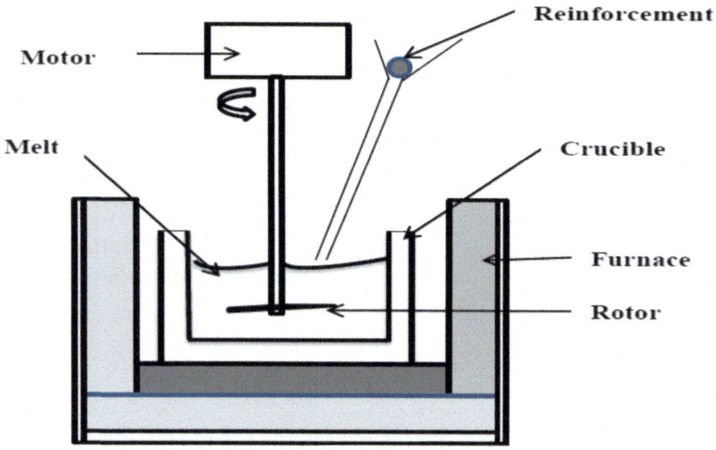

Fig. 3 Stir casting technique of AA6061/clay composites

2.3 Stir Casting Process of AA6061/Clay Composites

Firstly, 3 kg of AA6061 pieces are fed into the red hot heated electric furnace, and the temperature was elevated to 750 °C until the aluminium melts totally. Slag was extracted using scum powder to generate a good standard of the melt. The essential quantity of clay is preheated to 600 °C in the isolated oven and maintained for 20 min to eradicate moisture content. The melted material was poured into a crucible in the separate stirring set-up. The picture of the stir cast set-up is shown in Fig. 3. The mechanical stirrer was inserted into the melt and slowly stirred to maintain a rotary motion. Then, the preheated clay was introduced into

Table 2 Reinforcement weight parameters

Sample designation	Wt of clay (%)	Wt of AA6061 (%)	Particle size (μm)
AA 6061	0	100	–
AA 6061 + 2% clay	2	98	75
AA 6061 + 8% clay	8	92	75
AA 6061 + 2% clay	2	98	150
AA 6061 + 8% clay	8	92	150

the molten alloy at 720 °C. Maintain stirring for another 5 min until the even dispersal of reinforcing agent in the molten matrix was completed. The mould was preheated to prevent contraction of the cast sample, and the developed molten sample was removed and transferred at 680 °C into the moulds of 250 mm × 25 mm size cylindrical hole instantly. After total solidification, the product was extracted from the mould. The same process was reiterated for outstanding cast samples with diverse reinforcement. It is appropriate for the production of composites having 0–30% volume fractions of reinforcement and uses stirrer to stir the molten metal. The graphite stirrer is the main component in stir casting, and it consists of a substance, which can be used for a superior melting temperature than the matrix (Pal et al., 2015).

2.4 *XRD Analysis of AA6061/Clay Composites*

The XRD was used to classify single crystals, and to disclose its structure. The reinforcement of AA6061 with clay for all produced composites and the weight percentage composition is shown in Table 2.

2.5 *Microhardness Tester of AA6061/Clay Composites*

The technique computes the hardness by calculating the height of an indenter's insertion under a bulky load compared to the preload penetration. The detained surfaces happen in ranges, which are precisely monitored to four controlled axes. The microhardness was valued at 100 g load for 15 s. The impression gotten is calculated with a Brinell microscope between two diameters, mainly perpendicular to each other, and the results are averaged (d) (Hossain et al., 2017).

2.6 *Ultimate Tensile Strength of AA6061/Clay Composites*

The INSTRON 3369 universal testing capacity of 100 KN was used to compute the developed material's capacity. The diameter and gauge length of the sample were 10 mm and 170 mm, respectively, sourced from the cast samples. This machine operates when the sample is situated in between the grips of the machine. The machine records the force after the tensile test and occurs as the load is applied to the specimen. The elongation is computed for each sample using the capacity of 100 kN.

3 Results and Discussions

3.1 *Mechanical Properties of AA6061/Clay Composites*

Figure 4 represents the microhardness (BHN) for AA6061/clay composites in both 75 μm and 150 μm particle sizes. It showed the highest hardness value of 175 BHN as a result of reinforcement under different working condition for performance evaluation. This increase in sample hardness was due to change in silicon morphology and existence of aluminium silicates from clay particles compared to other samples such as plastics and fibre-reinforced plastics that melt after working long hours. The hardness values of the developed composites increased as the %wt clay addition improves in the alloy. From the result, the fabricated alloys increased substantially in hardness from 152 BHN to around 175 BHN at 15.1%. The existence of strong ceramic phase in the matrix has increased the hardness of the composite. It increases from 167 to 175 BHN at 75 μm particle size and from 168 to 173 BHN at 150 μm particle size for 2% wt and 8% wt clay particles, respectively. These increases were

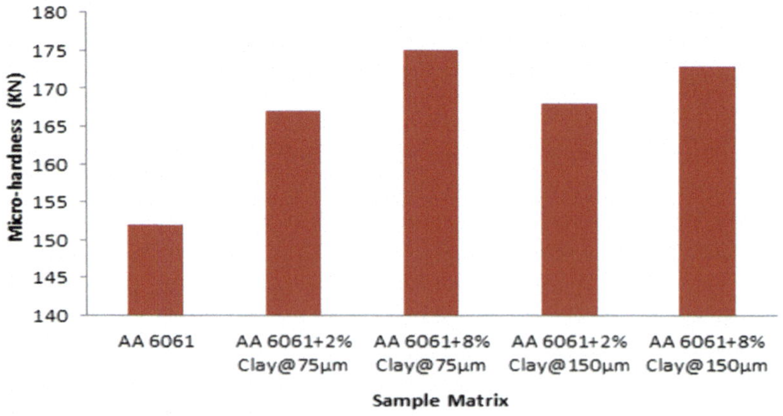

Fig. 4 Microhardness chart for AA6061/clay composites

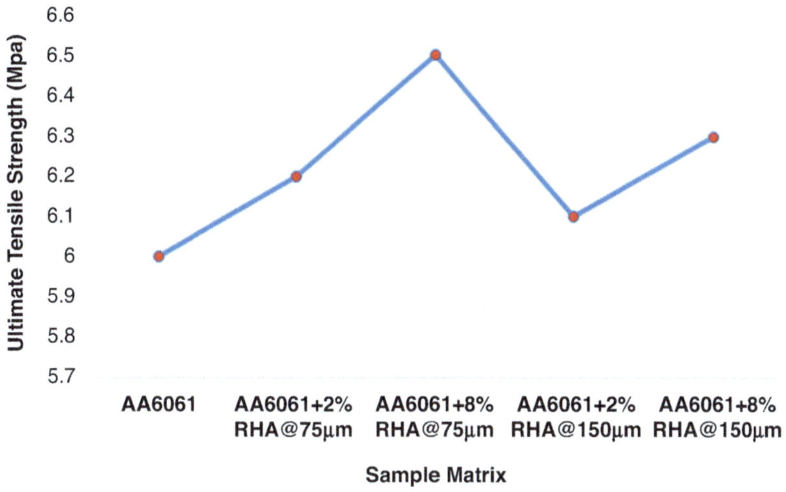

Fig. 5 Ultimate tensile strength chart for AA6061/clay composites

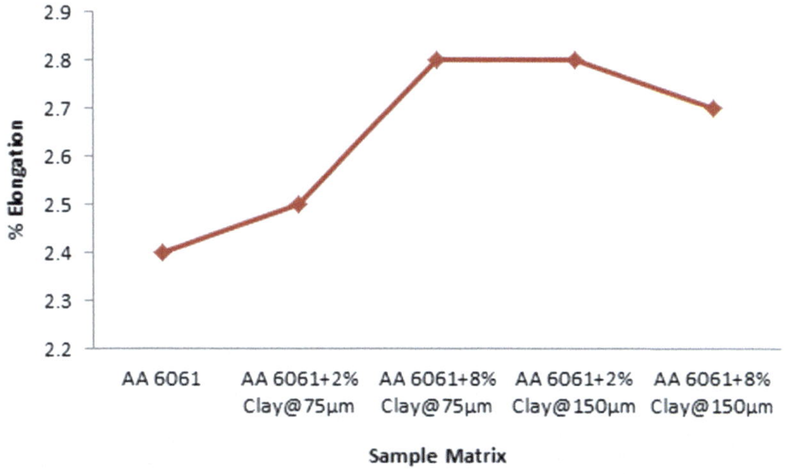

Fig. 6 Effect of clay particles on percentage elongation

due to enhancement in weight percentage of the solid phase of the clay particles in the aluminium alloy.

Figure 5 shows the UTS of the reinforced alloys for AA6061/clay composites in both 75 µm and 150 µm particle sizes. The UTS of the reinforced aluminium alloy increased considerably from 6 MPa starting aluminium alloy to almost 6.5 MPa for the developed alloy at 8.3%. The crystal phase presence in the ductile matrix has given rise to the increase in the strength of the composite. It shifts from 6 MPa to 6.5 MPa at 75 µm particle size and from 6.1 to 6.3 MPa at 150 µm particle size for

2%wt and 8%wt clay particles, respectively. AA6061/8% clay at 75 μm amidst all others gave a superior ultimate tensile strength.

Figure 6 shows the percentage elongation analysis of the reinforced alloys for AA6061/clay composites in both 75 μm and 150 μm particle sizes. From the result, the reinforced aluminium metal matrix composites increased intensely from 2.4% starting aluminium alloy to around 2.8% for the developed alloy at 16.7%. The percentage elongation values of the developed composites increased as the percentage weight clay inclusion rises in the alloy. It shifts from 2.5% to 2.8% at 75 μm particle size and from 2.8% to 2.7% at 150 μm particle size for 2 %wt and 8 %wt clay particles, respectively. For exact analysis of the reinforced matrixes, AA6061+ 8% clay at 75 μm among the others gave a better superior percentage elongation. This further implies that the perfection in percentage elongation generally improves the ductility of the starting material.

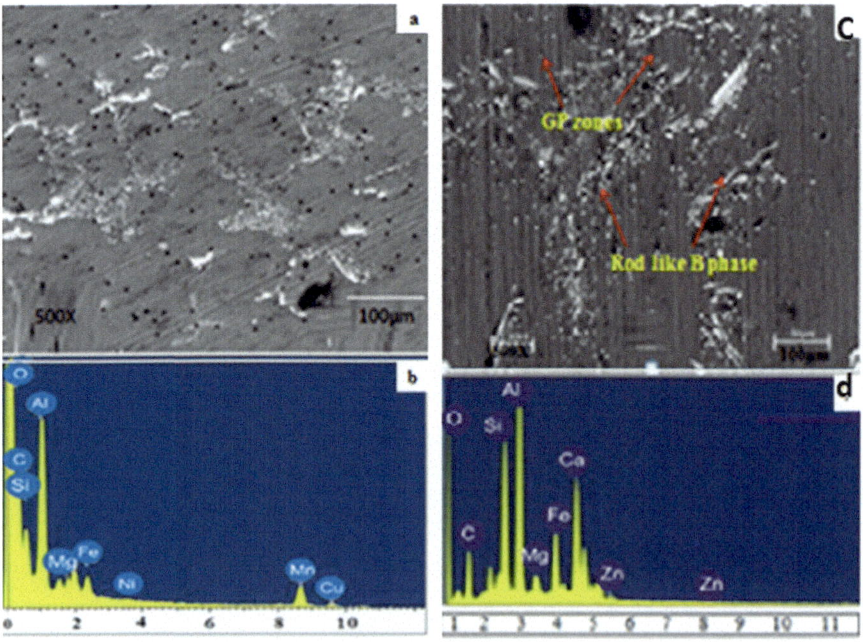

Fig. 7 SEM/EDS of AA6061/clay at 75 μm (**a**) and (**b**) 2% clay (**c**) and (**d**) 8% clay

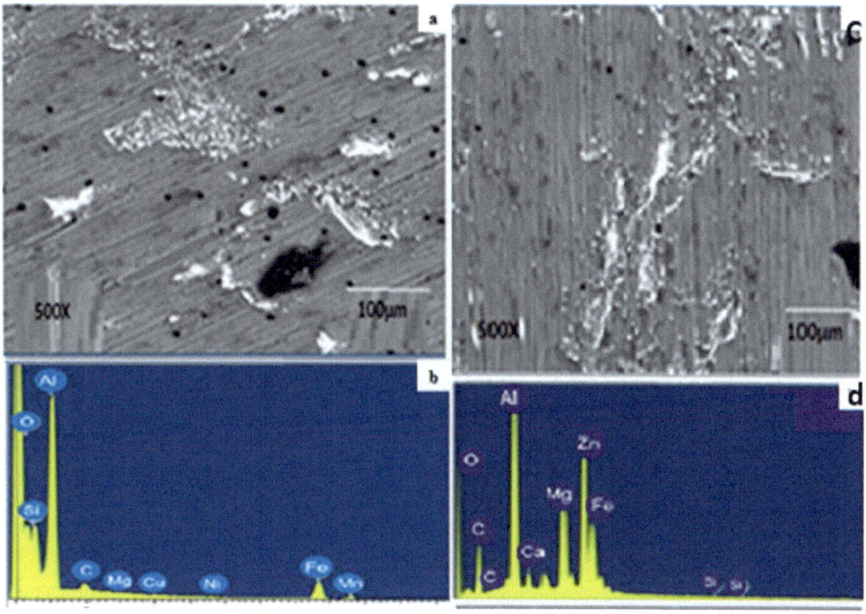

Fig. 8 SEM/EDS of AA6061/clay at 150 μm (**a**) and (**b**) 2% clay (**c**) and (**d**) 8% clay

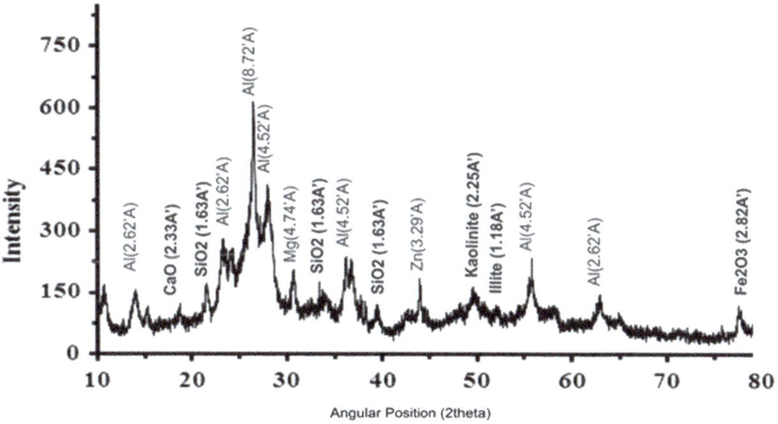

Fig. 9 Diffractogram of AA6061+2% clay at 75 μm reinforced composite

3.2 *Microstructural Analysis of AA6061/Clay*

Figure 7 (a and b) displays the SEM image of the composites' fractured surface, and it discloses that the reinforcements are an ingredient in composites. It was a rise in the particles' collection equal to an improvement in the percentage of reinforcement. The EDS image shown in Fig. 4 shows the peaks generated by various constituents of the composite. The effect of particulate of 8% clay is presented in Fig. **7** (c and d). It was noticed that there is better wetting of reinforcements into the matrix shown by the dark patches of clay. Figure **8** (a and b) revealed the structural propagation of AA6061/2% clay at 150 μm composite. SEM study shows the occurrence of eutectic silicon α-Al phase precipitation and rough distribution of particulates' particle refiners. Figure **8** (c and d) shows the structural propagation of AA6061/8% clay at 150 μm composite. SEM study depicts the presence of magnesium, iron, zinc, silicon and homogenous distribution of particulates' particle refiners.

3.3 X-Ray Diffraction (XRD) Analysis of AA6061/Clay + RHA Composites

Figure 9 depicts the XRD forms of the developed alloy and AA6061+2% clay at 75 μm. The XRD pattern confirms the presence of particulates of agro-based alloying element metallic materials used in the study. It was discovered that silicon IV oxide, aluminium silicates and kaolinites existed within the intermetallic phases at different peaks. The intermetallic phases of kaolinites, aluminium silicates, silicon IV oxide and iron III oxide reinforcement increase with agro-based waste content, while the peaks of AA6061 decrease. The presence of aluminium, magnesium, calcium oxide, silicon IV oxide, iron III oxide and aluminium silicate intermetallic phases at 25°, 30°, 18°, 32°, 78° and 50°, respectively, were found within the intensity line with notable peaks. Iron from iron III oxide helps to enhance the ductility and malleability of the developed composites. Figure 10 illustrates the XRD analysis for the AA6061/8% clay at 75 μm and its crystal constituents. The presence of

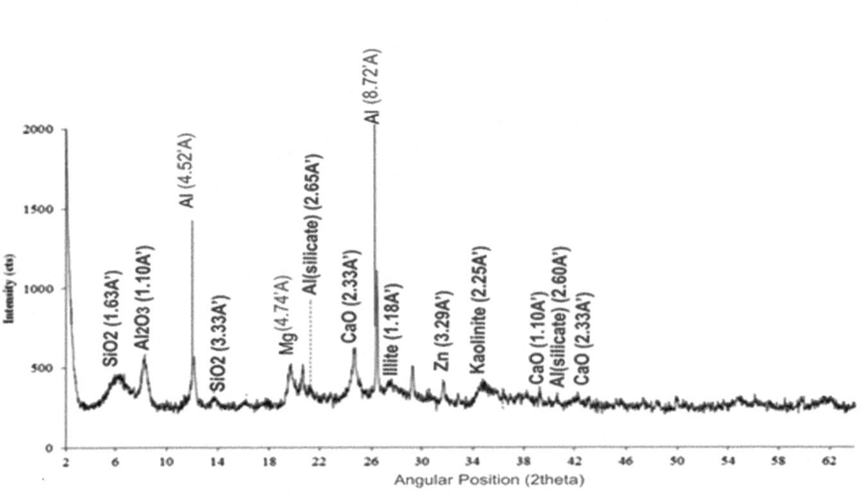

Fig. 10 Diffractogram of AA6061+8% clay at 75 μm

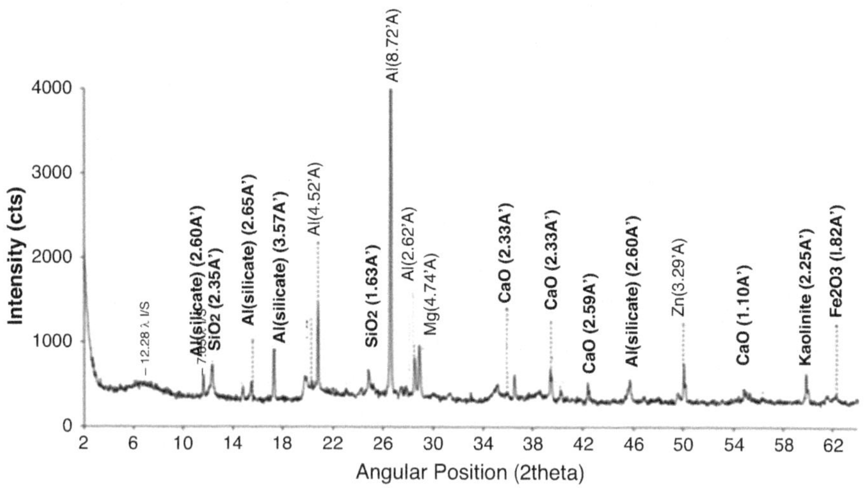

Fig. 11 Diffractogram of AA6061+2% clay at 150 μm reinforced composite

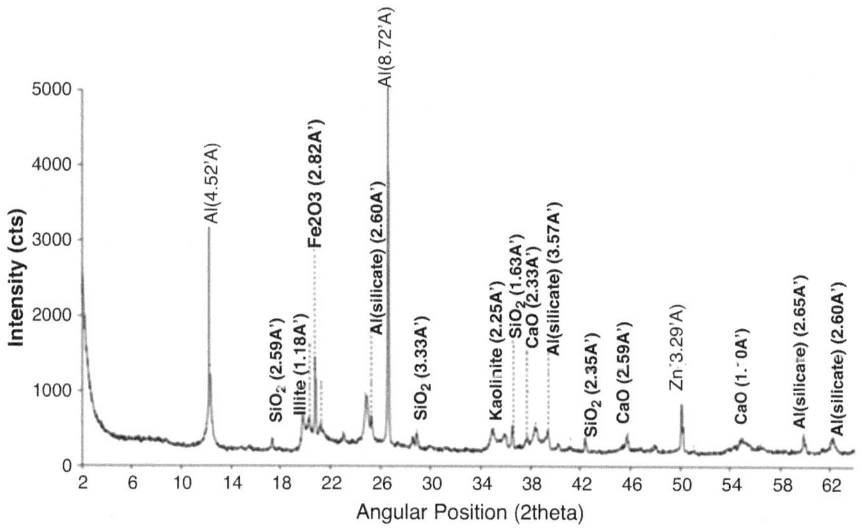

Fig. 12 Diffractogram of AA6061+8% clay at 150 μm reinforced composite

aluminium, magnesium, calcium oxide, silicon oxide, alumina and aluminium silicates at 26°, 17°, 25°, 6°, 8° and 20°, respectively, was found within the intensity line with notable peaks. Calcium is the principal agent used to increase the apparent viscosity of molten aluminium. It is also known as an effective silicon modifier in Al-Si alloys (Kumari et al., 2005). Kaolinites are obtained from the clay used as the reinforcing agent in the development of aluminium matrix composites. Figure 11 revealed the XRD analysis for the AA6061+2% clay at 150 μm and its crystal constituents. The patterns established the presence of composite particle reinforcements at different peaks. The presence of aluminium, calcium oxide, silicon IV oxide and aluminium silicate intermetallic phases at 26°, 35°, 12° and 15°, respectively, was found within the intensity line with notable peaks. Silicon is a component of high-strength ceramics that he lps in reducing the melting temperature of aluminium alloy (Fayomi et al., 2017). Figure 12 illustrates the XRD analysis for the AA6061+8% clay at 150 μm and its crystal constituents. The presence of aluminium, calcium oxide, silicon IV oxide, iron III oxide and aluminium silicates at 27°, 36°, 18°, 20° and 25°, respectively, was found within the intensity line with notable peaks. The existence of a trace element of magnesium in the intermetallic phase helps increase the strength and enhances the developed composites' corrosion resistance.

4 Conclusions

The following conclusions were deduced from the research work:

- AA6061/clay matrix composites enriched with calcium oxide, silicon IV oxide and aluminium silicate intermetallic phases have a greater wear resistance than the aluminium alloy.
- The surface morphology analyses showed equal dispersal of intermetallic phase in the AA6061 aluminium matrix.
- The materials reinforced with clay showed improved mechanical properties due to particle clusters' formation resulting from the reinforcing agents.

Acknowledgement The authors will like to acknowledge the support of Covenant University for the publication fund.

References

Ahamed, A. A., Ahmed, R., Hossain, M. B., & Billah, M. (2016). Fabrication and characterization of aluminium-rice husk ash composite prepared by stir casting method. *Rajshahi University Journal of Science and Engineering, 44*, 9–18.

Balakrishnan, M., Dinaharan, I., Kalaiselvan, K., & Palanivel, R. (2020). Friction stir processing of Al_3Ni intermetallic particulate reinforced cast aluminum matrix composites: Microstructure and tensile properties. *Journal of Materials Research and Technology, 9*(3), 1–12.

Basil Quent, O., Nwobi-Okoye, C. C., Ochieze, P. U., & Ochieze, I. A. (2018). Microstructural and properties evaluation of A356 alloy/cow horn particulate composites produced by spark plasma sintering. *Journal of the Chinese Advanced Materials Society, 6*(1), 30–43.

Baskaran, J., Prashad, A., & Kumar, R. (2019). Investigation on the influence of heat treatment of mechanical behavior enhancement of A356 alloy. *International Journal of Mechanical Engineering and Technology, 10*(3), 1088–1093.

Fayomi, O. S. I., Loto, C. A., & Adedokun, S. T. (2016). Evaluation of composition, microstructure characterization and interfacial properties of Zn—SnO_2 metal matrix composite coating. *Protection of Metals and Phys Chem Surfaces, 52*(2), 238–243.

Fayomi, O. S. I., Popoola, A. P. I., & Udoye, N. E. (2017). Effect of alloying element on the integrity and functionality of aluminium-based alloy. *Intechopen, 13*, 243–244.

Gupta, R., Chaudhari, G. P., & Daniel, B. S. S. (2018). Strengthening mechanisms in ultrasonically processed aluminium matrix composite with in-situ Al_3Ti by salt addition. *Composites Part B: Engineering, 140*, 27–34.

Hossain, R., Ali, H., Amin, A., Kibria, G., & Ferdous, S. (2017). Fabrication and performance test of aluminium alloy-rice husk ash hybrid metal matrix composite as industrial and construction material. *International Journal of Engineering Materials and Manufacture, 2*(4), 94–102.

Joseph, O. O., & Babaremu, K. O. (2019). Agricultural waste as a reinforcement particulate for aluminum metal matrix composite (AMMCs): A Review. *Fibers, 7*(4), 33.

Kumar, P. S. R., Smart, D. S. B., & Alexis, S. J. (2018). Evaluation of mechanical properties of mwcnt/nanoclay reinforced aluminium alloy metal matrix composite. *Journal of Material Sciences and Engineering, 346*.

Kumari, S. S., Pillai, R. M., & Pai, B. C. (2005). Role of calcium in aluminium based alloys and composites. *International Materials Reviews, 50*(4), 216–238.

Ogunyemi, S. A., Ajileye, O. O., Muibi, K. H., Alaga, A. T., Eguaroje, O. E., Samson, S. A., & Omisore, O. O. (2017). Geo-information and distribution pattern of petrol service station in Sango–Ota Metropolis in Ado–Odo Ota Local Government Area, Ogun State, Nigeria. *Asian Research Journal of Arts & Social Sciences, 2*(1), 1–11.

Ossowski, M., Moskalewicz, T., Czyrska-Filemonowicz, A., & Wierzchoń, T. (2008). Modification of the microstructure and properties of laminate titanium alloy-TiAl intermetallic phases composites by a duplex surface engineering treatment. *Advances in Materials Sciences, 8*(3), 23–31.

Pal, A., Saxena, S. S., Verma, A., & Kandpal, B. C. (2015). Stir casting of metal matrix composites – A review. *International Journal of Computer and Mathematical Science, 4*, 2347–8527.

Rao, R. N., Das, S., Mondal, D. P., & Dixit, G. (2009). Dry sliding wear behaviour of cast high strength aluminium alloy (Al–Zn–Mg) and hard particle composites. *Wear, 267*, 1688–1695.

Ufoegbune, G. C., & Fabiyi, S. (2016). Identification of land areas suitable for fadama farming at Federal University of Agriculture, Abeokuta, Nigeria using GIS. *Journal of Applied Sciences & Environmental Management, 20*(4), 1027–1032.

A Short Overview on the Role of Nanotechnology in Different Sectors of Energy System

O. Agboola, A. O. Ayeni, O. S. I. Fayomi, O. Oladokun, A. A. Ayoola, E. D. Babatunde, V. E. Efeovbokhan, O. A. Odunlami, A. Adeniyi, and E. R. Sadiku

1 Introduction

Energy is practically needed in most of our daily life. The practical need for energy includes transportation, hospital laboratory, agriculture, telecommunication, and other industrial activities that have an impact on the growth of the economy. As a result, the efficiency of energy stays low and a huge quantity of energy is lost in the form of heat in the course of industrial processes. The thermal power plants' efficiency ranges between 35% and 49%; this shows that 51% to 65% of the energy input is lost (Zhang, 2020). In the course of the thermal power plant operation, the fuel becomes kindled in the boiler; this produces vapors that inflate in the turbine and aiding its operation; then, the turbine would power the generator to produce electricity (Debnath, 2019). The transitions of energy in this process would result in substantial energy loss (Zhang, 2020), which will, in turn, affect the growth of the economy. The growth of the economy is an estimation of gross domestic product (GDP). GDP is a measure of the entire economic activities of a terrain, which

O. Agboola · A. O. Ayeni · O. Oladokun (✉) · A. A. Ayoola · E. D. Babatunde · V. E. Efeovbokhan · O. A. Odunlami
Department of Chemical Engineering, Covenant University, Ota, Nigeria
e-mail: oluranti.agboola@covenantuniversity.edu.ng; augustine.ayeni@covenantuniversity.edu.ng; olagoke.oladokun@covenantuniversity.edu.ng; ayodeji.ayoola@covenantuniversity.edu.ng; vincent.efeovbokhan@covenantuniversity.edu.ng; olayemi.odunlami@covenantuniversity.edu.ng

O. S. I. Fayomi
Department of Mechanical and Biomedical Engineering, Bells University of Technology, Ota, Ogun State, Nigeria
e-mail: osfayomi@bellstech.edu.ng

A. Adeniyi · E. R. Sadiku
Depatment of Chemical, Metallurgical and Materials Engineering, Tshwane University of Technology, Pretoria, South Africa

A. O. Ayeni et al. (eds.), *Advanced Manufacturing in Biological, Petroleum, and Nanotechnology Processing*, Green Energy and Technology,
https://doi.org/10.1007/978-3-030-95820-6_9

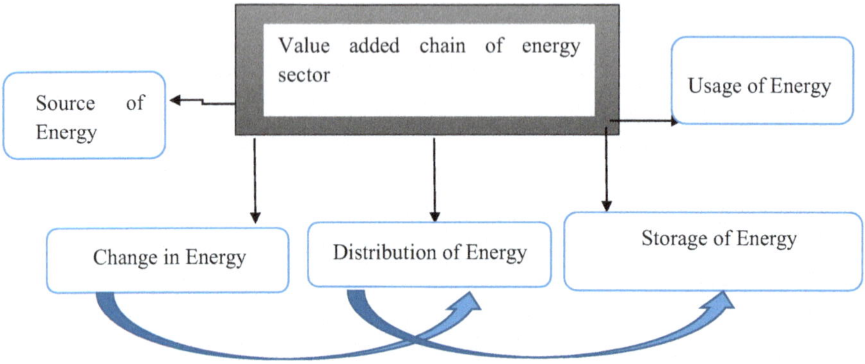

Fig. 1 The value-added chain of the energy sector

includes all public and private consumption, product inventory, investment, government expenditure, and net exports that occur within a well-defined terrain (gross domestic product – energy education). Most countries correlate GDP with the consumption of energy in the country (Shafie et al., 2011). Studies have shown that for all countries, GDP has a positive effect on energy consumption (Campo & Sarmiento, 2013; Lee, 2005; Lee & Chang, 2007).

Due to the continuous growth of the world's energy demand, the advancement in the use of more efficient and sustainable technologies for the generation, conservation, and storing of energy has found great importance and attention. According to the International Energy Agency forecasts, energy is anticipated to upswing by roughly 50% prior to 2030 (Rhiel, 2008). At the moment, the concentration of mankind in the community is to produce energy using low carbon base materials and instituting sustainable green technology (Singh, 2018) and nanotechnologies. Hence, provision has now been made by nanotechnologies to produce and improve the efficiency of energy all through the branches of industry. Nanotechnologies now economically provide controlled renewable energy production via novel technological solutions which will subsequently advance production technologies (Nanowerk, 2019). Nanotechnology advancements can influence every part of the energy sequence as shown in Fig. 1. This review will discuss the role of nanotechnologies in energy production. The following section will discuss the application of nanotechnologies for all the value-added chains of the energy sector depicted in Fig. 1.

2 Review on the Value-Added Chain of Energy Sector

The institute of industry value chain is one of the foundations of industry structure. This value chain helps in following the process of value discovery and recreation and completely incorporation of each enterprise's value chain in the industry chain. This will assist in conducting unceasing design and in redesigning the value system of the industry chain (Wenfeng, 2012). The energy value-added chain of energy sector comprises all events essential for the production, distribution, storage, and use of energy. The five major sectors are the source of energy (fuel procurement), change in energy, distribution of energy, and the end-market or service location (energy usage). The efficiency of energy can be associated with much more than the energy used in production. As a matter of reality, many processes in a value chain unswervingly or indirectly have an impact on the usage of energy (Rex et al., 2015). The necessity for a very effective materials supply chain could have dramatically swayed the company's bottom line, via productivity upsurge, risk reduction, and alleviating downtime (DHL, 2021). It is, hence, important to know the impact of nanotechnologies on the parts of the value-added chain in energy sector. This section will provide an understanding of the impact of the role nanotechnology has played on the value-added chain in different energy sectors.

2.1 *Sources of Energy*

The increasing demand for energy together with the increase in the price of crude oil, global warming as a result of emission of greenhouse gases, pollution from the environment, and the fast-falling supply of fossil fuels is the major crucial factor that results in the search for alternative sources of energy. Some of the most prominent alternative sources of energy that have the capacity of replacing fossil fuels are solar energy, water, wind energy, and biofuels. At present, 86% of the energy being consumed worldwide and almost 100% of the same being sought after in the transportation sector are gotten from nonrenewable fossil fuels (Abbaszaadeh et al., 2012; Azeta et al., 2021). Nanotechnologies offer vital enhancement capacities for the development of traditional sources of energy, such as fossil and nuclear fuels, and renewable sources of energy, like wind, sun, tides, water, geothermal energy, or biomass. As a means of illustration, nano-coated, wear-resistant, and drill probes give room for optimizing the duration and efficacy of systems for the advancement of oil and natural gas deposits or geothermal energy, hence making the system cost-effective (Gross domestic product - Energy Education). Additional illustrations are functional nanomaterials for lighter and super rugged rotor blades of wind and tide power plants and use of nanotechnologies in intensifying solar energy via photovoltaic systems.

2.1.1 Nanotechnology in Electricity Generated by Windmills

Globally, wind energy is presently a rapid-growing source of generating electricity. Yearly, modern wind turbines increase with regard to wind turbine blades due to the empirical super energy output demand and low carbon footprint. The upsurge in the length of blade and needs of enhancing useful properties have resulted in increased request for nano-enabled components in wind energy value chain (Patel & Mahajan,

2017). Hence, nanotechnology could be employed to crinkle the generated electricity by means of wind turbines. An epoxy comprising carbon nanotubes has being utilized for making stronger windmill blades. There are great possibilities in making stronger and lower weight blades through the utilization of nanotube-filled epoxy. The resultant longer blades upsurge the quantity of generated electricity by each windmill (Energy, 2019). The wind turbine blades' lifespan cycle could be upsurged through the use of nanocoatings and nanopaints. In addition, reduction of weight can be attained with the utilization of prepregs that are based on nanomaterials, and increase in efficiency can also be attained by employing nanofluids, nanolubricants, and nano-enabled wires and cables (Patel & Mahajan, 2017).

2.1.2 Nanotechnology in Electricity Generation from Waste Heat

With regard to the selection of technologies for generating power, some projects finally compare the technologies of generating power for an identical source. The generation of electricity through waste heat seems to be the most comprehensive project that deals with the perception of waste heat recovery. There are lots of waste heat sources; some are hot surfaces, exhaust gases, or liquid streams (Blanquart, 2017). Electricity generation from waste heat, known as waste heat to power (WHP), is the process that recovers waste heat by capturing waste heat by a current operating industrial process and utilizing the heat to generate power with no combustion and no emission. Recovering of waste heat is a novel prospect that provides the opportunity of decreasing gas emissions and at the same time cutting energy costs. The process is shown in Fig. 2, the energy demanding industrial operations like those in glass furnaces, steel mills, and refinery; discharge hot exhaust gases and waste streams that could be coupled with firm technologies to generate electricity. The waste heat recovery from industries that are used for electricity generation is

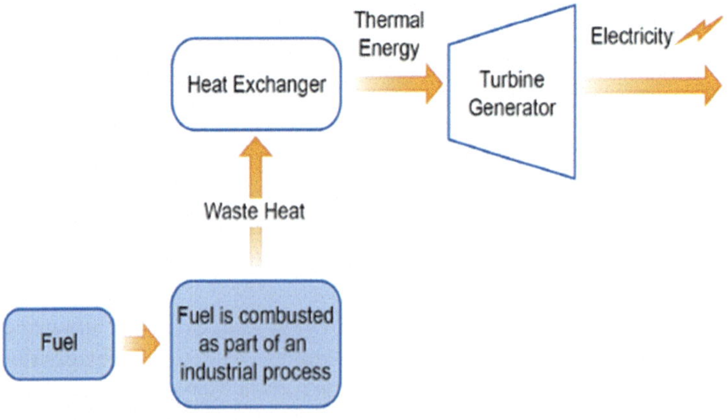

Fig. 2 Illustration of waste heat to power. (EPA, 2012)

basically an unexploited type of combined heat and power, and it uses a single fuel source for generating thermal energy and electricity (EPA, 2012).

Capturing waste heat has turned out to be a critical concern; as a result, numerous technologies have recently come up to address this concern. These technologies need to determine the utilization of the recovered heat as a source of heat or an energy input that will produce power. Depending on the source of heat which is peculiar to the need of certain industries, several technologies are available, and some are under development (Blanquart, 2017). An example of such available technologies that are used to transform waste heat into direct power is nanotechnology. Nanotechnology has recently been used for the generation of electricity from waste heat.

Literature has it that researchers have employed nanotube sheets to fabricate thermo-cells for electricity generation with the sides of the cell being operated at diverse temperatures. The nanotube sheets can be enfolded all over the hot pipes, for electricity generation using waste heat (Energy, 2019). Im et al. (2016) fabricated carbon nanotube aerogel-based thermo-electrochemical cells. These carbon nanotubes are inexpensive and are moderately high-efficiency materials for thermo-electrochemical cells. An optimum power output of $6.6Wm^{-2}$ was attained for an inter-electrode temperature differentiation of 51 °C, having a Carnot relative efficacy of 3.95%. The significance of the pureness of electrode, engineered porosity, and catalytic surfaces is for the performance improvement of the thermo-cell to be demonstrated. Some investigators from Sandia National Laboratories constructed a minute silicon-based device with the capacity to harness the formerly known waste heat and convert it into DC power (Shank et al., 2018). They demonstrate an undeviating rectified power generation from a neutral large-region nanoantenna-coupled tunnel diode rectifier known as a rectenna. By utilizing a measurement procedure known as vacuum radiometric, with irradiation from a temperature-stabilized thermal source, they observed a generated power density of 8 nW/cm^2 at a source temperature of 450 °C for the neutral rectenna through an enhanced load resistance. However, utilizing an enhanced system and better conversion efficiency, the power output per unit area will upsurge. Mistewicz et al. (2019) investigated a simple fabrication of SbSeI pyroelectric nanogenerator for a low-temperature waste heat recovery. The process comprises synthesizing a sonochemical of SbSeI nanowires into a bulk sample at an extreme compression pressure of 100 MPa and room temperature. SbSeI PENG produced electric output up to 11 nA with a power density of 0.59(4) $\mu W/m^2$ when exposed to heat cool conditions from 324 K to 334 K.

2.1.3 Nanotechnology in the Generation of Steam from Sunlight

The technologies of solar harvesting can be grouped into two major types; they are solar thermal collectors and photovoltaic (PV) systems. This is based on the technique of converting energy from sunlight. Electric energy generated by PV panels occurs by absorbing solar energy through solar cells, having an average efficiency of 15–17% (Boldoo et al., 2020). Solar energy can be garnered by different forms of energy like chemical (fuels), electricity, thermal, and chemical (fuels), and photothermal energy conversion processes can be used to obtain heat (Gao et al., 2019). Electric energy generated using solar energy occurs by solar thermal collectors. In this technique, solar energy directly heats an operational fluid, which can collect the heat in the solar collector for space and water heating. The efficiency of solar collectors varies as it depends on the type of solar collector.

The technology for electricity generation using solar energy is globally utilized. The utilization of solar energy to generate steam gives the foundation for numerous sustainable desalination, process heating technologies, and sanitization. There are several studies on cost-effective floating structures that absorb solar radiation and transmission energy to water through thermal conduction and driving evaporation. Nonetheless, the studies showed that the contact amid water and the structure results in fouling and pins the vapor temperature close to the boiling point. For this reason, Cooper et al. (2018) established solar-aided evaporation utilizing a structure that is not in any form of interaction with water. This structure absorbs solar radiation and reradiates infrared photons, which are unswervingly absorbed by the water in the range of a sub-100 μm infiltration deepness. As a result of the physical separation from the water, fouling was totally circumvented. Again, as a result of the thermal separation, the structure was not anymore held at the boiling point but was utilized to superheat the generated steam. This group of authors generated steam having temperatures of about 133 °C, establishing superheated steam in a non-pressurized system under one sun illumination. Though they were able to overcome the two challenges, solar energy harvesting technologies are, however, less efficient than other technologies (Boldoo et al., 2020), hence the introduction of nanotechnology for generating steam from sunlight.

There are many studies on the use of nanomaterials because they possess numerous advantageous and functional qualities. They have the ability to improve and influence light absorption, heat transfer efficiency, and thermal conductivity, which has fascinated noteworthy research considerations (Boldoo et al., 2020). When it comes to the generation of energy, the greatest use of nanotechnology appears to be in the area of efficient binding of solar energy employing photovoltaic (PV) cells. An efficient PV system helps in overcoming the challenges of the supply of energy in developing nations. The approach is very convenient for powering streetlights and for charging inverters. These serve as an excellent substitute to the typical fossil fuel-powered electric generators (Echiegu, 2016). Literature has it that some investigators have shown that sunlight, concerted on nanoparticles, could produce steam that possesses great energy efficacy. The utilization of the "solar steam gadget" is envisioned in regions of developing countries that don't have electricity for applications like water purification or disinfecting dental instruments (Energy, 2019). Solar illumination of mostly absorbing carbon or metal nanoparticles distributed in a liquid has the capacity to produce vapor devoid of the need to heat the fluid volume. Neumann et al. (2013) employed different materials such as metallic and carbon nanoparticles. The main aim of using these nanomaterials was to be able to absorb light. These nanoparticles direct most of the energy into making steam rather than heating the water when dispersed into water.

Furthermore, subwavelength metallic particles are powerful absorbers of optical radiation, as a result of their joint oscillations of delocalized conduction electrons, called surface plasmons. During the time of placement on a resonance, the energy that is not reradiated by means of light scattering is dissipated via Landau (non-radiative) damping (Gao et al., 2011). This leads to an intense upsurge in temperature around the nanometer-scale surrounding area of the particle surface (Neumann et al., 2013). This process of heat generation has attracted pronounced attention for several uses.

Studies have shown that carbon-based particles also bring about very sturdy photothermal heating impacts (Han et al., 2011; Chen et al., 2017). Particle-based methodologies have also attracted lots of consideration in the field of solar energy usages (Otanicar et al., 2010; Ma et al., 2017); nonetheless, these studies primarily put their attention on improving the thermal conductivity of operational fluids. However, the energy benefits of unswervingly apprehending the latent heat of vaporization needed for liquid vapor phase transition are also required. Neumann et al. (2013) used the phenomenon of solar illumination of carbon nanoparticles distributed in a liquid to produce vapor devoid of the requisite of heating the fluid volume. Also, this phenomenon can significantly get the solar applications compacted, like sterilization of waste and surgical instruments in resource-poor locations. Hazra et al. (2019) reported carbon black-ethylene glycol (CB-EG)-based nanofluids as the operational fluids for direct absorption solar collector (DASC) applications.

Nanofluids were prepared by the two-step method. When compared with other base fluids, improved absorption features were observed in all cases toward incident irradiance. Measured data depicted an upsurge in direction of local photothermal efficacy having the thickness of the liquid stratum and the concentration of the suspended nanoparticles. Their studies confirmed that CB-EG-based nanofluids can be employed as potential operational fluids for DASCs. Articles discussed in this section have shown that the use of nanoparticles in the making of solar cells is very advantageous (NANOPINION, 2015). As a result of their exceptional electronic and optical characteristics of the nanostructures, they can lessen manufacturing costs, and they also possess the capacity that enables them to attain overall higher efficacy degrees than traditional ones. Also, they could be crystal clear and malleable permitting their utilizations in more places than just roofs.

2.1.4 Nanotechnology in Hydrogen Storage for Fuel Cell-Powered Cars

Hydrogen as a form of energy storage has tremendous prospects. Characteristically, it is not a renewable source of energy when compared to solar power, nor is it a fuel that occurs naturally such as coal. Nonetheless, it uses an efficient process to release its energy, and pure water is the only exhaust gas produced (Scouter, 2012). Hydrogen burns like any other fuel; it could hence be employed to directly produce electricity in a hydrogen fuel cell. Hydrogen fuel cell cars are powered by an electric motor; they are thus categorized as e-cars. Hence, they are called fuel cell electric vehicle (FCEV). Hydrogen cars produce electricity themselves, and the vehicle doesn't get its power from an in-built battery that could be charged from an external power source. By contrast, hydrogen cars efficiently inherently possess an effective power plant on board, which is "the fuel cell." Also, hydrogen-powered cars are considered electric vehicles (EVs) for the reason that oxygen and hydrogen are converted to electric energy, which in turns powers the electric motor with a battery. They can also recapture the energy that is lost in the course of braking and store it in a battery (BMW, 2019, 2020). However, hydrogen as a fuel cell has some disadvantages. The problems with the implementation of hydrogen come from the production and storage of hydrogen gas. The current processes for producing hydrogen need considerably more quantity of energy which is usually gotten from fossil fuels, refuting the ecological advantages of the otherwise fuel that is carbon- and pollution-free. Also, storing hydrogen is an issue, as it is highly flammable when in free gaseous form. Hence, the process of storing and safely transporting the fuel is a priority (Scouter, 2012). Nanotechnology has the ability to offer a solution to the challenges.

For hydrogen power generation development, it is essential to build safe and effective systems for the reversible storage of hydrogen that will possess high capacity and stability, and the likelihood of swift hydrogen evolution (Alekseeva et al., 2020). The use of novel nanomaterials to efficiently produce hydrogen from water is a promising development in hydrogen production and storage. In addition, novel nanomaterials are used to improve the performance of hydrogen storage technologies (Alekseeva et al., 2020). Graphene layers have been prepared by some researchers to upsurge the binding energy of hydrogen to the surface of graphene in a fuel tank. This has resulted in a higher quantity of hydrogen storage, hence a lighter-weight fuel tank (Energy, 2019). There has been great interest in the utilization of graphene-based nanomaterials. It is a 2D crystal made of a monolayer of sp2 hybridized carbon atoms that made a honeycomb structure. This structural alignment of carbon atoms could offer strong bonding with hydrogen atoms. With regard to hydrogen storage, graphene has other valuable characteristics like high thermal and electrical conductivities and high surface area with an excellent mechanical strength (Alekseeva et al., 2020). Some have also established the fact that sodium borohydride nanoparticles can effectively store hydrogen (Energy, 2019). Lai et al. (2019) explored the likelihood of fabricating a core-shell nanocomposite ($NaBH_4$-Ni) by using a metallic nickel catalyst to directly facilitate the hydrolysis as a support onto $NaBH_4$ nanoparticles. In the course of the hydrolysis, the closely joined Ni^0 and $NaBH_4$ permit the production of hydrogen at rates of 6.1 L min^{-1} g^{-1} at 39 °C which is quite high, when water is utilized in excess. The effective gravimetric hydrogen storage capacity of nanosized $NaBH_4$-Ni was optimized by regulating the needed quantity of water for hydrolysis, and an effective hydrogen capacity of 4.4 wt% was attained.

2.2 Nanotechnology in Change in Energy/Energy Conversion

Energy can transform from one form to another and it is conserved when it changes form. Energy conversion or change of primary energy sources into heat, electricity, and kinetic energy needs maximum efficacy. The upsurge in the efficacy, particularly in fossil-fired gas and steam power plants, could assist in avoiding substantial quantities of carbon dioxide emissions. Nonetheless, enhanced power plant efficacies need higher working temperatures and heat-resistant turbine materials too (Nanowerk, 2019). There could be possible improvements via nanoscale heat and corrosion shielding stratums for turbine blades in power plants to improve the efficacy by increasing working temperatures or by utilization of lightweight fabrication materials such as titanium aluminides (Nanowerk, 2019).

Nanotechnology possesses a wide use in the current process of membrane synthesis, putting into consideration their exceptional features as membranes (Delgado et al., 2014). Nano-optimized membranes could be extended to the possible use in the area of separation and climate-neutral storage of carbon dioxide for power generation in coal-fired power plants, to eventually render this imperative technique of generating ecologically pleasant power. The energy harvested from the conversion of chemical energy via fuel cells could be increased by nanostructured electrodes, catalysts, and membranes. This will lead to the possibilities of profitable utilization in building, automobiles, and the operation of mobile electronics (Nanowerk, 2019). The conversion of thermoelectric energy appears to have a great prospect. Nanostructured semiconductors, having enhanced at the border stratum design, are advantageous to an upsurge inefficacy that has the capacity of paving the way for extensive utilization of waste heat (Photonics, 2021). 2D "nanosheets" fabricated of bonds amid metal atoms and organic molecules have been found considerable attention for photoelectric conversion; however, they get corroded easily. Some researchers in Japan and Taiwan reported a novel nanosheet design by utilizing iron and benzene hexathiol that displays the record of stability to air exposure for 2 months, beckoning the future employment of these 2D materials for commercial optoelectronic (NANOWERK, 2021). Another set of researchers recently developed coordination nanosheets that have proven to possess maximum stability under their exposure to air. The work could find application in optoelectronic; coordination nanosheets have spawned attention as a result of their capability to absorb light at manifold ranges of wavelength and convert them into electrons with superior efficacy than other types of nanosheets (Photonics, 2021).

2.3 Nanotechnology in the Distribution of Energy

The distribution of energy is a technology that collects man-made systems that transport energy, which includes the primary energy material such as coal, crude oil, or energy currencies for end-users such as electricity or gasoline, together with electric networks and district heating/cooling networks. The distribution of energy allows the above-listed commodities to be globally transported to drive the economy. When energy is produced in the form of liquefied natural gas, electricity, gasoline, etc., it needs to be transported to where it could be made useful (Hanania et al., 2020). The transportation of energy systems is immensely in diverse ways depending on what is being transported. These systems need broad infrastructures such as the pipeline grid, the electricity grid, and a network of road, rail, and water shipping (Photonics, 2021). These infrastructures help to carry electricity from the transmission system to individual consumers. However, there are losses of energy in the existing transmission system which need to be reduced.

For the decrease in energy losses in existing transmission, it is envisaged that the uniqueness of electric conductivity of materials at nanoscale, such as carbon nanotubes, could be employed during the fabrication of electric cables and power lines. Additionally, orthodox metal wires suffer from an extreme degradation or result in a total electrical performance failure, when exposed to severe reacting surroundings. Nonetheless, wires fabricated from carbon nanotubes have, in point of fact, been discovered to boost their electrical performance when imperiled to any harsh reacting surroundings. These opposing reactions could offer novel and stimulating uses for CNT wires (Lepak-Kuc et al., 2018). In addition to that, nanotechnology can be used in optimizing superconductive materials to avoid loss of current conduction (Nanowerk, 2019). For example, carbon nanotube fiber conductors are newly attractive types of nanomaterials. They possess yarn-like, tubular, highly anisotropic, and nanostructured, carbon-based morphology that makes them essentially different from any current metallic conductors (Lepak-Kuc et al., 2018). These unique properties are hypothetically making ways for noteworthy developments in the performance of present wiring systems together with entirely innovative utilizations like power lines.

Apart from the broad infrastructure listed above, with time, there will be options for wireless energy transport, such as the utilization of microwaves, laser, or electromagnetic resonance. In the future, power distribution would need power systems that will provide dynamic load and failure management and highly request energy supply that will possess mechanisms that are not too expensive together with the likelihood of feeding across some decentralized renewable energy sources. Nanotechnologies have the capacity to decisively contribute to the fulfillment of such an idea. This can be attained via nano-sensory tools and power-electronically components that have the capability of coping with the extremely multifaceted control and monitoring of such grids (Nanowerk, 2019).

2.4 Nanotechnology in the Storage of Energy

Energy storage is simply the capturing of energy produced at a particular time for use in the future to decrease imbalances between the demand for energy and the production of energy. It is therefore the conversion of electrical energy from a power network into a form that can be stored until converted back to electrical energy (Price, 2011). There are different types of energy storage. Bulk energy storage is the energy storage that possesses a large energy capacity and charges or discharges over the duration of time; it is, hence, regarded as a significant contributor for the transition in the direction of a more flexible and sustainable electricity system (Hittinger & Azevedo, 2015). A rechargeable battery is a storage battery that can be charged and discharged into a load; it can, however, be recharged many times unlike the disposable battery, which is supplied fully charged and thrown away after use. As a result of high cell voltage and exceptional energy together with power density, the lithium-ion technology is considered the most promising electrical energy storage However, nanotechnologies have found promising interest in enhancing storage of electrical energy like batteries and super-capacitors.

Also, nanotechnologies can decisively enhance the capacity and safety of lithium-ion batteries; this can, for instance, be achieved via new ceramic, heat-resistant, and still flexible separators together with high-performance electrode materials. Hydrogen also appears to be prospective energy storage for ecologically friendly energy supply. Besides the adjustments of the necessary nanostructure, the effectual storage of hydrogen is considered as one of the serious factors of success concerning likely hydrogen management. Present materials used for chemical hydrogen storage do not meet the demands of the automotive industry, which needs a hydrogen storage capacity that is close to ten weight percent. However, numerous nanomaterials fabricated based on nanoporous metal-organic compounds now exist. These materials offer advanced capabilities, which, at least, appear to be economically attainable to the operation of fuel cells in portable electronic devices (Nanowerk, 2019).

Thermal energy storage is another important type of energy storage system. The energy required in buildings could be significantly reduced via the utilization of phase change materials like latent heat stores. The use of nanomaterial for thermal energy storage has also attracted great attention for economic reasons. Economically, adsorption stores that are based on nanoporous materials such as zeolites could be applied as heat stores in local heating grid or industries. The adsorption of water in zeolite permits the reversible storage and release of heat (Nanowerk, 2019). Furthermore, nanoparticles can be distributed into base fluids for thermal energy storage. The utilization of nanoparticles distributed into base fluids for thermal energy storage is a relatively new research area. The quantity of nanoparticles needed to be distributed into base fluids to suit a particular application for precise thermal energy storage is an ongoing study. Currently, carbon nanotubes are the most extensively employed as additives because they possess unique thermophysical properties and their benefits have been proven to be the best when compared to other nanoparticles (Al-Kayiem et al., 2013).

2.5 *Usage of Energy*

There is a great tendency to associate energy usage with evident applications like transport, lighting, and cooking. Economy-wise, energy needs are typically classified into industrial, domestic, commercial, transportation, and power generation. Energy is needed in the home for our domestic use; it is also needed at our business place for commercial operations and in the air and on the roads, rail, and sea for our transportation (Singh, 2018). However, we rarely raise the value and the important roles nanotechnology plays in energy usage. The following subsections will discuss the important roles of nanotechnology in energy usage.

2.5.1 Nanotechnology for Producing High-Efficiency Light Bulbs

The use of incandescent lighting and its warmth is fast changing as guidelines pointed to the direction of enhancing energy efficacy, hence slowly getting rid of the old bulbs for the utilization of more effective compact fluorescent bulbs (CFLs) and recently developed light-emitting diode bulbs (LEDs) (LEDinside, 2016). In the past few decades, advances in efficiency light bulbs have been obtained due to the utilization of nanotechnology. Nanotechnologies are introducing novel means to lots of new forms of light bulbs; an instance is a nano-engineered polymer matrix employed in a novel fabrication of high-efficiency light bulbs undergoing study (NANOPINION, 2015). These bulbs produce white light, comparable to sunlight. These new bulbs possess the precedence to become shatterproof and possess efficiency that is twice of compact fluorescent light bulbs. This shows that nanostructured materials possess tunable physicochemical features like color change, light absorption, and electrical and thermal properties. Updating incandescent light bulbs by means of encompassing the traditional filament with a crystalline material that will convert some of the waste infrared radiation into visible light is a novel technology that is under development (Energy, 2019).

In the last two decades, developments in understanding physical phenomena and synthesis/construction techniques have driven and guided novel research development in the plasmonic nanostructure. As a result of current developments in fabrication and techniques of characterization, plasmonic modes and their interaction with nano-architectures can be extensively studied (Kasani et al., 2019). This new development involves the manipulation of light at the nanoscale. Some scientists are also developing high-efficiency LEDs using arrays of nano-sized structures called plasmonic cavities (Energy, 2019). For example, the use of nanotechnology has been developed to make surface structures that will boost light extraction efficiency (Huang et al., 2006). Yi et al. (2017) numerically studied the local electromagnetic field improvement and the optical characteristics of a many-layer structure having square lattice nano-hole arrays in Au-SiO_2-Au many-layer films by employing the finite-difference time-domain technique. The results they obtained reveal that the multiple surface plasmon (SP) resonances are made of SP on the air/Au boundary (the upper stratum), SiO_2/Au boundaries (the intermediate strata), and Au/SiO_2 boundary (the lower stratum) together with coupling modes on the Au film. The result of the distributions of electromagnetic field depicts that the location of the local electromagnetic field boost could stipulate the diverse SP resonances patterns. The study could see dipole, quadrupole, and 12-pole SP resonances modes in the many-layer nanostructure. This shows that support to the metal strata are metal nanostructures that reinforced localized optical resonances.

2.5.2 Nanotechnology for Transportation Energy Usage

Close to 15% of the fuel energy is spent as a result of the friction of the touching mechanical parts of the engine in present-day automobiles. Cylinder wall, piston cranking in the air condition, connecting rod with bearings, and valve drive system, including the valves and the camshaft, are all vehicle parts that are the main friction parts of the engine. Among these parts, cylinder wall aggregate and piston are the main parts of mechanical frictional loss. Another automobile problem is the generation of heat via the engine and its cooling. In order to mitigate these problems, engine oils have been used in between the moving parts to reduce friction and to ensure that engine friction and heat stayed checked (Mathew et al., 2019). Furthermore, radiator, coolant, and water reservoir should be topped off to lessen the heat generated in the engines. The mixture is pumped through tubes in the engine to absorb heat. The functions of the coolant, radiator, and engine oils could be enhanced via the utilization of nanotechnology (Srinivasan & Kumar, 2016). Furthermore, radiator has been utilized for the removal of heat from the engine; hence, traditional heat transfer fluids such as ethylene glycol, water, and mineral oil have been used for this reason. However, these fluids are not very effective for the removal of all the heat generated. The addition of nanofluids into the coolant significantly enhances its heat transfer rate (Mathew et al., 2019).

Abrasion and friction in vehicles can be reduced via coating of the cylinder wall with nanocrystalline materials which will consequently influence fuel consumption. Some researchers are currently working on directly coating tracks of the aluminum crankcase with nanomaterials. Iron carbide and boride nanocrystals that have a size range between 50 nm to 120 nm are potential friction modifying agents that have been employed to coat engine parts (Mourya, 2018). These tribological coatings can reduce friction coefficients and enhance wear resistance, which will subsequently improve the efficiency of the engine and reduced fuel consumption (Pathak et al., 2021). Hence, new smart high-speed and efficient vehicles in present-day transportation cannot be attained without the application of nanotechnologies like lightweight nanomaterials, nanocrystalline materials for coating, and nanofluids.

2.5.3 Nanotechnology for Energy Usage in Vehicle Battery

Electric vehicles possess the potential to radically upsurge the energy efficacy of the transport industry. They are much more effective at transforming stored energy into motion when compared with internal combustion engines. The batteries employed in electric vehicles are currently centered on nickel-metal hydride or lithium-ion technology. Contemporary instances almost completely make use of lithium-ion batteries because they provide many advantages over nickel-metal hydride. Nevertheless, no battery technology is presently satisfactory for making electric vehicles that will be feasible enough to substitute internal combustion. The span and the time used in recharging electric cars put boundaries to their utilization to not too far trips where recharging points are usually close by. Nanotechnology has the capacity to embrace the key to extensively manufacture electric vehicles as a result of their ability to increase battery performance. Better-quality electrolytes that employ nanoparticles and nanocomposite materials have been revealed to significantly boost the precise potentials of lithium batteries and other more novel battery technologies (Soutter, 2013).

Nanotechnology can play an important role in attaining precise performance objectives in batteries. Customarily, graphite powder has been employed as an intercalation material on the negative electrode for lithium-ion batteries. The removal or insertion rate of lithium and the capacity of the battery could be boosted by substituting micrometer-sized powder with carbon nanomaterials like carbon nanotubes. As a result of the high surface area carbon nanotubes possess, they can bind much higher concentrations of lithium. Nanowires made of nanoparticles such as vanadium oxide (V_2O_5), titanium dioxide (TiO_2), and tin oxide (SnO) are also promising materials to be used as negative electrode materials (Soutter, 2013). For example, studies have shown that lithium titanate ($Li_4Ti_5O_{12}$, LTO) spinel has proven to be a feasible substitute to graphite as anode material due to its exceptional safety features (Ohzuku et al., 1995). Several nanotechnologies have therefore been vigorously used in moving these nanomaterials for better electrochemical performance in lithium batteries.

3 Conclusion

The energy demand is growing as the world population upsurges. This has made the production of energy one of the cogent global problems that need to be solved. Hence, there is a need to develop alternative technologies that will be financially viable and at the same time lessen reliance on fossil fuels (which pollute the environment). Nanotechnology offers quite a number of energy applications which are constantly increasing daily. As far as technologies are concerned, nanotechnology has found vast attention in the different energy sectors, such as sources of energy, change in energy, and distribution of energy, storage of energy, and uses of energy.

Acknowledgment The authors wish to acknowledge the financial support provided by Covenant University in the actualization of the publication for this review.

References

Abbaszaadeh, A., Ghobadian, B., Omidkhah, M. R., & Najafi, G. (2012). Current biodiesel production technologies: A comparative review. *Energy Conversion and Management, 63*, 138–148.

Al-Kayiem, H. H., Lin, S. C., & Lukmon, A. (2013). Review on nanomaterials for thermal energy storage technologies. *Nanoscience & Nanotechnology-Asia, 3*(1), 60–71.

Alekseeva, O., Pushkareva, I., Pushkarev, A., & Fateev, V. (2020). Graphene and graphene-like materials for hydrogen energy. *Nanotechnologies in Russia, 15*(3), 273–300.

Azeta, O., Ayeni, A. O., Agboola, O., & Elehinafe, F. B. (2021). A review on the sustainable energy generation from the pyrolysis of coconut biomass. *Scientific African, 13*, e00909.

Blanquart, F. (2017). *Perspectives for power generation from industrial waste heat recovery*. Master of Science Thesis, KTH School of Industrial Engineering and Management Energy Technology, Stockholm.

BMW. (2019). *Plug-in hybrid and other electric vehicles* | BMW.com. https://www.bmw.com/en/innovation/Plug-in-hybrid-and-other-kinds-of-electric-cars.html

BMW. (2020). *Hydrogen fuel cell cars: What you need to know* | BMW.com. https://www.bmw.com/en/innovation/how-hydrogen-fuel-cell-cars-work.html

Boldoo, T., Ham, J., Kim, E., & Cho, H. (2020). Review of the Photothermal Energy Conversion Performance of Nanofluids, Their Applications, and Recent Advances. *Energies, 13*(21), 5748.

Campo, J., & Sarmiento, V. (2013). The relationship between energy consumption and GDP: Evidence from a panel of 10 Latin American countries. *Latin american journal of economics, 50*(2), 233–255.

Chen, L., Liu, J., Fang, X., & Zhang, Z. (2017). Reduced graphene oxide dispersed nanofluids with improved photo-thermal conversion performance for direct absorption solar collectors. *Solar Energy Materials and Solar Cells, 163*, 125–133.

Cooper, T. A., Zandavi, S. H., Ni, G. W., Tsurimaki, Y., Huang, Y., Boriskina, S. V., & Chen, G. (2018). Contactless steam generation and superheating under one sun illumination. *Nature Communications, 9*(1), 1–10.

Debnath, A. K. (2019). *Power plant instrumentation and control handbook: A guide to thermal power plants*. Academic.

Delgado, J., Öchsner, A., & de Lima, A. (2014). *Nanotechnology for energy and environment*. Hindawi.

DHL. (2021). *The power behind the energy supply chain.*

Echiegu, E. A. (2016). Nanotechnology as a tool for enhanced renewable energy application in developing countries. *Journal of Fundamental of Renewable Energy and Applications, 6*(6), 01–09.

Energy Na. (2019) *Hawk's perch technical writing*. https://www.understandingnano.com/nanotechnology-energy

EPA C. (2012) *Waste heat to power systems.*

Gao, M., Zhu, L., Peh, C. K., & Ho, G. W. (2019). Solar absorber material and system designs for photothermal water vaporization towards clean water and energy production. *Energy & Environmental Science, 12*(3), 841–864.

Gao, Y., Yuan, Z., & Gao, S. (2011). Semiclassical approach to plasmon–electron coupling and Landau damping of surface plasmons. *The Journal of Chemical Physics, 134*(13), 134702.

Gross domestic product – Energy Education (2021). https://energyeducation.ca/encyclopedia/Gross_domestic_product

Han, D., Meng, Z., Wu, D., Zhang, C., & Zhu, H. (2011). Thermal properties of carbon black aqueous nanofluids for solar absorption. *Nanoscale Research Letters, 6*(1), 1–7.

Hanania, J., Stenhouse, K., Strange, M., & Donev, J. (2020). *Energy distribution technology – Energy Education*. https://energyeducation.ca/encyclopedia/Energy_distribution_technology

Hazra, S., Ghosh, S., & Nandi, T. (2019). Photo-thermal conversion characteristics of carbon black-ethylene glycol nanofluids for applications in direct absorption solar collectors. *Applied Thermal Engineering, 163*, 114402.

Hittinger, E. S., & Azevedo, I. M. (2015). Bulk energy storage increases United States electricity system emissions. *Environmental Science & Technology, 49*(5), 3203–3210.

Huang, H.-W., Kao, C., Chu, J., Liang, W., Kuo, H.-C., Wang, S., & Yu, C. (2006). Improvement of InGaN/GaN light emitting diode performance with a nano-roughened p-GaN surface by excimer laser-irradiation. *Materials Chemistry and Physics, 99*(2-3), 414–417.

Im, H., Kim, T., Song, H., Choi, J., Park, J. S., Ovalle-Robles, R., Yang, H. D., Kihm, K. D., Baughman, R. H., & Lee, H. H. (2016). High-efficiency electrochemical thermal energy harvester using carbon nanotube aerogel sheet electrodes. *Nature Communications, 7*(1), 1–9.

Kasani, S., Curtin, K., & Wu, N. (2019). A review of 2D and 3D plasmonic nanostructure array patterns: Fabrication, light management and sensing applications. *Nanophotonics, 8*(12), 2065–2089.

Lai, Q., Alligier, D., Aguey-Zinsou, K.-F., & Demirci, U. B. (2019). Hydrogen generation from a sodium borohydride–nickel core@ shell structure under hydrolytic conditions. *Nanoscale Advances, 1*(7), 2707–2717.

LEDinside. (2016). *Nanotechnology raises incandescent bulb efficiency to same level as LEDs – LEDinside*. https://www.ledinside.com/news/2016/1/nanotechnology_raises_incandescent_bulb_efficiency_to_same_level_as_leds

Lee, C.-C. (2005). Energy consumption and GDP in developing countries: A cointegrated panel analysis. *Energy Economics, 27*(3), 415–427.

Lee, C.-C., & Chang, C.-P. (2007). Energy consumption and GDP revisited: A panel analysis of developed and developing countries. *Energy Economics, 29*(6), 1206–1223.

Lepak-Kuc, S., Boncel, S., Szybowicz, M., Nowicka, A., Jozwik, I., Orlinski, K., Gizewski, T., Koziol, K., Jakubowska, M., & Lekawa-Raus, A. (2018). The operational window of carbon nanotube electrical wires treated with strong acids and oxidants. *Scientific Reports, 8*(1), 1–12.

Ma, Z., Zhang, R., & Sawaged, F. (2017). Design of particle-based thermal energy storage for a concentrating solar power system. In *Energy sustainability* (p. V001T005A003). American Society of Mechanical Engineers.

Mathew, J., Joy, J., & George, S. C. (2019). Potential applications of nanotechnology in transportation: A review. *Journal of King Saud University-Science, 31*(4), 586–594.

Mistewicz, K., Jesionek, M., Nowak, M., & Kozioł, M. (2019). SbSeI pyroelectric nanogenerator for a low temperature waste heat recovery. *Nano Energy, 64*, 103906.

Mourya, S. (2018). *Recent advances of nanotechnology in transportation*. https://www.azonano.com/article.aspx?ArticleID=4826

NANOPINION. (2015). *Energy | nanopinion | Nanotechnologies: Where should they take us? Have your say*. https://nanopinion.archiv.zsi.at/en/about-nano/energy.html

NANOWERK. (2021). *Unlocking efficient light-energy conversion with stable coordination nanosheets – NewsBreak*. https://www.newsbreak.com/news/2310270620685/unlocking-efficient-light-energy-conversion-with-stable-coordination-nanosheets

Nanowerk. (2019). *Nanotechnology in energy*. https://www.nanowerk.com/nanotechnology-in-energy.php

Neumann, O., Urban, A. S., Day, J., Lal, S., Nordlander, P., & Halas, N. J. (2013). Solar vapor generation enabled by nanoparticles. *ACS Nano, 7*(1), 42–49.

Ohzuku, T., Ueda, A., & Yamamoto, N. (1995). Zero-strain insertion material of Li [Li1/3Ti5/3] O 4 for rechargeable lithium cells. *Journal of the Electrochemical Society, 142*(5), 1431.

Otanicar, T. P., Phelan, P. E., Prasher, R. S., Rosengarten, G., & Taylor, R. A. (2010). Nanofluid-based direct absorption solar collector. *Journal of Renewable and Sustainable Energy, 2*(3), 033102.

Patel, V., & Mahajan, Y. (2017). Techno-commercial opportunities of nanotechnology in wind energy. In *Nanotechnology for energy sustainability* (pp. 1079–1106).

Pathak, S., Saha, G. C., Abdul Hadi, M. B., & Jain, N. K. (2021). Engineered nanomaterials for aviation industry in COVID-19 context: A time-sensitive review. *Coatings, 11*(4), 382.

Photonics. (2021). *Stable coordination nanosheets enable efficient light-energy conversion* | Research & technology | Jul 2021 | Photonics.com. https://www.photonics.com/Articles/Stable_Coordination_Nanosheets_Enable_Efficient/a67173

Price, A. (2011). Energy storage for small and micro combined heat and power (CHP) systems. In *Small and micro combined heat and power (CHP) systems* (pp. 307–322). Elsevier.

Rex E, Brunklaus B, Lorentzon K (2015) Energy efficiency along the value chain ways of working for increased competitiveness.

Rhiel, A. (2008). *Application of nanotechnologies in the energy sector*. Available fron: www hessen-nanotech de

Scouter, W. (2012). *Production and storage of hydrogen using nanotechnology*. https://www.azonano.com/article.aspx?ArticleID=3067

Shafie, S. M., Mahlia, T. M. I., Masjuki, H. H., & Andriyana, A. (2011). Current energy usage and sustainable energy in Malaysia: A review. *Renewable and Sustainable Energy Reviews, 15*(9), 4370–4377.

Shank, J., Kadlec, E. A., Jarecki, R. L., Starbuck, A., Howell, S., Peters, D. W., & Davids, P. S. (2018). Power generation from a radiative thermal source using a large-area infrared rectenna. *Physical Review Applied, 9*(5), 054040.

Singh, A. (2018). *Talking renewables: A renewable energy primer for everyone*. Morgan & Claypool Publishers.

Soutter, W. (2013) *Nanotechnology in electric vehicle batteries*. https://www.azonano.com/article.aspx?ArticleID=3157%20

Srinivasan, V., & Kumar, P. (2016). *Review on nanoparticles in CI engines with a new and better proposal on stabilisation* (pp. 1656–1668). IJIRSET.

Wenfeng, Z. (2012). *The manufacturing value chain of power generation equipment: A case study*. ISCTE-Instituto Universitario de Lisboa (Portugal).

Yi, Z., Liu, M., Luo, J., Zhao, Y., Zhang, W., Yi, Y., Yi, Y., Duan, T., Wang, C., & Tang, Y. (2017). Multiple surface plasmon resonances of square lattice nanohole arrays in Au-SiO2-Au multilayer films. *Optics Communications, 390*, 1–6.

Zhang, T. (2020). Methods of improving the efficiency of thermal power plants. *Journal of Physics: Conference Series, 1*, 012001.

Experimental Study of Enhanced Oil Recovery Potential of Nanoparticle (Silicon Dioxide) Coated with Guar Gum

O. O. Olabode, O. Okafor, B. Oni, P. Alonge-Niyi, and V. Abraham

1 Introduction

Ongoing research areas in enhanced oil recovery are to formulate and invent low-cost methods to increase oil recovery from hydrocarbon reservoirs. Special reservoirs like those producing heavy oil have undergone processes such as carbon dioxide flooding, various forms of thermal injections, and chemical EOR. These processes have incurred challenges such as low oil recovery, difficulty in deploying the processes and scales formation especially for thermal, high cost of the chemicals, and its adverse effect on the environment (Al-Campbell, 1981).

The primary and secondary oil recovery strategies regularly produce just 15–30% of the first untapped oil setup (Nikolova & Gutirrez, 2020). Among all the chemical-enhanced oil recovery methods, polymer flooding is a direct technique with a commercial history and proven outcomes. This technology by far outstrips other chemical technologies because the risk of polymer flooding is indeed very minimal, and the envelope of use has greatly widened over the years. EOR projects involving polymer flooding recorded in literatures have been technically and economically successful with incremental oil recoveries between 12% and 15% (Wang et al., 2002).

Nanofluids are essentially a fluid with nanoparticles of explicit capacities as steady colloidal suspension. One cogent issue for the two-stage nanofluid is the dependability of the nanoparticles. The nanoparticles are so little in size, thus light in weight that regularly they can stay suspended in the base fluid, paying little attention to the gravity impact (Kewen Li et al., 2018).

O. O. Olabode (✉) · O. Okafor · P. Alonge-Niyi · V. Abraham
Department of Petroleum Engineering, Covenant University, Ota, Ogun State, Nigeria
e-mail: babalola.oni@covenantuniversity.edu.ng

B. Oni
Department of Chemical Engineering, Covenant University, Ota, Ogun State, Nigeria

A. O. Ayeni et al. (eds.), *Advanced Manufacturing in Biological, Petroleum, and Nanotechnology Processing*, Green Energy and Technology,
https://doi.org/10.1007/978-3-030-95820-6_10

Current studies on nanoparticles are seen as potential solutions to majority of the challenges faced during the deployment of traditional EOR techniques (Sun et al., 2017).

Nanofluid can be balanced out by changing charge thickness and zeta capability of the nanoparticles (Jazeyi et al., 2014). As of late, analysts have discovered that nanoparticles could be used as rheology control operators in polymer flooding. This paper intends to show that using silicon dioxide coated with guar gum as an enhanced oil recovery technique, oil production can still be continued after primary and secondary recovery methods have been depleted with guar gum acting as the polymer and silicon dioxide acting as a nanoparticle.

2 Experimental Procedure

2.1 Materials

The core plugs employed for this experiment were Berea sandstone cores. Four core samples were used (three cores for GCNP flooding and one core for polymer guar gum flooding). From a crude oil reserve from the Niger Delta (zombe Field of OML 124), having an API of 34.97° at 29 °C (medium crude), guar gum and silicon dioxide (SO_2) (purity >99%, average particle size 40–1500 nm) were employed for this experiment.

The equipment engaged in this study includes Soxhlet extractor (for cleaning the core samples), Reservoir Permeability Tester (for core flooding process and permeability measurements), manual saturator, Ofite Model 800 Viscometer (for estimating the rheological properties of liquids), vernier caliper (for mensuration), and pycnometer.

2.2 Characterization of the Core Samples

The core samples are initially flooded with a base fluid (water) to estimate the absolute permeability. The flooding process is continuous to attain complete saturation and steady-state condition of the fluid flow in the core sample. The permeability and porosity values recorded for each core sample as shown in Table 1 are estimated in

Table 1 Characteristics of all the cores

Core samples	Weight before saturation	Weight after saturation	Pore volume	Bulk volume	Porosity	Absolute K (mD)
Core B1	84.3	96.7	12.2	52.66	0.2316	294.3
Core B2	60.6	66.3	5.7	26.87	0.2121	286.07
Core B3	53.9	59.9	6	26.15	0.2294	288.91

conjunction with porosity and relative permeability equations. The vernier caliper assisted in obtaining estimates for each core sample diameters and lengths. With the utilization of the gauging balance, the core samples (B1, B2, and B3) were weighed after it was dried, and at that point immersed with brine by utilizing the manual saturator, the (soaked) samples were weighed once more.

2.3 *Core Flooding (Reservoir Permeability Tester)*

Permeability is the proportion of the capacity of a rock to transmit hydrocarbon liquid. The reservoir permeability analyzer (RPT) is gear initially implied for testing core samples to quantify their permeability during flow hydrocarbons and injected fluids. A further amplification of its capacity is its utilization to test core samples with different fluid samples to observe oil recovery. Core samples can be injected with water, gas, and other liquid mixtures depending on the nature of the experiment. Parameters, for example, water and oil saturations, lingering oil and water saturations, oil recuperations, and permeability changes, might be estimated and determined while utilizing the RPT (www.ofite.com, n.d.). For this work, oil recuperations after water, polymer, and GCNP flooding were estimated, and permeability weakness, water and oil saturations, and leftover oil saturations were additionally estimated. The setup, schematics, and different parts and functions of the equipment are described by Olabode et al. (2020).

The core flood setup comprises three distinct aggregators, topped off with water, raw petroleum, and nanofluid. Every one of the vessels had valves on the deltas and outlets so as to control the liquid stream. Estimations were taken each 3 min and were proceeded for all the flooding phases. Core flooding tests were run on three different cores (B1, B2, and B3). Water flooding analysis was performed on core B1, polymer flooding at 1.5% wt. and 2.5% wt. on core B2, and BpNp with a 0.1% wt. nanoparticles mixed with 1.5% wt. and 2.5% wt. of biopolymer on core B3. The % oil recovered at the different %wt. conc is compared with % oil recovered after water flooding.

2.4 *Characterization of NPs and Polymer*

The stability of nanofluids relies upon the technique for preparation, nanoparticle (NP) attributes, and sort of base liquids, pH, ultra-sonication, and so forth. An attractive stirrer which is a closer option in contrast to the ultrasonic vibration was utilized in this examination to scatter the NPs in the base liquid which was later increased with manual hand blending.

The nano-liquid was set up by weakening the much-focused nano-suspensions with a scattering specialist (deionized water). The nanoparticle being used was SO2I (40–150 nm, virtue more noteworthy than 99%). 50 g of SO2 was scattered in 1 liter of deionized water to make nano-liquid suspensions, making a 5 wt% blend.

Table 2 Density and viscosity of fluids for experiments

Fluid sample	Density(g/ml)	Viscosity (cP)
Nanofluid (silicon dioxide)	1.089	0.891
Oil	0.884	4.623
Water	1.088	0.894
Polymer (guar gum)	1.201	1.942

Table 3 Viscosity determination of 0.1%wt nanoparticles +0.5%wt dissolved in 500 ml of water at room temperature

RPM	Dial readings	Shear rate (S-1)	Shear stress (Pa)	Viscosity
600	6.5	1021.8	3.315	0.070
300	4.5	510.9	2.295	0.097
200	3.5	340.46	1.785	0.113
100	3.5	170.23	1.785	0.227
60	3	102.14	1.53	0.325
30	3	51.07	1.53	0.650
6	2.2	10.21	1.122	2.383
Yield point	2.2	5.11	1.122	

Table 4 Viscosity determination of 0.1%wt nanoparticles +1.5%wt dissolved in 500 ml of water at room temperature

RPM	Dial readings	Shear rate (S-1)	Shear stress (Pa)	Viscosity
600	11	1021.8	5.61	0.201
300	8.2	510.9	4.182	0.300
200	5.9	340.46	3.009	0.324
100	5.1	170.23	2.601	0.561
60	5.2	102.14	2.652	0.953
30	3.7	51.07	1.887	1.356
6	4.6	10.21	2.346	8.433
Yield point	4.6	5.11	2.346	

Table 2 shows the initial density and viscosity of the fluids used during the experiment. The guar gum (polymer) was blended with deionized water at a concentration of 10 wt% yet before it was utilized.

2.5 *Rheological Properties of BpNp*

The rheological properties of the biopolymer to be used for the core flooding procedure were measured. The viscosity, yield point, shear stress, and rate values of the BpNp mixtures measured at room temperatures are shown in Tables 3, 4, and 5.

The yield point at 0.1%wt nanoparticles +0.5%wt compared with the others in Tables 4 and 5 is relatively low, indicating a rapid degradation of the polymer

Table 5 Viscosity determination of 0.1%wt nanoparticles +2.5%wt dissolved in 500 ml of water at room temperature

RPM	Dial readings	Shear rate (S-1)	Shear stress (Pa)	Viscosity
600	9.3	1021.8	4.743	0.144
300	6.5	510.9	3.315	0.201
200	3.5	340.46	1.785	0.162
100	3.5	170.23	1.785	0.325
60	6.9	102.14	3.519	1.069
30	4.8	51.07	2.448	1.488
6	5.9	10.21	3.009	9.145
Yield point	5.9	5.11	3.009	

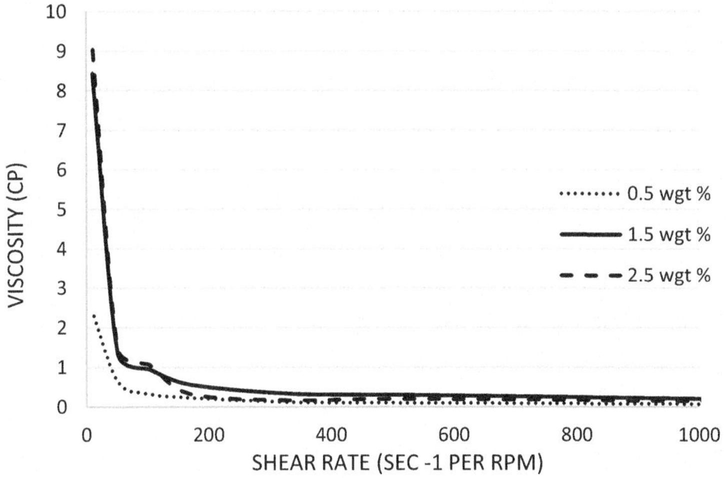

Fig. 1 Rheology summary

mixture during injection. The viscosities of the mixture is inversely proportional to the shear rates observed (Fig. 1) thus, mixtures with lower weight percentages will experience a drastic decline in viscosities as the share rates increases. The shear stress vs shear rate curves in Fig. 2 shows a more stable mixture of higher weight percentages under higher rates. For optimum oil recoveries, the polymer mixtures must improve the viscosities on the injected fluids and maintain its rheological properties prior to its injection.

3 Results

Core sample B1 was injected with three pore volumes of water at a rate of 3 cc/min. To obtain a uniform trend of results, core samples B2 and B3 are desaturated and cleaned using the Soxhlet extractor and re-saturated with oil again

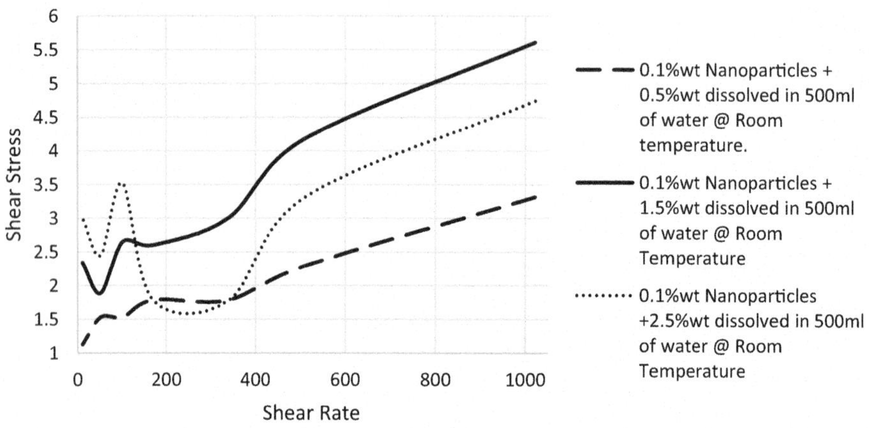

Fig. 2 Shear stress and shear rate relationship at 0.5wt%, 1.5wt%, 2.5wt% BpNp

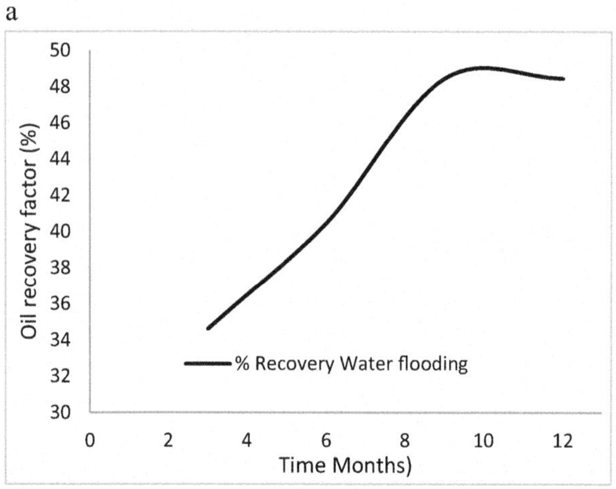

Fig. 3a Oil recovery (water flooding)

after the first injection processes. Then these cores are simultaneously injected with either biopolymers for core B2 or BpNps for B3. Thus, core B2 was first injected with a 1.5% wt. of biopolymer and then a 2.5% wt., while C is injected with a 0.1%wt nanoparticles +1.5%wt biopolymer and 0.1%wt nanoparticles +2.5%wt biopolymer. Figure 3a, 3b, 3c, 3d, and 3e shows the individual oil recoveries from the core samples. From the charts, an oil recovery estimate of 48.62% (figure 3a) is recorded from water flooding, while an incremental oil

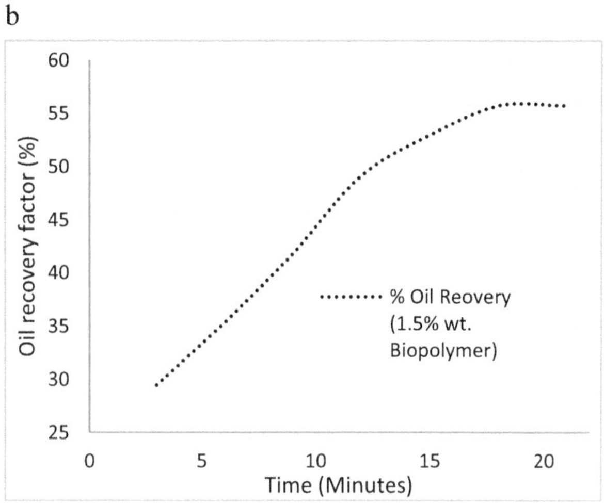

Fig. 3b Oil recovery (at1.5% wt. polymer)

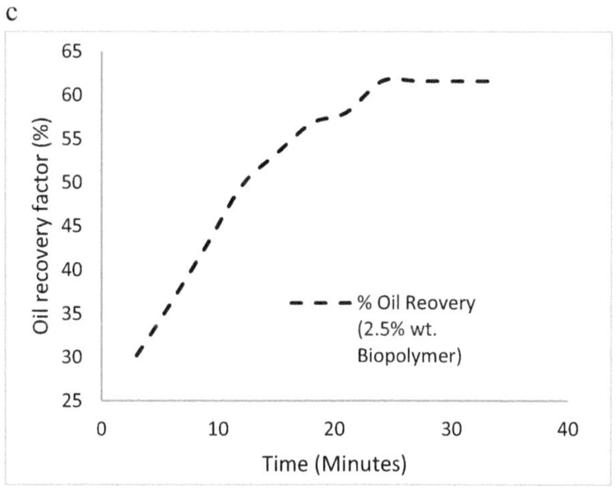

Fig. 3c Oil recovery (at 2.5% wt. polymer)

recovery of 7.1%, 13%, 23.7%, and 28.5% is recorded for polymer flooding at 1.5% wt. and (Fig. 3b) 2.5% wt. and (Fig. 3c) then BpNp at 0.1% wt. and 1.5% wt. (Fig. 3d) and 0.1% wt. and 2.5% wt. (Fig. 3e), respectively. The summary of the oil recovery trends is progressive/incremental as shown in Figs. 4 and 5.

This shows that a BpNp combination is more effective when compared to water flooding and normal polymer flooding in EOR because of increase in

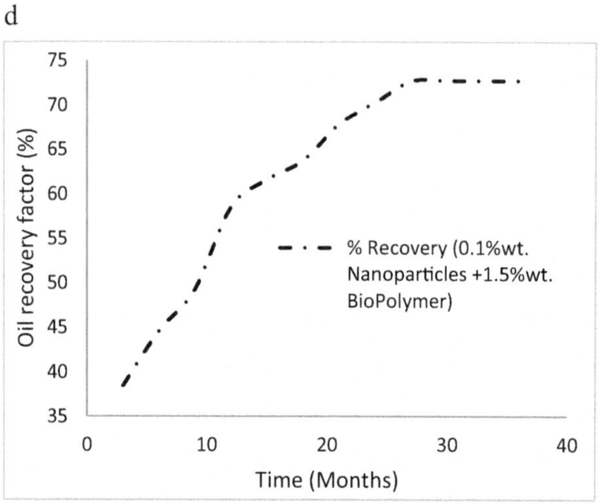

Fig. 3d Oil recovery (at 1.5% wt. BpNp)

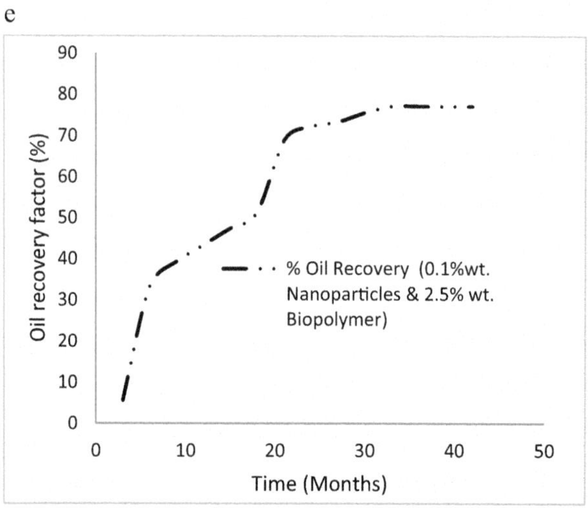

Fig. 3e Oil recovery (at 2.5% wt. BpNp)

mobility of the displacing EOR fluids and the reduction in the rate of polymer adsorption on rock surfaces. This results in a longer time of reaction between BpNp and rock surfaces leading to increased oil recovery. Also, a higher %wt. concentration of BpNp will recover more oil when compared to a lower %wt. BpNp concentration.

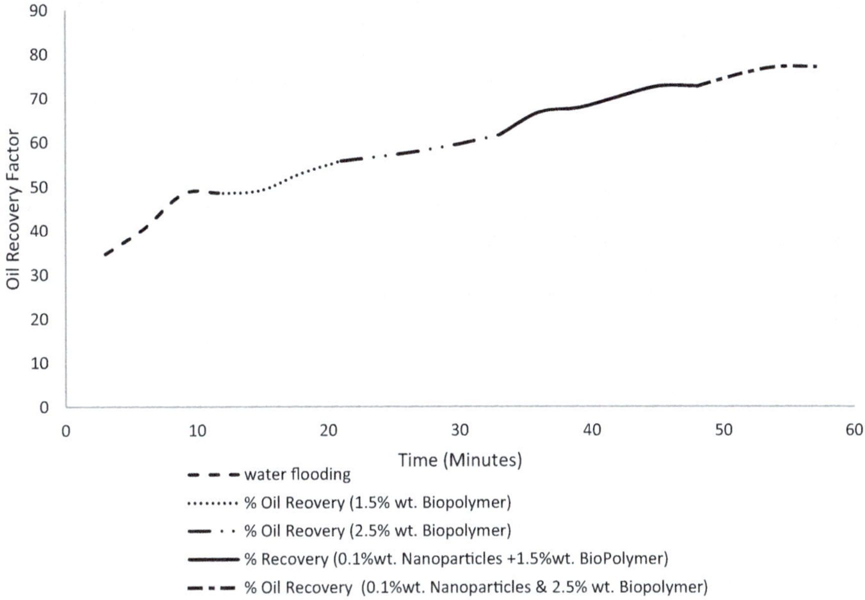

Fig. 4 Oil recovery performances during water flooding and injected BpNp (flooding analysis of core sample (A1), (A2), (A3))

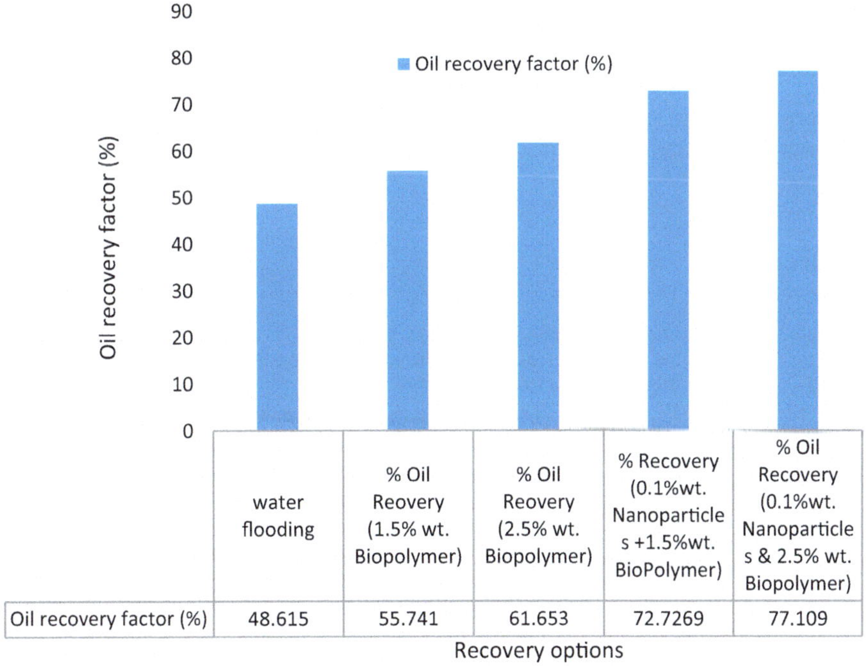

Fig. 5 Oil recovered in % of water, BpNp conc 1 and conc 2, respectively

4 Conclusion

Formulation of biopolymer coated with nanoparticles for oil recovery processes is essential for maximizing oil recovery in medium to heavy oil reservoirs. This formulation helps to increase the viscosity of the injection fluid to recover more oil, and the presence of the nanoparticles aids in reducing the adsorption rates and potentials of the polymer on rock surfaces. This study has successfully shown that utilizing a mixture of biopolymer and nanoparticle can improve oil recovery from reservoirs producing medium crude oil. By infusing BP-NPs a tertiary recovery for medium crude reservoirs, oil recovery can be increased up to 70%.

5 Recommendation

It is recommended that sensitivity analysis be done with respect to the effect of different mixture polymers and nanoparticles during core flooding analysis to ascertain which is best suitable for enhanced oil recovery. These results can be compared with other enhanced oil recovery techniques such as surfactant flooding as described by Abraham et al. (2020) and low salinity flooding options by Olabode et al. (2020b). Before a proper field implantation of an EOR scheme, results from core flooding analysis can be compared with those simulation studies described by Olabode et al. (2021). Lastly, silicon nanoparticle as utilized in this work can be joined with different nanoparticles to exploit their effects on oil recovery.

Acknowledgment The authors thank Covenant University management for creating an enabling environment to carry out this research and sponsorship.

References

Abraham, V., Orodu, O., Efeovbokhan, V., Olabode, O., & Ojo, T. (2020). The influence of surfactant concentration and surfactant type on the interfacial tension of heavy crude oil/brine/surfactant system. *Petroleum and Coal, 62*(2), 292–298.

Campbell, D. A. (1981). Enhanced oil recovery and its environmental and economic implications in the United States. *Environmental Considerations, 8*(1), 5.

Jazeyi, S., Miller, C., Wong, M., Tour, J., & Verduzco, R. (2014). Polymer coated nanoparticles for enhanced oil recovery. *Journal of Applied Polymer Science, 131*(15).

Li, K., Wang, D., & Jiang, S. (2018). Review on enhanced oil recovery by nano fluids. *Oil and Gas Science and Technology, 73*. https://doi.org/10.2516/ogst/2018025

Modelling Low Salinity Water Flooding as a Tertiary Oil Recovery Technique, Oluwasanmi Olabode, David Alaigba, Daniel Oramabo, Oreofeoluwa Bamigboye (2020b). Journal of Modelling and Simulation in Engineering, https://doi.org/10.1155/2019/6485826, Article ID-6485826.

Nikolova, C., & Gutierrez, T. (2020). Use of Microorganisms in the Recovery of Oil From Recalcitrant Oil Reservoirs: Current State of Knowledge, Technological Advances and Future Perspectives. *Frontiers in Microbiology, 10*, 2996. https://doi.org/10.3389/fmicb.2019.02996

Olabode, O., Ojo, T., Oguntade, T., & Oduwole, D. (2020). Recovery Potential of biopolymer (B-P) formulation from solanum tuberosum (waste) starch for enhancing from oil reservoirs. *Energy Reports, 2020*(6), 1448–1455.

Olabode, O., Ogbebor, O., Onyeka, E., & Belema, F. (2021). The effect of chemically enhanced oil recovery on thin oil rim reservoirs. *Journal of Petroleum Exploration and Production Technology*. https://doi.org/10.1007/s13202-021-01090-9

Sun, X., Zhang, Y., Chen, G., & Gai, Z. (2017). Application of nanoparticles in enhanced oil recovery: A critical review of recent progress. *Energies, 10*(3), article no 345.

Wang, D., Cheng, J., Wang, G., & Wu, J. (2002). Experiences learned after production of more than 300 million Barrels of oil by polymer flooding in daqing oil field. In *SPE annual technical conference and exhibition* (p. 10). Society of Petroleum Engineers.

Part II
Pollution Control and Management

Public-Private Partnership: A Veritable Tool for Handling Environmental Pollution and Infrastructural Degeneration in Nigeria

O. J. Oyebode

1 Introduction

Sustainable processes in science and engineering are vital for the protection of our environment, pollution removal, energy provision and enhancement of public health. It is therefore crucial for private and government agencies to work together for robust plan and implementation strategy. Infrastructural decadence is a complicated issue in many countries of the world, and we have to contend with it globally in diverse field of endeavour especially engineering discipline. Health, aviation, road, energy, transportation and every sector can derive maximum benefits from this initiative. Infrastructure development has gained an important significance in Nigeria and it's vital for fighting and attaining social stability. Federal and state governments start infrastructural development because of the main intention during their administrations. The Infrastructure Concession Regulatory Commission (ICRC) was established to assist the federal government's obligation towards the PPP model to finance much-needed infrastructural projects.

2 Public-Private Partnership in Nigeria

Public-private partnership was established in order to achieve quality project with private participation and for speedy completion of projects and adequate financing. With this arrangement, private establishment can be motivated to play active roles in addressing infrastructural deficit and escalating cost of construction, especially in

O. J. Oyebode (✉)
Civil and Environmental Engineering Department, Afe Babalola University Ado-Ekiti (ABUAD), Ado-Ekiti, Ekiti State, Nigeria

A. O. Ayeni et al. (eds.), *Advanced Manufacturing in Biological, Petroleum, and Nanotechnology Processing*, Green Energy and Technology,
https://doi.org/10.1007/978-3-030-95820-6_11

urban areas of the country (Oladokun & Aluko, 2012; Aduwo et al., 2017; Ibem & Aduwo, 2012).

Transportation, energy, public amenities, education, telecommunications, building structures and other important infrastructures have remarkable impact on sustainable development, national growth and environmental sustainability. PPP is an agreement between private establishment and government for the provision of assets and facilities for the populace with special financing system. The current infrastructure scenario in Nigeria needs improvement in all ramifications. The Federal Ministry of Environment is saddled with responsibilities of protecting the environment, conserving natural resources and implementing sustainable development, and Nigeria has a lot of work to do to mitigate the impact of illegal construction and poor partnership between private and government. There are threats of flooding, environmental degeneration, climate change and failure of projects because of various issues emanating from construction projects and financial aspects of public-private initiatives. Public-private partnership (PPP) procurement method and implementation process will definitely assist governments to develop infrastructures and provide facilities and opportunities for the reduction of the debt of governments. Feedbacks of projects, leadership skills, special focus, economic policy, excellent governance, risk allocation and political support will give favourable socio-economic factors and will deliver public infrastructures in a decent way. The achievement of any public-private partnership (PPP) construction works is largely nation's maturity, economic growth, financial schemes and drive for sustainable development. Various forms of public-private partnerships for project execution include build-transfer-operate (BTO), design-build-finance-operate (DBFO), build-operate-transfer (BOT), build-own-operate (BOO) and design-build-operate-maintain (DBOM), among others (Sanni, 2016).

Poverty level, increase in the cost of accommodation and failure of several development schemes, environmental degradation, failure of infrastructures and safety issues in Nigeria necessitated the need for adequate and robust public-private partnership. The government cannot provide affordable housing at low cost to its citizens. It had been the expectation of several experts that the adoption of PPPs can facilitate in addressing housing affordability challenge effort made by many low-income households in Nigerian cities (Ibem, 2010, 2011). This shows that no vital progress has been made in reasonable housing for the low-income urban residents of Nigeria underneath the present PPP arrangement. Among various reasons adduced for this, embrace over reliance on the venture model of surgery in housing; inadequate offer of land; high rate of interest on housing finance, and high price of building materials; high building standards and thus the act of presidency authorities and not-for-profit private-sector organizations in surgery housing comes. Other areas include the shortage of correct definition, monitoring, and incidence of corruption inside the implementation of surgery housing comes in Nigeria. Figure 1 indicates several aspects of project financial options and portfolio representation for government.

Adoption of PPP approach in Nigeria was predicated for the private sector to play a lot of active roles in addressing infrastructural deficit and escalating the value

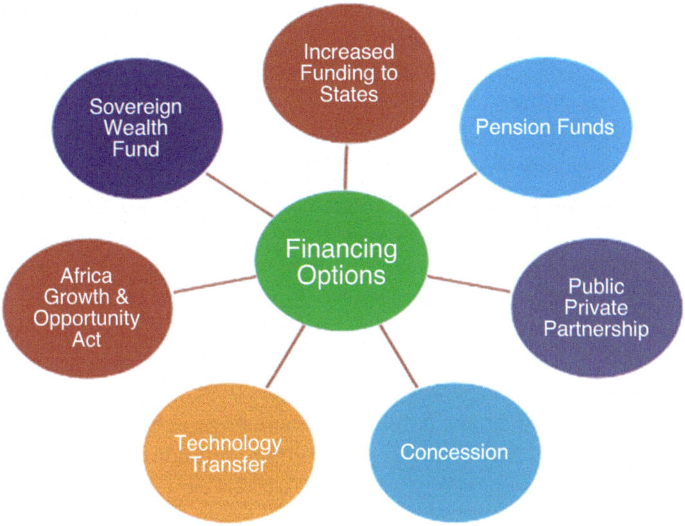

Fig. 1 Project financial options and portfolio representation for government. (Source: Effiom & Ubi, 2016)

of construction, particularly in urban areas of Nigeria (Oladokun & Aluko, 2012; Aduwo et al., 2017; Ibem & Aduwo, 2012). The current infrastructure situation in Nigeria should be improved. The high level of poverty, high cost of housing units and failure of the past development schemes to adequately cater for the needs of this populace in Nigeria should also addressed. There is poor performance of government-led housing delivery ways in providing smart housing at cheap worth to most Nigerians ((Ibem, 2010; Taiwo et al., 2014), it had been the expectation of the many specialists that the adoption of PPPs will facilitate address housing affordability challenge effort several low-income households in Nigerian cities Ibem (2010), Ibem (2011), Ukoje and Kanu (2014), Olofa and Nwosu (2015).

This shows that no important progress has been created in cheap housing for the low-income urban residents throughout this country beneath the present PPPs arrangement. Among the various reasons adduced for this, embrace over reliance on the venture model of surgery in housing; inadequate provision of land; high rate of interest on housing finance, and high worth of building materials; high building standards then the act of presidency authorities and not-for-profit private-sector organizations in surgery housing comes. Others unit the shortage of correct definition, monitoring, and incidence of corruption within the implementation of surgery housing comes in Nigeria. Figure 1 indicates illustration of a portfolio of funding choices on the marketplace for state.

Regarding the transparency measures at acquisition stage of the comes, the study conjointly found that competitive and clear bidding method was the key live enforced by the operators interviewed. This finding is not a surprise as a result of in line with Chan et al. (2010), clear acquisition is one in each of the five main parts of success in PPPs (Chan, 2010). Moreover, proof at intervals the literature (FRN, 2009) shows that clear and competitive bidding technique ensures that every tender submitted is evaluated victimization uniform standards and criteria; then, reducing the use of bribes and differing types of corruption at intervals the award of surgery contracts. This might be collectively in agreement with the essence of transparency as a tool for fighting corruption in surgery comes as given at intervals the literature (Nelson, 2003; FRN, 2009; Greve & Hodge, 2010).

Public-private initiatives and concession activities are very beneficial to construction projects as it improves facility development, public infrastructure, community service, operation and maintenance of engineering projects (Aderibigbe, 2008; Batley, 1996; Akintoye et al., 2003; Murtala, 2007; IP3, 2009). Infrastructure parts like sewerage systems can affect the environment, residences, hospitals, communities and government corporations (Oyebode, 2019a, b). The government must improve housing for low-income public servants in Nigeria by handling the issue of housing holistically. All infrastructures should be well constructed and managed (Taiwo, 2015).

3 Benefits of PPPs for State and Taxpayers

PPP consider creating cooperation between public authorities and the private sector for optimal performance. These initiatives have taken many forms like the outright privatization of antecedent nation-owned industries, obtaining out services to non-public organizations and conjointly the use of private finance at intervals the supply of social infrastructures and services. Public-private partnership (PPP) provides effective service to the public expeditiously by specializing publicly service output, they supply a tons of refined and cost-efficient approach to the management of risk by public sector than is typically achieved by ancient input-based on public sector acquisition (Carr, 1998; Babalakin, 2013).

Table 1 provides the share of the transport sector in Nigeria's public sector (1962–1998).

Table 2 presents examples of PPP projects in various professionals in the country. Figure 2 indicates percentage of implementation, development and procurement in notable professionals in Nigeria. Figure 3 presented energy, industrial, transport and other sectors of PPPs in Nigeria.

Table 1 Percentage share of the transport sector in Nigeria's public sector (1962–1998)

Plan	Plan period	% share of TS
First Countrywide Development Plan	1962–1968	21.3%
Second Countrywide Development Plan	1970–1974	23.7%
Third Countrywide Development Plan	1975–1980	22.2%
Fourth Countrywide Development Plan National Development Plan	1981–1985	15.2%
National Development Plan	1993–1995	11.0%
National Development Plan	1994–1996	8.6%
	1996–1998	10.1%

Source: National Development Plans

Table 2 Example of PPP projects in Nigeria

Name	Sector	State	Sponsoring agency
Lagos-Ibadan expressway	Transport	Lagos	Federal Ministry of Works
Kuto-Bagana bridge	Transport	Nasarawa, Kogi	Federal Ministry of Works
Lekki-Epe	Transport	Lagos	Lagos State
Maervis Management of airport	Transport	All international airports in Nigeria	Federal Ministry of Works
Dagbolu inland container depot	Transport	Osun	Nigerian Shippers' Council
Lolo inland container depot	Transport	Kebbi	Nigerian Shippers' Council
Onitsha inland container depot	Transport	Anambra	Nigerian Shippers' Council
Apapa port terminal c	Transport	Lagos	Nigerian Ports Authority
Lekki deep water port	Transport	Lagos	Nigerian Ports Authority
Warri new terminal a	Transport	Delta	Nigerian Ports Authority
Bakolori dam	Energy	Zamfara	Federal Ministry of Power, Works and Housing
Badagry deep water port	Transport	Lagos	Nigerian Ports Authority

Source: ICRC (2020)

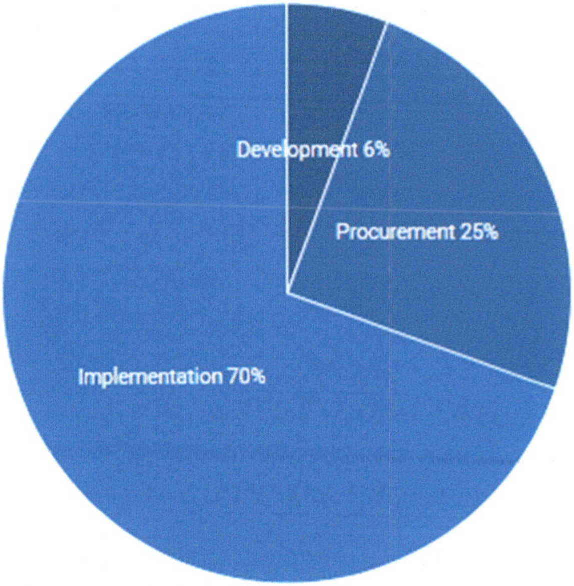

Fig. 2 Percentage of implementation, development and procurement in Nigeria

Figure 4 presents public-private partnership benefits, Fig. 5 presents PPP models and relationship between private sector involvement and risks and Fig. 6 presents PPP participation and various concessions in government works.

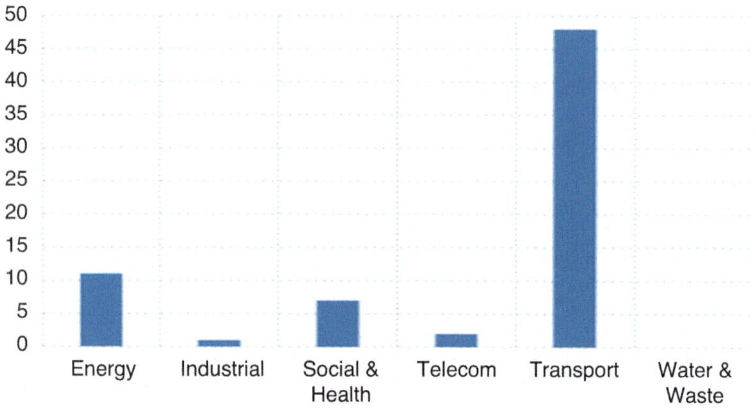

Fig. 3 Energy, industrial, transport and other sectors of PPPs in Nigeria

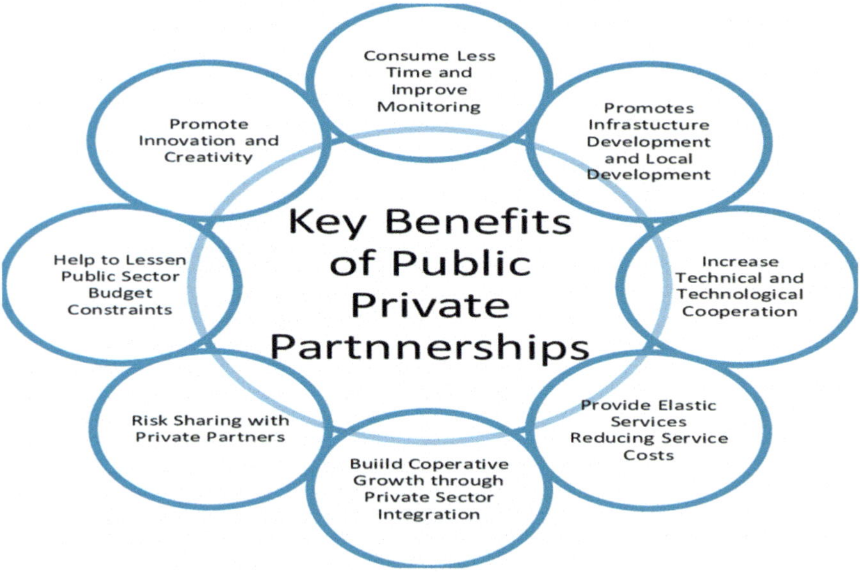

Fig. 4 Public-private partnership benefits

3.1 Challenges Facing PPP in Nigeria

1. The public sees the immediate burden but they do not see deferred benefits.
2. Change in stakeholders' behaviour and shift in existing political and economic power network.
3. Government employees erroneously believe that PPP means an end to traditional procurement.

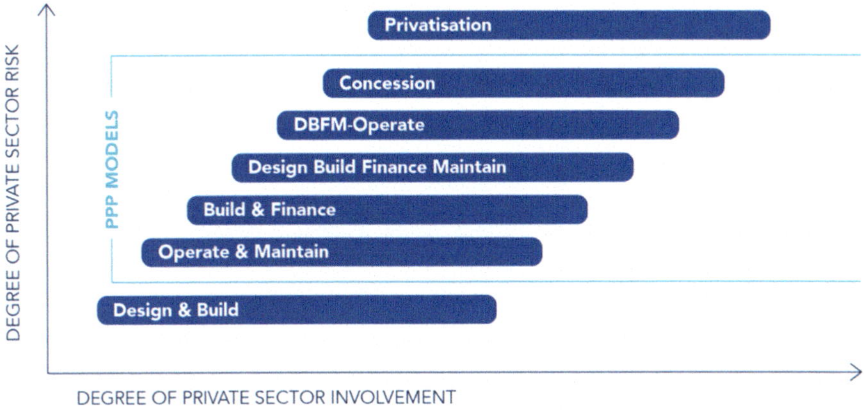

Fig. 5 PPP models and relationship between private sector involvement and risks

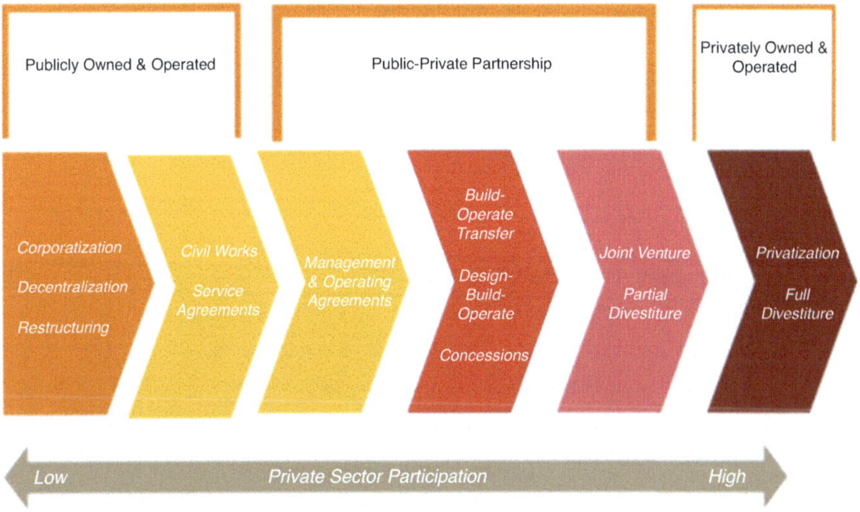

Fig. 6 PPP participation and various concessions in government works. (Source: Jamnadass et al. 2014)

4. Lack of drive for safety, quality control and construction integrity.
5. The public is always suspicious of the motive of the private sector because of the belief that only profit drives the private sector.
6. Illicit acts, corruption and poor financing system.
7. Lack of innovative policies or frameworks that can assist environmental engineers.
8. Lack of positive mindset towards PPP is a major challenge. PPP transfer significant risk to private sector in most cases.

4 Remedy for Civil Infrastructure Degeneration in Nigeria

Plates 1 and 2 indicate infrastructural construction as a remedy for infrastructural decay in Nigeria. Plate 3 presented precast concrete for a typical construction project.

Remedies for engineering infrastructural degeneration include:

(i) Excellent delivery of projects in terms of scope and quality
(ii) Precision of budgets and completion of project within scheduled time
(iii) Robust management plan and implementation strategies
(iv) Periodic maintenance of civil infrastructure
(v) Appropriate manpower development and staff remuneration
(vi) Transparency in planning and procurement of materials for project works
(vii) Formulation of policies, regulations and rules for public-private partnership
(viii) Occupation health and safety measures for workers and management staff
(ix) Reduction of environmental pollution through engineering interventions
(x) Innovative approach in contractual works
(xi) Clear understanding of contract agreement and payment terms
(xii) Comprehensive feasibility studies, design, supervision, implementation and maintenance strategies

Plate 1 Planned road network for sustainable infrastructures. (Source: Field Study 2021)

Plate 2 Ongoing construction works with emphasis on sustainable infrastructures. (Source: Field Study 2021)

Plate 3 Precast concrete for a typical construction project

5 Conclusions

It has been established from this paper that:

(i) Corruption and issues linked with budgeting and financing and unhealthy partnership need to be overcome in all facets of PPP administration.
(ii) The availability of infrastructures is an incentive to increasing economic efficiency, national development and productivity.

(iii) The government has not done enough by establishing mechanism which can help in fighting corruption among PPP arrangements.
(iv) Many public-private partnerships have not been designed and monitored properly. This experience has eroded support for PPP concept.
(v) The intensity of commitment from the public and private sector changes from contract to contract.
(vi) Public-private partnership is essential for economic development.

6 Recommendations

Based on this paper, the following were recommended:

(i) Public-private partnership should be set up and designed for positive benefits to government. Implementation of viable policies for its monitoring will yield greater benefits.
(ii) Proper laws, regulations and institutions should be encouraged to facilitate improvement of PPP projects.
(iii) Enabling environment is required for effective management of infrastructures.
(iv) There is a need for cooperation and understanding among partners and other stakeholders; this might reduce the frequency of litigation.
(v) There is need for a well-organized system, human resource development and in-depth assessment of facilities for achieving infrastructural developments.

References

Aderibigbe, A. (2008). Procurement methods – Procurement of goods/works. In *Basic procurement course for procurement officers* (pp. 11–15).

Aduwo, E. B., Ibem, E. O., & Onyemaechi, P. D. (2017). Challenges and opportunities in public-private partnerships (PPPs) for Housing low-income earners in Nigeria. *Advances in Business and Management, 12*, 145–168. (eprints.covenantuniversity.edu.ng).

Akintoye, A., Beck, M., & Hardcastle, C. (2003). *Public private partnership, management, risk and opportunities*. Wiley–Blackwell.

Babalakin, B. O. (2013). *Public private partnership: "Infrastructure development as a vehicle for economic development*" (being a paper presented at the 2nd Mustapha Akanbi Public Lecture organized by the Faculty of Law, University of Ilorin, p. 6.)

Batley, R. (1996). Public private relationships and performance in service provision. *Urban Studies, 33*(4–5), 723–751.

Carr, G. (1998). The Canadian council for public private partnership (online). *The Canadian Council for Public Private Partnerships*.

Chan, A. P. (2010). Critical success factors for PPPs in infrastructure developments: Chinese perspective. *Journal of Construction Engineering and Management, 136*(5), 484–494.

Chan, A. P., Lam, P. T., Chan, D. W., Cheung, E., & Ke, Y. (2010). Critical success factors for PPPs in infrastructure developments: Chinese perspective. *Journal of Construction Engineering and Management, 136*(5), 484–494.

Effiom, L., & Ubi, P. (2016). Deficit, decay and deprioritization of transport infrastructure in Nigeria: Policy options for sustainability. *International Journal of Economics and Finance, 8*(3), 55–68.

Federal Republic of Nigeria. (2009). *National policy on public private partnerships* (dspace.ut.ee). (dspace.ut.ee). Infrastructure Concession Regulatory Commission (ICRC).

Greve, C., & Hodge, G. (2010). Public-private partnerships and public governance challenges. In *The new public governance: Emerging perspectives on the theory and practice of public governance* (pp. 149–162). Routledge.

Ibem, E. O. (2010). 'An assessment of the role of government agencies in public-private partnerships in housing delivery in Nigeria', Journal of construction in developing countries. *Penerbit Universiti Sains Malaysia, 15*(2), 23–48.

Ibem, E. O. (2011). The contribution of Public–Private Partnerships (PPPs) to improving accessibility of low-income earners to housing in southern Nigeria. *Journal of Housing and the Built Environment, 26*(2), 201–217.

Ibem, E. O., & Aduwo, B. E. (2012). Public-Private Partnerships (PPPs) in urban housing in Nigeria: Evidence from Ogun State. *International Journal of Architecture and Urban Development (IJAUD), 2*(2), 2–15.

IP3. (2009). Contract compliance and performing monitoring for public private partnership (online). A training workshop, October 2–13, 2000. The institute for public private partnership (IP3) Washington, DC. http://www.ip3.org/contract2000.htm. Assessed 21 Dec 2000.

Jamnadass, R., Langford, K., Anjarwalla, P., & Mithöfer, D. (2014). Public–private partnerships in agroforestry. pp. 544–564.

Murtala, S. S. (2007). *Public private dialogue, myth or reality*. Department of Business Administration, Bayero University, Kano, Nigeria. Dec 13 2007.

Nelson, P. J. (2003). Multilateral development banks, transparency and corporate clients: "public–private partnerships" and public access to information. *Public Administration and Development, 23*(3), 249–257.

Oladokun, T. T., & Aluko, B. T. (2012). Public-private partnership in housing delivery in Lagos State, Nigeria. In *International conference on construction and real estate management* (pp. 1–2).

Olofa, S., & Nwosu, A. (2015). Investigating the problems associated with public private partnership in the process of housing delivery in Nigeria. *International Journal of Education and Research, 3*(1), 130–133.

Oyebode, O.J. (2019a). Impacts of civil engineering infrastructures in the sustainability of the environment. In *Proceeding of 2019 international conference of Nigerian Institution of Civil Engineers (NICE) held in Maiduguri in Nigeria* (pp. 75–85).

Oyebode, O. J. (2019b). Strategies for effective delivery of structural engineering projects in Nigeria through manpower development and adequate remuneration. In *Proceeding of 2019 International Conference and 32nd AGM of Nigerian Institution of Structural Engineering Eba Odan 2019 held between 22nd and 23rd October 2019 at the University of Ibadan*, Conference Center, Ibadan, Oyo State.

Sanni, A. O. (2016). Factors determining the success of public private partnership projects in Nigeria. *Construction Economics and Building, 16*(2), 42–55.

Taiwo, A. (2015). The need for government to embrace public–private partnership Initiative in Housing delivery to low-income public servants in Nigeria. *Urban Design International, 20*(1), 56–65.

Taiwo, A. A. T., Adeboye, A., & Adewuyi Aderonmu, P. (2014). Housing finance: The role of the private sector in public-private partnership in housing delivery for the low-income in Nigeria. *International Journal of Architecture and Urban Development, 4*(3), 5–10.

Ukoje, J. E., & Kanu, K. U. (2014). Implementation and the challenges of the mass housing scheme in Abuja, Nigeria. *American International Journal of Contemporary Research, 4*(4), 209–218.

Smart Toilets and Toilet Gadgets in Sustainable Smart Cities: An Overview of Personal Health Monitoring

C. C. Mbonu, O. Kilanko, M. B. Kilanko, and P. O. Babalola

1 Introduction

Various economic and social crises have resulted in the degradation and collapse of the environment and the cities we live in (Yigitcanlar & Lee, 2014). These cities and communities are a concern, which impacts negatively on economic growth and liveability. Hence, the offspring of sustainable cities and communities was shortlisted as the 11th sustainable development goal among the 17 slated sustainable development goals of the 2030 sustainable development agenda by the United Nations (UN) in 2015 (Mustafa et al., 2021). The aim to attain improved economic condition, social condition, environmental condition, infrastructure, and services drives the concept of a sustainable city (De Jong et al., 2015).

2 Smart Cities

2.1 *Smart Sustainable Cities and Services*

A sustainable city in this frame is addressed as one, which makes the offspring of new business and career possible and safe, provides a resilient building that has great positive social and economic impact, and provides affordable and safe shelter

C. C. Mbonu · O. Kilanko (✉) · P. O. Babalola
Department of Mechanical Engineering, Covenant University, Ota, Ogun State, Nigeria
e-mail: oluwaseun.kilanko@covenantuniversity.edu.ng;
phillip.babalola@covenantuniversity.edu.ng

M. B. Kilanko
General Hospital Idiroko, Ipokia Local Government Area, Idiroko, Ogun State, Nigeria

A. O. Ayeni et al. (eds.), *Advanced Manufacturing in Biological, Petroleum, and Nanotechnology Processing*, Green Energy and Technology,
https://doi.org/10.1007/978-3-030-95820-6_12

for occupants. All these are to be provided without creating any negative impact on lives and the environment. For a city to attain the state of sustenance, it must make available sustainable transportation, green energy generation and supply, services, infrastructures, green environment, improved city planning, and improved management or governance (Townsend, 2013). The development of a sustainable city is the first step in successfully developing a smart city.

Yigitcanlar (2016) views that the name smart city is a substitute for sustainable digital or information city, owing to the point of view that each city name is because of a change of name, technology, or author's view. The building of smart cities and the innovation of already existing cities to attain smartness is an attempt to improve the efficiency of the operation and operating condition of a city. Smart cities employ the use of technology, which targets the improvement of the services rendered to occupants, living experience, and process of operation of business and business owners (Reichental, 2020).

As viewed by various authors, a smart city is an outcome of smart-oriented governance, management, services, citizen, environment, and concept of living (Deakin, 2013; Halegoua, 2020). Dey et al. (2018) give a representation of the

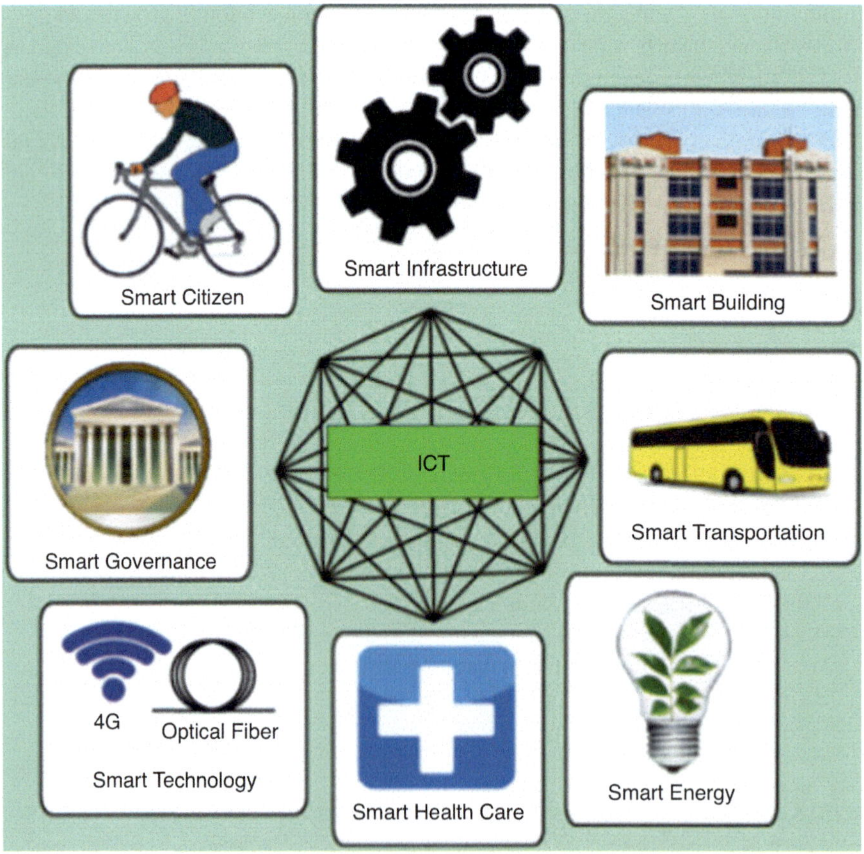

Fig. 1 Components of a smart city. (Dey et al., 2018)

component of a smart city as highlighted by the authors (Fig. 1). The word "smart" in this context has its foundation on the concept of data acquisition, data processing and analysis, data management, and implementation (Fayomi et al., 2019). Employing the Internet of things (IoT) and information and communication technology (ICT) has made the development of smart cities possible over the years (Lara et al., 2016).

Hence, it is permissible to say that a city is termed smart if data acquisition, data processing, and analysis, data management, and implementation are key factors applied by all sectors in its day-to-day activities (Da Silva & Flauzino, 2016; Oyedepo et al., 2020; Reichental, 2020). Novotný et al. (2014) highlighted that smart cities involve the attainment of sustainable cities by the use of technologies. This involves the collection, analysis, and implementation of data using sensor and actuator technology, software, servers and server subsystems, digital and analogue devices, and other networking systems. As listed above, this technology employs the use of ICT and IoT, which connects the building, management or governance, structures, services, and occupants together, exchanging and transferring data in real time.

Smart service is one of the indicators of a smart city. Some of the other smart city indicators include smart energy and energy grid, smart buildings, smart metering, smart water, smart public services, and smart water management. Smart services involve the use of automation in real-time data collection, data analysis, data implementation, etc. when using various services. The data obtained is used for remote monitoring, improvement of user's experience, and personalised user experience or for maintenance and managerial purposes. Novotný et al. (2014) show the collaboration of various components and services in a smart city with ICT in Fig. 2.

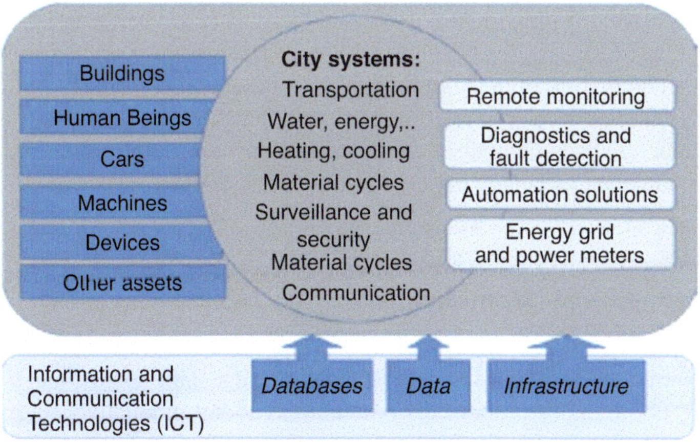

Fig. 2 Collaboration of various components and services in a smart city with ICT. (Novotný et al., 2014)

Smart services are interconnected to various systems and infrastructures in a smart city; this is because the infrastructures and amenities provide services. Hence, smart services involve smart approaches in rendering services by various infrastructures. This spreads across various sectors in the smart city such as the transportation sector, education sector, health sector, power and energy sector, utility sector, security, etc. (Georgakopoulos & Jayaraman, 2016). The provision of these services makes smart energy grid, smart security, smart assist and well-being, smart home and automated home, smart healthcare and remote health monitoring, machine-to-machine interaction, machine-to-human interaction, and vice versa effective and reliable (Mishra & Chakraborty, 2020).

From the aspect of rendering services, XaaS (which denotes anything as a service) cuts across all possible service that can be rendered. This ranges from infrastructural services or infrastructure as a service, security as a service, housing as a service, healthcare as a service, health monitoring as a service, metering as a service, and anything that can be rendered as a service (Duan et al., 2015). Healthcare as a service in most developed countries has advanced more than just electronic medical records (EMR), automated hospital experience, hospital information systems, and the like of it (John & Shenoy, 2014).

Healthcare services have attained the height of remote surgery, computer-aided surgery, and remote patient monitoring and care due to the advancements in information and communication technology coupled with the advancement in artificial intelligence (AI) ability on the Internet of things (Pradhan et al., 2021). The offspring of digital healthcare services and the advancement in the present obtainable healthcare services is attributed to the synergetic and collaborative approach by various fields and professions. This strives to attain more digitised or smart healthcare that is sustainable in process and operation (Kabadayi et al., 2019).

The broad application of sensors in healthcare services has improved the quality and efficiency of medical services received by individuals. This provides accurate and real-time data acquisition on an individual's health status by detecting and analysing the function of individual organ or body vitals (e.g. heartbeat, colour of the tongue, colour of the eye, liver function, kidney function, etc.) or the measurement of the constituents from or in the individual (e.g. urine colour, urine turbidity, blood sugar, etc.). These senses also acquire or obtain valuable medical data by monitoring the characteristic behaviour of the individual such as eating pattern, activeness of the individual, and sleeping pattern (Yilmaz et al., 2010).

2.2 Smart Health Monitoring

With the introduction of personal health monitoring, data acquisition for health monitoring traditionally processed by humans is today gradually being replaced with the use of senses and other technologies. Brent (2018) defined personal health monitoring as the use of a system or a device with the ability of data collection, data storage, and transmission of data on health-related issues of an individual outside the hospital environment. The author highlighted that personal health monitoring involved the use of sensors that could be implanted, worn on the body of the individual, or strategically placed in the environment of the person. This sensor technology provides a cheap, non-invasive, non-destructive, and continuous approach to health monitoring which updates the individual's medical record in real time.

Various devices and equipment have been equipped with various sensor technologies capable of monitoring and detecting the health status of the user. This ranges from wearable devices, smartphones, and other devices that have the capacity or ability to house and operate wireless sensors. Baig and Gholamhosseini (2013) expressed that health monitoring involves the analysis of a user's vitals or organs with the purpose of detecting variations from original functionality, characteristics, or contents of the body. Health monitoring systems are aimed at reducing the number of hospital patients, reducing the burden on hospital staff, conserving time, reducing queuing, saving money, and eliminating unnecessary trips to the hospital. The author expresses that the health monitoring system can be divided into three groups.

- Remote health monitoring system (RHMS)
- Mobile health monitoring system (MHMS)
- Wearable health monitoring system (WHMS)

2.2.1 Remote Health Monitoring System (RHMS)

The significant characteristic of the remote health monitoring system is the ability of hospitals to receive health data from a remote location using remote access networking. This type of health monitoring system can be strategically placed at home, officers, parks, or any other facility away from the hospital environment to gather multiple or single types of sensing parameters. The processor of information acquired from the sensor in this type of health monitoring system can be located with the device or in hospitals for the analysis of data. This health monitoring system and stationery can detect one location at a time (Mohammed et al., 2019).

2.2.2 Mobile Health Monitoring System (MHMS)

The mobile health monitoring system involves the use of easy-to carry devices capable of collecting medical data from the user and is able to process the data acquired. Hence, it must consist of the ability to process data with a processor. The mobile phone, smart gadget, pocket computer, digital assistant, and any other pick-to-use device are examples of mobile health monitoring systems. The user is able to receive health monitoring services and medical care at the same time as scheduled by the user (Agrafioti et al., 2011; Mohammadzadeh & Safdari, 2016).

2.2.3 Wearable Health Monitoring System (WHMS)

The third health monitoring system is the wearable health monitoring system. The wearable health monitoring system involves all types of health monitoring device that is worn on the body in order to collect medical data from the user, which could be in the form of a vest, a band, or an implant. This provides a ground to conduct real-time health monitoring throughout the period the health monitoring device is worn or is in contact with the body (Yilmaz et al., 2010).

As highlighted by John and Shenoy (2014), the level of development in healthcare service is dependent on the level of collaboration, patient-centredness, and real-time data acquisition employed by health service providers. Smart healthcare service employs the use of ICT, which involves the collection, or acquisition, of tons of data. This size of data input of the present healthcare services in the developed nations is overwhelming for the traditional techniques used in analysing and storing data. Using the new healthcare technology, an individual is capable of producing millions of data in half a minute (Holzinger et al., 2015). This brings into play the use of cloud computing in managing and processing of data.

Cloud computing provides easier access and storage of data having tools and applications capable of performing networking, database provision, software provision, and data storage rather than the use of local storage devices, bookkeeping which involves documentation recording, and filing using traditional approach or the use of hard drive and other portable storage devices. Cloud computing involves the storage of information or data to a remote database that is easily accessible. The data stored can be easily assessed, processed, and analysed when proper identification is provided. The incorporation of cloud computing in the health sector for the rendering of health services gives room for a new type of cloud database referred to as health cloud. This provides an improved medical recording process, which can be easily shared and stored in a secure manner (Kumar et al., 2020).

2.3 *Automated Toilets and Toilet Gadgets*

The initial automation in toilets involved the implementation of sensor and actuator technology, which provides comfort to the user. Common technology found in such modern toilets involves the use of push to control functions such as warming of the toilet seat, wiping system, opening and closing of the toilet lid, music, and night light. It also comprises sense-controlled features such as automatic toilet lid control, auto flush, auto disinfectant, etc. Today, smart toilets have advanced as a device not only for just providing luxury and automated services but also for health monitoring purposes by targeting and analysing the use of urine and stool samples to detect disease markers (Bhatia et al., 2020; Lokman & Kanesaraj Ramasamy, 2019; Shaikh et al., 2019).

2.4 Impact of Smart Toilets on the SDGs

The SDGs of interest in this study involve SDG 3, SDG 6, and SDG 11. Health is the centroid of SDG 3, which is focused on attaining a healthy lifestyle and well-being of all irrespective of age, gender, and physical or mental state (Filho et al., 2019). SDG 6 which is about clean water and sanitation for all is targeted at providing substantial systems and methods for the availability of water and improved sanitation for all by 2030 (Hutton & Chase, 2018). As expressed by Ban Ki-Moon (former Secretary-General of the United Nations), the city is the determining factor if sustainable development is attainable or not since the city accounts for the higher population and GDP globally. The 11th SDG is focused on attaining a state of safety, inclusion, sustenance, and resilience for cities and communities (Al-Zu'bi & Radovic, 2018). The impact of smart toilets in attaining the SDGs mentioned is summarised below.

2.4.1 Impact on SDG 3 (Good Health and Well-Being)

The aspect of providing improved healthy living includes a provision for sustainable healthcare service; this satisfies some aspects of SDG 3. As earlier discussed, the provision of sustainable healthcare service includes a substantial healthcare management system (NU CEPAL, 2019). Various automated or smart toilets and smart toilet gadgets are equipped with the capacity to monitor the functionality of a human body. This provides substantial information that is of great importance and gives an advantage in providing the best medical services to the user. The use of order automation and sense of technology in the toilet such as mentioned above provides a serene toilet environment that is safe, stress-free, and comfortable (Park et al., 2020).

2.4.2 Impact on SDG 6 (Clean Water and Sanitation)

The concept of clean water in SDG 6 addresses the availability of clean water, which is achievable by the production of a new water supply or the management of already existing water supply and the rate of water usage. The introduction to different alternatives in the use of water in performing various functions makes water available for other functions. Various smart toilets incorporate various methods and techniques to reduce the use of water in the toilet. The reduction in the usage of water in the toilet has been achieved by proper metering of water usage, structural innovation to reduce the volume of water used in the toilet, or the provision of alternatives to the use of water such as the use of a suction pump, etc. (Hauber-Davidson & Idris, 2006; Hashemi et al., 2015).

Improved sanitation is one of the sixth SDG, which is directed to provide a clean and safe environment for all. The smart toilet is focused on providing good sanitation in the toilet by providing auto-flush capacity. By this, it is assured that the toilet remains clean from biowaste at all times. The use of this technology prevents the event of an unkempt toilet in public and private toilets, which can cause irritations and infections (Wath, 2016). The use of various sterilisation technologies and other technologies for disinfecting the toilet such as the use of ultraviolet steriliser and the automated spray of disinfectant prevents the occurrence of infection.

2.4.3 Impact on SDG 11 (Sustainable Cities and Communities)

Sustainable cities are cities that have attained a state of sustenance in all the sectors of a city which include the education sector, healthcare sector, energy distribution sector, agricultural sector, social sector, economic sector, etc. (Filho et al., 2020). The use of smart toilets in a city gives room for networking, intelligent systems, machine-to-machine communication, artificial intelligence-based analysis, statistics, and algorithms. This provides job opportunities for IT inclined and associated graduates (since the development of smart toilets involved a synergetic approach), improves data collection, provides comfort for the user, saves time and money, and provides a means to generate revenue.

2.5 Smart Toilet as a Personal Health Monitoring System

In a smart health system, which provides smart health services, personal health monitoring is key in properly rendering healthcare services. This is aimed at detecting health-related issues in the individual, managing the condition of the individual's health, or preventing the occurrence of any health-related issue by the use of sensors, which measure and observe various functions of the individual's body. The essence of personal health monitoring is to provide medical care in the most convenient, cheap, efficient, and accurate way outside the traditional health monitoring system and hospital environment (Mittelstadt, 2013; Mittelstadt et al., 2014; Nordgren, 2015).

The use of toilets at housing to embed various sensors in them is aimed at conducting health monitoring on human excreta and human activity in the toilet. The use of smart gadgets or toilets for data collection relevant to healthcare makes the gadget or a smart toilet a personal health monitoring device, which is a supplement to traditional health monitoring (Mittelstadt, 2013). The information obtained from sensors embedded in smart toilets and toilet gadgets is valuable data employable in the monitoring of a person's health status. This is also a method of health monitoring, which was expressed in Jiuping and Lei (2017) as a process that conducts active monitoring of a system to prevent failure or reduce maintenance cost without causing destruction of the system.

Ghosh et al. (2020) develop a smart toilet system which has the capacity to estimate the level of blood sugar in the body through the urine in a non-invasive way, with an aim of controlling type 2 diabetes, which is a chronic disease. The method employed by the author involved the measurement of the sugar in the blood using the urine. This was used to determine the drug dosage to be administered to the patient in order to control type 2 diabetes. The author measured the amount of blood sugar by employing the use of Benedict's quantitative solution as the reagent to the urine sample. The urine reacts with Benedict's quantitative solution to provide a colour, which is detected using an RGB colour sensor and analysed using a fuzzy logic-based colour estimation technique.

Bhatia et al. (2020) designed an IoT-based smart toilet system for the prediction of urinary tract infection. The author employed urine acquisition, urine analysis, temporal extraction, and temporal prediction layer as the four-layer architecture for the monitoring and prediction of urinary tract infection. Artificial neural networks (ANN) were employed in the determination of the degree of infection and infection index value. The results of the monitoring and prediction conducted were displayed to the user using self-organised mapping technology.

Bae and Lee (2018) portrayed an example of a smart toilet in Fig. 3 which has the ability to impact SDGs 6 and 7 by reducing the use of clean water for flushing stool and urine by 90%, which addresses the affordability of clean water in SDG 6. When the amount of water usage can be supplemented by various alternative techniques, then there will be more water for other use. The aspect of sanitation was implemented by providing a contact safe environment in the toilet. The concept of affordable energy was implemented by the conservation of toilet waste.

Since it is imperative to separate stool waste and toilet waste in the production of effective biofuel, the illustrated smart toilet has the capacity to separate urine and stool (Hotta & Funamizu, 2007). This feature in this smart toilet is due to the technical alternatives embedded in the toilet. The structure of the toilet provides different channels, which separate urine from stool, and stores them differently. In place of the standard water to flush mechanism embedded in traditional toilets, this toilet utilises the use of a vacuum suction pump as its flushing mechanism. The smart toilet consists of an ultraviolet steriliser, which is positioned to sterilise the sit before use.

Fig. 3 Smart toilet with the ability of urine storage and separation from the stool. (Bae & Lee, 2018)

Park et al. (2020) designed a personalised health monitoring device, which was applied in the toilet shown in Fig. 4. The system acquires medical data from analysing the urine and stool of the user. Using the dipstick method, the author was able to conduct ten key urine analyses, which are useful in the determination of diabetes, urinary tract infection, kidney disease, prostatic hypertrophy, and neurogenic bladder issues. The urine analysis conducted by the mounted toilet system involves the measurement and detection of glucose, protein, blood, and several other disease markers detectable in the urine. This urine analysis was carried out using an automatic dipstick loading mechanism.

Using two high-speed cameras labelled (VI) in Fig. 4, computer vision flow metering was conducted on the urine to detect the urine flow rate and the urodynamic of the urine during voiding. This serves the purpose of uroflowmetry. The analysis of the solid waste streamlined to the detection of the stool type with reference to the Bristol stool form scale (BSFS) which gives a range of three types of stool: the first range is BS 1–2 (for constipation), the second range is BS 3–5 (a normal stool) and the third range is BS 6–7 (for diarrhoea stool). The image of the stool is taken with a camera labelled v in Fig. 4, and the type of stool is analysed using a deep convolutional neural network (CNN) (Park et al., 2020).

The duration in the use of the toilet was detected using a pressure sensor labelled (I) in Fig. 4. This gives substantial information on the medical status of the user, obtained from the analysis carried out. The science Walden smart toilet introduced by Park et al. (2020) is a personal health monitoring system with the capacity of

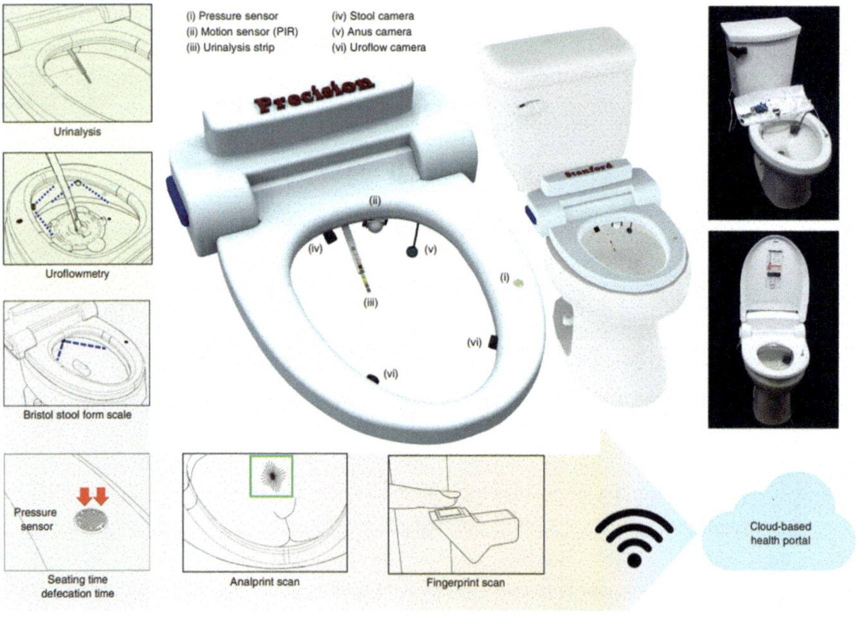

Fig. 4 Mountable smart toilet. (Park et al., 2020)

conducting analysis on both urine and stool. The smart toilet is embedded with two infrared sensors to monitor the behavioural pattern of the user while using the toilet. After the separation of the urine and the stool, the separated urine is then analysed by introducing a dipstick to the urine sample.

The dipstick method is able to determine the pH, specific gravity, and presence of protein, haemoglobin, bilirubin, glucose, ketone, nitrite, urobilinogen, acetone, and leukocytes in the urine. The colour obtained from the process was detected by the use of an RGB colour sensor. The data acquired from the sensor are sent to the cloud database using a BLE and LoRa for short-range and long-range communication, respectively.

3 Conclusion

The essence of a smart city is to incorporate data collection and data analysis in the day-to-day activity in the act, which helps ensure more effective decision-making, saves time and money, reduces stress, and improves the living condition of the occupants of the city. When the value and use of data are appreciated by all sectors of a city and implementation of acquired data is part of the city's operation, this will help the development and sustainability of the city. The smart approach in the provision of services in a smart city gives the service provider access to large data which, when analysed, provides a foundation for better decision-making and efficiency. Smart healthcare services involve the use of IoT and ICT in the normal operation carried out in the healthcare environment. Healthcare monitoring is one of the regular operations carried out in the healthcare environment.

A smart approach to health monitoring is a non-invasive data collection, collected in the comfort of the user outside the hospital environment. The use of a toilet for the acquisition of medical data is an effective way of addressing health issues detectable from the urine, behavioural pattern, or stool of the user. The summary of the contribution of various authors in the design and approach to attain a smart personal health monitoring system have been highlighted in this article. The review covers various methods of detection and various means of data transfer and communication. The approach employed by the authors discussed above provides a smart way to address health services. Concentration should be drawn towards the creation of smart toilet gadgets for personal health monitoring. This serves as a support system or an innovative attachment to the standard toilet system without inquiring about any other additional cost, such as the cost of purchasing a new smart toilet with personal health monitoring and other smart abilities.

Acknowledgement The author acknowledges the management of Covenant University for the conference support given to this paper.

References

Agrafioti, F., Bui, F. M., & Hatzinakos, D. (2011). Medical biometrics in mobile health monitoring. *Security and Communication Networks, 4*(5), 525–539. https://doi.org/10.1002/sec.227

Al-Zu'bi, M., & Radovic, V. (2018). *SDG11 – Sustainable cities and communities: Towards inclusive, safe, and resilient settlements*. Emerald Group Publishing.

Bae, J. H., & Lee, H. K. (2018). User health information analysis with a urine and faeces separable smart toilet system. *IEEE Access, 6*, 78751–78765.

Baig, M. M., & Gholamhosseini, H. (2013). Smart health monitoring systems: An overview of design and modeling. *Journal of Medical Systems, 37*(2), 9898.

Bhatia, M., Kaur, S., & Sood, S. K. (2020). IoT-inspired smart toilet system for home-based urine infection prediction. *ACM Transactions on Computing for Healthcare, 1*(3), 1–25.

Brent, M. (2018). An ethical analysis of personal health monitoring in the UK: A case study. *The Orbit Journal, 1*(3), 1–16.

Da Silva, I. N., & Flauzino, R. A. (2016). *Smart cities technologies*. InTechOpen. ISBN-10:9789535128076.

De Jong, M., Joss, S., Schraven, D., Zhan, C., & Weijnen, M. (2015). Sustainable–smart–resilient–low carbon–eco–knowledge cities; making sense of a multitude of concepts promoting sustainable urbanization. *Journal of Cleaner Production, 109*, 25–38.

Deakin, M. (2013). *Smart cities: Governing, modelling and analysing the transition*. Routledge.

Dey, K., Fries, R., & Ahmed, S. (2018). Future of transportation cyber-physical systems–Smart cities/regions. In *Transportation cyber-physical systems* (pp. 267–307).

Duan, Y., Fu, G., Zhou, N., Sun, X., Narendra, N. C., & Hu, B. (2015) Everything as a service (XaaS) on the cloud: Origins, current and future trends. In *2015 IEEE 8th international conference on cloud computing, IEEE* (pp. 621–628)

Fayomi, G. U., Mini, S. E., Fayomi, O. S. I., Mbonu, C. C., & Udoye, N. E. (2019). Sustainable smart city and its promising urban value: An overview. *IOP Conference Series: Earth and Environmental Science, 331*(1), 012046.

Filho, W. L., Wall, T., Azul, A. M., Brandli, L., & Özuyar, P. G. (2019). *Good health and well-being*. Springer.

Filho, W. L., Azul, A. M., Brandli, L., Özuyar, P. G., & Wall, T. (2020). *Sustainable cities and communities*. Springer.

Georgakopoulos, D., & Jayaraman, P. P. (2016). Internet of things: From internet scale sensing to smart services. *Computing, 98*(10), 1041–1058. https://doi.org/10.1007/s00607-016-0510-0

Ghosh, P., Bhattacharjee, D., & Nasipuri, M. (2020). intelligent toilet system for non-invasive estimation of blood-sugar level from urine. *IRBM, 41*(2), 94–105.

Halegoua, Germaine R. (2020). *Smart Cities*. Londres: Cambridge, MA – MIT Press, [e-book].

Hashemi, S., Han, M., Kim, T., & Kim, Y. (2015). Innovative toilet technologies for smart and green cities. In *Proceedings of the 8th conference International Forum on Urbanism (IFoU), Incheon, Korea* (pp. 22–24).

Hauber-Davidson, G., & Idris, E. (2006). Smart water metering. *Water, 33*(3), 38–41.

Holzinger, A., Röcker, C., & Ziefle, M. (2015). *Smart health: Open problems and future challenges*. Springer.

Hotta, S., & Funamizu, N. (2007). Biodegradability of fecal nitrogen in composting process. *Bio Tech, 98*(17), 3412–3414.

Hutton, G., & Chase, C. (2018). Water supply, sanitation, and hygiene. In C. N. Mock, R. Nugent, O. Kobusingye, & K. R. Smith (Eds.), *Injury prevention and environmental health*. The International Bank for Reconstruction and Development/The World Bank.

Jiuping, X., & Lei, X. (2017). *Sensor system and health monitoring. Integrated system health management, perspectives on systems engineering techniques* (pp. 55–99). Academic. https://doi.org/10.1016/B978-0-12-812207-5.00002-X

John, N., & Shenoy, S. (2014). Health cloud-Healthcare as a Service (HaaS). In *2014 International conference on advances in computing* (pp. 1963–1966). Communications and Informatics (ICACCI), IEEE.

Kabadayi, S., Ali, F., Choi, H., Joosten, H., & Lu, C. (2019). Smart service experience in hospitality and tourism services. *Journal of Service Management, 30*(3), 326–348. https://doi.org/10.1108/josm-11-2018-0377

Lara, A. P., Da Costa, E. M., Furlani, T. Z., & Yigitcanla, T. (2016). Smartness that matters: towards a comprehensive and human-centred characterisation of smart cities. *Journal of Open Innovation: Technology, Market, and Complexity, 2*(2), 8.

Lokman, A., & Kanesaraj Ramasamy, R. (2019). Smart toilet. In *Proceedings of the 3rd international conference on big data and internet of things – BDIOT 2019*. https://doi.org/10.1145/3361758.3361765

Mishra, K. N., & Chakraborty, C. (2020). A novel approach toward enhancing the quality of life in smart cities using clouds and IoT-based technologies. In *Internet of Things* (pp. 19–35). https://doi.org/10.1007/978-3-030-18732-3_2

Mittelstadt, Brent, D. (2013). *On the ethical implications of personal health monitoring*. A Conceptual Framework for Emerging Discourses. A PhD Thesis, submitted to Centre for Computing and Social Responsibility, De Montfort University, Leicester, Uk.

Mittelstadt, B., Fairweather, B., Shaw, M., & McBride, N. (2014). The ethical implications of personal health monitoring. *International Journal of Technoethics, 5*(2), 37–60. https://doi.org/10.4018/ijt.2014070104

Mohammadzadeh, N., & Safdari, R. (2016). Mobile health monitoring. *Mobile Health Technologies – Theories and Applications*. https://doi.org/10.5772/64704

Mohammed, K. I., Zaidan, A. A., Zaidan, B. B., Albahri, O. S., Alsalem, M. A., Albahri, A. S., Hadi, A., & Hashim, M. (2019). Real-time remote-health monitoring systems: A review on patients prioritisation for multiple-chronic diseases, taxonomy analysis, concerns and solution procedure. *Journal of Medical Systems, 43*(7), 223.

Mustafa, M. A., Mabhaudhi, T., Avvari, M. V., & Massawe, F. (2021). Transition toward sustainable food systems: A holistic pathway toward sustainable development. In *Food security and nutrition* (pp. 33–56). https://doi.org/10.1016/b978-0-12-820521-1.00002-2

Nordgren, A. (2015). Privacy by design in personal health monitoring. *Health Care Analysis: An International Journal of Health, Philosophy and Policy, 23*(2), 148–164.

Novotný, R., Kuchta, R., & Kadlec, J. (2014). Smart city concept, applications and services. *Journal of Telecommunications System & Management, 3*(2), 117.

NU CEPAL. (2019). The 2030 agenda for sustainable development and the challenges of financing for development. (2019). In *Economic Survey of Latin America and the Caribbean 2016*. https://doi.org/10.18356/d839133a-en

Oyedepo, S. O., Anifowose, E. G., Obembe, E. O., Dirisu, J. O., Khanmohamadi, S., Kilanko, O., Babalola, P. O., Ohunakin, O. S., Leramo, R. O., & Olawole, O. C. (2020). Assessment of economic and environmental impacts of energy conservation strategies in a university campus. In *Green Energy: Solar Energy, Photovoltaics, and Smart Cities* (pp. 441–468). Scrivener Publishing LLC, Wiley Inc.

Park, S.-M., Won, D. D., Lee, B. J., Escobedo, D., Esteva, A., Aalipour, A., Ge, T. J., Kim, J. H., Suh, S., Choi, E. H., Lozano, A. X., Yao, C., Bodapati, S., Achterberg, F. B., Kim, J., Park, H., Choi, Y., Kim, W. J., Yu, J. H., & Gambhir, S. S. (2020). A mountable toilet system for personalized health monitoring via the analysis of excreta. *Nature Biomedical Engineering, 4*(6), 624–635.

Pradhan, B., Bhattacharyya, S., & Pal, K. (2021). IoT-based applications in healthcare devices. *Journal of Healthcare Engineering, 2021*, 6632599.

Reichental, J. (2020). *Smart cities for dummies*. Wiley.

Shaikh, F., Shaikh, F., Sayed, K., Mittha, N., & Khan, N. (2019). Smart toilet based On IoT. In *3rd international conference on computing methodologies and communication (ICCMC)*. https://doi.org/10.1109/iccmc.2019.8819606

Townsend, A. M. (2013). *Smart cities: Big data, civic hackers, and the quest for a New Utopia*. W. W. Norton & Company.
Kumar, Vinoth, B., Maheswari Uma, J., Karpagam G. R., and Sivakumar P. (2020). *Computing paradigms for smart healthcare*. Nova Science Publishers Inc.
Wath, S. B. (2016). A mechanical automatic urinal-toilet flusher for Swach Bharat mission. *Procedia Environmental Sciences, 35*, 185–192. https://doi.org/10.1016/j.proenv.2016.07.075
Yigitcanlar, T. (2016). *Technology and the city: Systems, applications and implications*. Routledge.
Yigitcanlar, T., & Lee, S. H. (2014). Korean ubiquitous-eco-city: A smart-sustainable urban form or a branding hoax? *Technological Forecasting and Social Change, 89*, 100–114.
Yilmaz, T., Foster, R., & Hao, Y. (2010). Detecting vital signs with wearable wireless sensors. *Sensors, 10*(12), 10837–10862.

Biomedical Engineering Education: Equipment, Prospect and Challenges for Environmental Healthcare in Nigeria

O. J. Oyebode

1 Introduction

Engineers across the globe can bring BME into limelight and sustainable development in health sector. The African Union needs to ensure monitoring of national educational policies to identify gaps and challenges and strategize a way forward to cover gaps in mitigate challenges to improve quality education in Africa.

In the growing West African population, healthcare becomes a necessity that expands with it, and with the expansion of healthcare technology, a demand for constant maintenance and repair arises.

Biomedical engineering is identified as one of the niche engineering branches in the country which deals with the study of engineering principles. These principles are further combined with the principles of medical sciences aiming to streamline the healthcare services in the country.

The study of biomedical engineering includes analytical and curative applications. Many students have not developed interest in biomedical engineering in Nigeria and other African countries. Health technology innovation needs more attention especially on the material and equipment selection for better healthcare delivery.

The healthcare expenditure is likely to witness an increase due to the rise of awareness and population growth. Furthermore, spreading awareness of medical advancements has led to an increase in the number of people seeking biomedical solutions for their health issues. This will eventually witness a rise in the employment graph of biomedical engineers. These facts are evident to prove that the growth

O. J. Oyebode (✉)
Civil and Environmental Engineering Department, Afe Babalola University, Ado-Ekiti, Ekiti State, Nigeria

A. O. Ayeni et al. (eds.), *Advanced Manufacturing in Biological, Petroleum, and Nanotechnology Processing*, Green Energy and Technology,
https://doi.org/10.1007/978-3-030-95820-6_13

rate of biomedical engineers is likely to be much faster than the average pace for all the existing occupations.

Nigerian Institute for Biomedical Engineering (NIBE) is the professional body for biomedical education in Nigeria, and they represent biomedical engineering in international organizations (Nkuma-Udah et al., 2015).

Nigerian universities offering biomedical engineering:

(i) Afe Babalola University, Ado-Ekiti (ABUAD)
(ii) Achievers University, Owo (AC)
(iii) Bells University of Technology, Ota (BUT)
(iv) University of Ilorin (UNILORIN)
(v) University of Lagos (UNILAG)
(vi) Oyo State Technical University, Ibadan (TECH-U)
(vii) Federal University of Health Science, Otukpo
(viii) Federal University of Technology Owerri, Nigeria
(ix) Department of Food Science and Technology, Abia State University, Uturu, Nigeria

2 Biomedical Engineering and Healthcare Delivery System

There is a need to improve healthcare delivery system in Nigeria across various aspects, and every tier of government has a definite role to play (Coker et al., 2009).

Biomedical engineering (BME) programmes are relatively few in Africa. It was reported in 2008 that only 12 African universities in only six countries offered biomedical engineering compared with 229 universities in North America (Abu-Faraj, 2008).

As of 2015, there was less than 0.01 biomedical engineer per 10,000 population in Ghana compared to 0.49 in the United States (Mohedas et al., 2015).

Biomedical engineering can be defined as the application of the principles and problem-solving techniques of engineering and design concepts to medicine and biology for healthcare purposes (e.g. diagnostic or therapeutic).

Biomedical engineering, also known as bioengineering, refers to the field of study that combined biology and engineering. It solves medical and biological problems. Biomedical engineers are involved in making newer and better instruments.

In the growing West African population, healthcare becomes a necessity that expands with it, and with the expansion of healthcare technology, a demand for constant maintenance and repair arises. The field of bioengineering, as the term implies, includes the mathematical modelling of the biological systems, design and computation of the algorithms which help to analyse biological signals, bioinformatics, biomechanics, applications of micro-electromechanical systems, molecular engineering, nanotechnology and development of signal processing and control algorithms of artificial parts of the body.

It is a diverse range of study that can encompass different studies, e.g. biomaterials, bioinstrumentation, biomechanics, biological systems modelling, biotechnology, biomechatronics, biophysics, rehabilitation, neural engineering, bionics, etc.

Its importance cannot be understated as it helps to drive advancements to the field of medicine in an ever-changing world of technology and provide a fresh new source of innovation for the field, e.g. with the installation of various medical devices and machines like X-rays, MRIs, ultrasound, electrocardiograph (ECG) and dialysis machines, as well as high-tech surgical machines and equipment like an endoscope, etc.

Biomedical engineering is one of the emerging fields which combines engineering expertise with the needs in the medical industry for the growth and development of the healthcare sector. It is the unique branch of engineering in which the concepts, knowledge, expertise and skills are designated and applied to the field of biology and medicine in order to meet the daily challenges.

There are various areas of specialization within the field of biomedical engineering such as clinical engineering, tissue engineering, rehabilitation engineering, biomechanics and bioinstrumentation.

Biomechanics involves applying mechanics to biological or medical problems. They study the flow of body fluids such as blood.

Bioinstrumentation is the application of electronics and measurement techniques to develop devices used in diagnosis and treatment of diseases.

Biochemical engineering is very essential in hospital and health care delivery generally. The use of advanced technology to deliver healthcare services at a distance has established to be one of the defining medical revolutions of the twenty first century. With the unwieldy healthcare system under scrutiny, telemedicine can be one of the answers for increasing access and at the same time decreasing the cost of healthcare service delivery. Many managers and policymakers in health and medicine are not fully familiar with the dimensions, affected areas, advantages and added values caused by these new technologies and therefore do not consider telemedicine in strategic plans for health and medical systems. The pace of development was slow, most efforts were in the area of training - short courses, continuing education or professional development (Adeyemo and Onikoyi, 2012; Nkuma-Udah et al., 2015; Ajala et al., 2015). Inadequate of biomedical engineering training has grossly affected facilities in hospitals and this has resulted to short life expectancy, poor economic growth and poor management of resources in Nigerian health sector (Bamigboye and Bello, 2021). The biomedical engineering program is designed to integrate engineering and medical research spanning from the mechanics of man-made materials investigations using engineering methods to explore fundamental physiological processes. The biomedical engineers occupies a special role, having a back ground in a branch of engineering, combined with a knowledge of the physical structure and function of the human body systems and an understanding of how there were engineering principles and methods can be applied to technical innovations and problem solving in the field of medicine. Clinical engineering is the branch of biomedical engineering dealing with the actual implementation of medical equipment and technologies in hospitals or other clinical settings (Okorie, 2015; Kybartaite, 2010).

3 Education of Biochemical Engineering in Nigeria

This paper examined BME education in Nigeria for better healthcare delivery, sophisticated equipment in hospitals and environmental health. Challenges facing biomedical engineers in the medical industry were examined as well as probable prospects in the medical field documented.

Literature review, interview and secondary data were collected for this research.

Literatures and secondary data were used to check the viability, extent and growth of biomedical engineering education in Nigeria. It has been discovered that Nigeria is one of the few African countries with reasonably organized biomedical engineering practice.

Only few universities such as Bells, UNILAG, UI, UNILORIN, and ABUAD are offering biomedical engineering in Nigeria.

Various factors surrounding the field of the biomedical engineers were checked for technological expansion, equipment, maintenance, installation, upkeep and the challenges they face in the field or as undergraduates plus their duties and speciality in the healthcare sector in Nigeria. It is worthwhile to note that biomedical engineering closes the gap between engineering and medicine to create advance healthcare treatment such as diagnosis, monitoring and therapy.

Figure 1 shows a hospital equipment that can be handled by biomedical engineers.

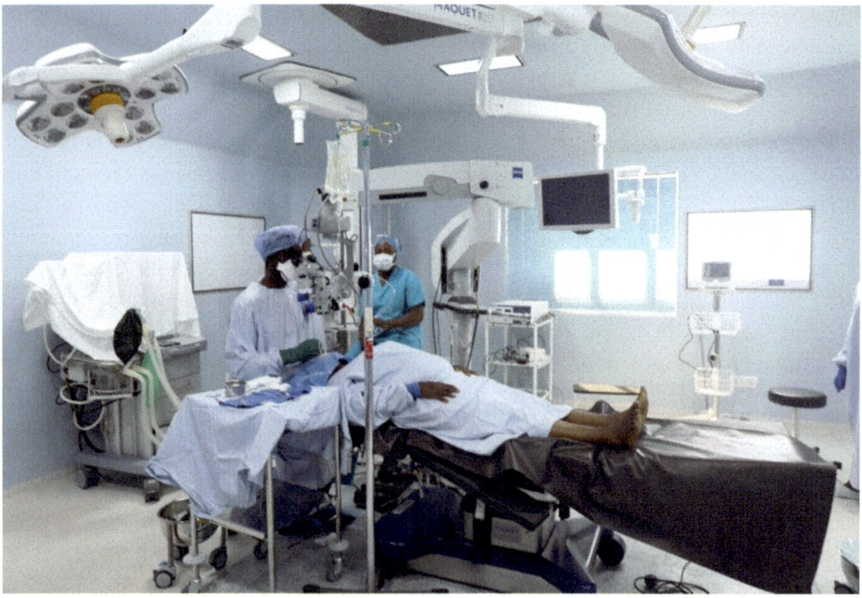

Fig. 1 Equipment designed by biomedical engineers

3.1 Responsibilities of Biomedical Engineers

Since biomedical engineering is an interdisciplinary field based on both engineering and the life sciences, it is important for biomedical engineers to have knowledge about and be able to communicate in both areas. Biomedical engineers must understand the basic components of the body and how they function well enough to exchange ideas and information with physicians and life scientists. Biomedical engineers are also involved in the designing of electrical circuits, software to run this medical equipment or computer simulations in order to test the new drug therapies. In addition to these, they also develop the materials that are required to make the replacement of certain body parts. In our body, we have mechanical, electrical, chemical, thermal, pneumatic and hydraulic and many other types of system. Each system communicates internally with other systems of the body and externally it communicates with surroundings. In medical terms, a study of the structure of the body and the relationship of its constituent's parts to each other is known as 'anatomy', while the study of function of these parts as a system is known as physiology.

The key responsibilities of a biomedical engineer:

(i) Designing equipment like artificial internal organs for replacement of various body parts and machines which can diagnose medical problems.
(ii) Installation, adjustment, maintenance, repair and providing technical support for biomedical equipment.
(iii) Evaluation of the safety, effectiveness and efficiency of all the biomedical equipment.
(iv) Training of the clinicians and other personnel involved in the operation of this equipment, on the proper use of this equipment.
(v) Working in coordination with the life scientists, chemists and medical scientists in order to conduct research to find out the engineering aspects of the biological systems of humans and animals.
(vi) Preparing procedures and technical reports, publishing research papers and also making recommendations based on the conducted researches and their outcomes.
(vii) Presenting research outcomes to the scientists, nonscientist executives, hospital management, clinicians, engineers and others involved.
(viii) Biomedical engineers are also involved in the designing of electrical circuits, software to run this medical equipment or computer simulations in order to test the new drug therapies.
(ix) In addition to these, they also develop the materials that are required to make the replacement of certain body parts.

4 Challenges Biomedical Engineering Facing in Nigeria

Challenges facing biomedical engineering in Nigeria include:

(i) Over-reliance on importation of medical equipment produced in the western world
(ii) Poor understanding of concept of systematic management of medical equipment by hospital clinicians
(iii) Epileptic power supply,
(iv) Poor government funding of research related activities
(v) Exorbitant cost of maintaining purchased equipment
(vi) Corruption and nepotism
(vii) Dearth of skilled personnel to effectively maintain purchased or donated equipment

4.1 Prospect of Biomedical Engineering

(i) They can become successful as freelancers by giving consultancy to companies or managing teams of doctors and engineers while profiting greatly as well as filling the demands in the economy for biomedical engineers and repairing several company machines at once.
(ii) It becomes much easier for current biomedical engineers to gain employment as there is a high demand for their services.
(iii) Because biomedical engineering has such a broad range of specialties, it makes them versatile in the field.
(iv) Bioinstrumentation is a type of prospect that involves creating and developing devices that help to treat illnesses.

5 Conclusions

In conclusion, biomedical engineering must be embraced by government, health practitioners and engineering leaders for diagnostic or therapeutic healthcare purposes. Lack of proper design and poor installation of hospital facilities and disregard of health policy should be handled holistically.

It is very clear that a biomedical engineer plays a major role in the healthcare system. Starting from minor equipment like bandages to more complex devices like implants, a biomedical engineer makes sure to fight the challenges in the healthcare system.

6 Recommendations

It has been recommended that:

(i) Government should provide more funding for research and training of clinical engineers, biomedical engineers, technicians and scientists on procurement, handling and maintenance of equipment in the hospital for better environmental health.
(ii) There should be a general training on how to use a medical device once it's bought and installed.
(iii) The government should create more awareness in schools, health centres, etc. on biomedical engineering and how useful they are to the healthcare sector.
(iv) Government should reduce the cost of raw materials. By so doing, the rate of importation would be reduced as biomedical engineers would be able to manufacture some of these devices themselves.

References

Abu-Faraj, Z. O. (2008). Bioengineering/biomedical engineering education and career development: literature review, definitions, and constructive recommendations. *International Journal of Engineering Education, 24*(5), 990.
Adeyemo, A. A., & Onikoyi, M. P. (2012). Prospects and challenges of large scale commercial poultry production in Nigeria. *Agricultural Journal, 7*(6), 388–393.
Ajala, F. A., Adetunji, A. B., & Akande, N. O. (2015). Telemedicine acceptability in south western Nigeria: Its prospects and challenges. *International Journal of Advanced Computer Technology, 4.9*, 1970–1976.
Bamigboye, A. A., & Bello, K. A. (2021). Biomedical engineering in Nigeria: The genesis, present and the future. *Clinical Reviews and Opinions, 10*(1), 1–4.
Coker, A., Sangodoyin, A., Sridhar, M., Booth, C., Olomolaiye, P., & Hammond, F. (2009). Medical waste management in Ibadan, Nigeria: Obstacles and prospects. *Waste Management, 29*(2), 804–811.
Kybartaite, A. (2010). Impact of modern educational technologies on learning outcomes application for e-learning in biomedical engineering (Unpublished doctoral dissertation). Tampere University of Technology, Tampere.
Nkuma-Udah, K. I., Agoha, E. E. C., Ejeta, K., & Ndubuka, G. I. (2015). Biomedical engineering in Nigeria: A developmental overview. In *World congress on medical physics and biomedical engineering, June 7–12, 2015, Toronto, Canada* (pp. 1643–1648). Springer, Cham.
Okorie, P. U. (2015). The significant of biomedical engineering to medical field in Nigeria. *America Journal of Biomedical Science and Engineering, 1*(2), 20–24.

Inhibition Performance of Admixed Grapefruit and Lemongrass Oil Extracts on Low Carbon Steel in Weak Acid Formulation

R. T. Loto, P. Okpaleke, and U. Udoh

1 Introduction

The universal utilization of carbon steels worldwide for structural components and platforms coupled with their inherent mechanical, physical, and metallurgical properties establishes them as the standard wherewith the properties of other metallic alloys can be compared (Dwivedi et al., 2017). Carbon steels are applied in construction, petrochemical crude distillation unit, mining, energy generation, automobile, and desalination plants. However, the weak resistance of the steel to corrosion severely shortens their operational life span because of the reaction effect of SO_4^{2-}, Cl^-, $S_2O_3^{2-}$, NO_3-, etc. in aqueous environments on the steel surface (Zarras & Stenger-Smith, 2017; Loto & Babalola, 2017). The capacity of carbon steels to be inert is because of the absence of important passivating species that is responsible for their weak corrosion resistance. As a result, the cost of corrosion control, repair, replacement of damaged parts, and continual maintenance to prevent collapse of structure, industrial accidents, fluid leakages, etc. is shifted to consumers/end-users. Corrosion inhibitors are chemical derivatives which remain the most versatile and cost-effective method of controlling carbon steel degradation during service (Branko, 2015). The adequate corrosion inhibitors in service are inorganic and tend to be very poisonous (Singh & Bockris, 1996). Synthetic organic compounds which have seen limited service are also toxic to personnel and the environment (Brycki et al., 2017; Loto, 2017; Loto & Loto, 2012, 2013). Research on plant extracts for application as corrosion inhibitor is ongoing. The results gotten so far are appreciable; however, plant extracts are limited by short shelf life, weak adsorption, and general poor inhibition (Li et al., 2005; Quartarone et al., 2012; Ashassi-Sorkhabi et al., 2004; ÖZcan, 2008; Loto & Oghenerukewe, 2016). Essential oils extracts

R. T. Loto (✉) · P. Okpaleke · U. Udoh
Department of Mechanical Engineering, Covenant University, Ota, Ogun State, Nigeria

A. O. Ayeni et al. (eds.), *Advanced Manufacturing in Biological, Petroleum, and Nanotechnology Processing*, Green Energy and Technology,
https://doi.org/10.1007/978-3-030-95820-6_14

have phytochemical compounds with numerous research being done to assess their corrosion inhibition properties (Fu et al., 2010; Bouoidina et al., 2017; Hamdani et al., 2015; El Ouadi et al., 2014; El Ouariachi et al., 2015; Boumhara et al., 2015; Lahhit et al., 2011; Loto et al., 2018). Results have shown they tend to be effective at higher concentrations, and their performance can be strongly dependent on concentration and exposure time. Non-oil extract-based inhibitors have been proven to be effective. However, research on them is ongoing before commercialization (Ye et al., 2020a, b, 2021; Turcio-Ortega et al., 2007; Guadalupe et al., 2011; Hussin et al., 2016; Cruz et al., 2005). Synergistic combination effect of the oil extracts has given more effective results. However, data investigation of the inhibition output is important to establish the peak performance of the extracts with respect to experimental factors. This investigation targets the corrosion rate and performance output of admixed grapefruit and lemongrass oil extracts on low carbon steel in 0.5 M H_2SO_4 and HCl formulation, and statistical investigation of the data.

2 Experimental Methods

Grapefruit and lemongrass essential oils (GFLG) are produced by NOW Foods, United States, and obtained through an intermediary in the synthesized form with assured 100% purity which were admixed together in ratio 1:1 and formulated in volume concentrates of 1.0%, 2.0%, 3.0%, 4.0%, and 5.0% in 200ml of 0.5 M H_2SO_4 and HCl acid formulation. Low carbon steel (LCS) rods with diameter of 1.2 cm and average thickness of 1 cm have wt. % constituent of 0.06% S, 0.05% P, 0.9% Mn, and 98.99% Fe by confirmation with PhenomWorld scanning electron microscope. Low carbon steel (LCS) was prepared into six test subjects for coupon investigation. LCS samples were sandpapered with coarse SiC polishers (60–2500 grits) before burnishing with 6 μm Struers DiaDuo-2 polishing formulation and subsequently rinsed with acetone. Computed LCS samples were inserted into the acid formulation for 240 h. LCS samples were weighed at 24 h interludes with Ohaus scaling instrument. The weight difference was computed from the deduction between the starting weight of LCS (maintained for 240 h) and successive weight recorded at 24 h intermittently, amounting to 240 h in total. Corrosion rate of LCS was computed according to the equation below (Ali & Fulazzaky, 2020; Khaksar & Shirokoff, 2017):

$$R = \left[\frac{87.6W}{DAT}\right] \tag{1}$$

W represents weight loss (g), D represents density (g/cm^2), A represents area (cm^2), and T represents time of measurement (h). Inhibition output (η) was computed according to Eq. 2:

$$\eta = \left[\frac{\omega_1 - \omega_2}{\omega_1}\right] \times 100 \tag{2}$$

ω_1 represents weight loss of LCS free the plant extracts and ω_2 represents weight loss of LCS at set GL concentrations. Binary-part mono level ANOVA test (F-test) was utilized to investigate the statistical applicability of GFLG concentrations and exposure time on GFLG inhibition data. The assessment was executed at confidence factor of 95% (significance level of $\alpha = 0.05$) analogous to the following successive equations. The aggregation of squares of columns (exposure time) was computed as follows:

$$SS_c = \frac{\Sigma T_c^2}{nr} - \frac{T^2}{N} \tag{3}$$

The summation of squares between rows (extract concentrate) was computed with Eq. 4:

$$SS_r = \frac{\Sigma T_r^2}{nc} \quad \frac{T^2}{N} \tag{4}$$

Total computation of squares as:

$$SS_{\text{Total}} = \Sigma x^2 - \frac{T^2}{N} \tag{5}$$

3 Results and Discussion

3.1 Gravimetric Analysis

Table 1 presents the corrosion rate data for LCS from 0.5 M H_2SO_4 and HCl formulations, and Table 2 presents the analogous data for GFLG inhibition output from both acids. Inspection of Table 1 reveals the substantial variation in LCS corrosion rate from both acids at 0% GFLG and GFLG at 1–5% concentration. Figure 1 exhibits the comparative plots of GFLG inhibition output at 1% and 5% GFLG concentration. At 0% GFLG, the corrosion rate initiated at 102.24 and 62.071 mm/y (24 h) in H_2SO_4 and HCl formulations, and terminated at 34.59 and 17.53 mm/y (240 h). The higher corrosion rate is due to accelerated degradation of LCS due to SO_4^{2-} and Cl^- anions. The anions oxidize the steel surface causing the rapid discharge of ionized steel cations into the electrolyte. LCS corrosion rate value (0% GFLG) is significantly higher in H_2SO_4 than HCl by reason of the diprotic nature of H_2SO_4 when it ionizes in H_2O compared to HCl which is monoprotic. As a result,

Table 1 LCS corrosion rate data from H_2SO_4 and HCl solution after 240 h of exposure

	LCS corrosion rate (mm/y) in H_2SO_4 solution						LCS corrosion rate (mm/y) in HCl solution					
	GFLG conc. (%)											
Exp. time (h)	0% GL	1.0% GL	2.0% GL	3.0% GL	4.0% GL	5.0% GL	0% GL	1.0% GL	2.0% GL	3.0% GL	4.0% GL	5.0% GL
24	102.24	0.662	0.818	0.006	0.253	0.500	62.071	0.247	0.337	0.350	0.136	0.247
48	59.75	0.370	0.464	0.013	0.286	0.471	31.626	0.136	0.204	0.237	0.107	0.211
72	59.00	0.264	0.355	0.095	0.201	0.405	32.799	0.229	0.164	0.171	0.102	0.147
96	44.58	0.140	0.112	0.086	0.154	0.451	24.263	0.219	0.138	0.217	0.088	0.117
120	50.75	0.312	0.104	0.092	0.129	0.476	27.640	0.276	0.148	0.266	0.104	0.104
144	42.66	0.344	0.217	0.093	0.124	0.449	23.943	0.242	0.275	0.328	0.111	0.114
168	46.13	0.351	0.205	0.160	0.134	0.452	23.820	0.271	0.286	0.345	0.107	0.211
192	40.97	0.340	0.204	0.153	0.187	0.467	20.895	0.273	0.288	0.449	0.196	0.312
216	37.24	0.402	0.221	0.163	0.216	0.427	19.358	0.387	0.485	0.479	0.248	0.344
240	34.59	0.376	0.221	0.172	0.220	0.466	17.530	0.474	0.541	0.646	0.282	0.399

Table 2 GFLG inhibition efficiency data from H_2SO_4 and HCl solution after 240 h of exposure

	GFLG inhibition efficiency (%) in H_2SO_4					GFLG inhibition efficiency (%) in HCl				
	GFLG conc. (%)									
Exp. time (h)	1.0% GL	2.0% GL	3.0% GL	4.0% GL	5.0% GL	1.0% GL	2.0% GL	3.0% GL	4.0% GL	5.0% GL
24	99.35	99.20	99.99	99.75	99.51	99.60	99.46	99.44	99.78	99.60
48	99.38	99.22	99.98	99.52	99.21	99.57	99.35	99.25	99.66	99.33
72	99.55	99.40	99.84	99.66	99.31	99.30	99.50	99.48	99.69	99.55
96	99.69	99.75	99.81	99.65	98.99	99.10	99.43	99.10	99.64	99.52
120	99.39	99.80	99.82	99.75	99.06	99.00	99.46	99.04	99.62	99.62
144	99.19	99.49	99.78	99.71	98.95	98.99	98.85	98.63	99.53	99.53
168	99.24	99.56	99.65	99.71	99.02	98.86	98.80	98.55	99.55	99.11
192	99.17	99.50	99.63	99.54	98.86	98.70	98.62	97.85	99.06	98.51
216	98.92	99.41	99.56	99.42	98.85	98.00	97.49	97.53	98.72	98.22
240	98.91	99.36	99.50	99.36	98.65	97.29	96.91	96.31	98.39	97.72

degradation by SO_4^{2-} anions tends to be general over the steel exterior compared to degradation by Cl^- which tends to be localized at specific sites/regions of the steel (Loto et al., 2020). Moreover, LCS corrosion rate in both acids at 0% GFLG decreases significantly akin to exposure time by reason adulteration of the acid formulation with released corrosion precipitates. Nevertheless, the degradation rate of LCS at 0% GFLG remains significantly higher than values obtained higher GFLG concentration. The significant decrease in LCS corrosion from 1% to 5% GFLG concentration is because of the reaction effect of protonated GFLG molecules which obstructs the reduction-oxidation reaction processes inducing corrosion. As a result, discharge of Fe^{2+} cations from the steel exterior is effectively suppressed. The corrosion rate data for LCS at 1–5% GFLG concentration shows GFLG concentration

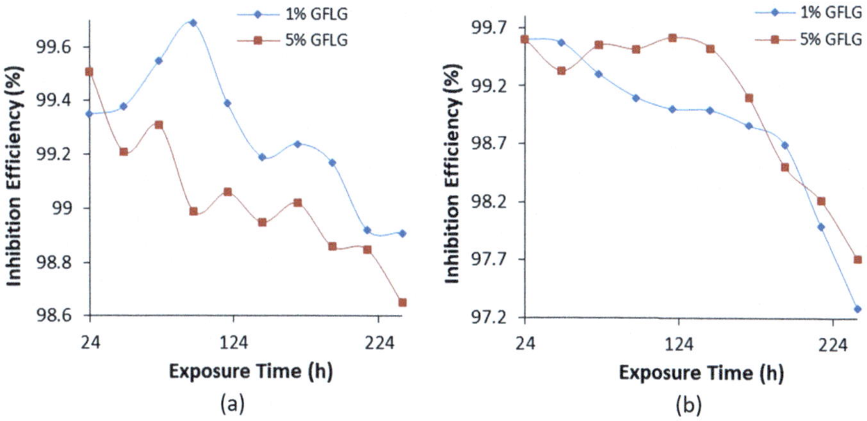

Fig. 1 Relative plot of GFLG inhibition output at 1% and 5% concentration in (**a**) H_2SO_4 formulation and (**b**) HCl formulation

has weak impact on the corrosion rate outputs of LCS. The rate of degradation of LCS in H_2SO_4 decreases analogous to exposure time, while in HCl the values increase. In H_2SO_4 solution, LCS corrosion rate values (1% and 5% GFLG concentration) initiated (24 h) at 0.662 and 0.5 mm/y. The correlating outcomes in HCl solution are 0.247 and 0.247 mm/y. At 240 h, the corrosion rate values in both acids declined to 0.376 and 0.466 mm/y in H_2SO_4, and 0.474 and 0.399 mm/y in HCl. These observations show GFLG improves in performance with time in H_2SO_4, while in HCl its performance marginally decreases with exposure time. Cl^- ions are significantly smaller than their SO_4^{2-} ion counterparts. As a result, limited transport of the Cl^- ion in the acid electrolyte diffuses through the barrier film formed by the oil extract on the steel (Cabrini et al., 2015; Loto, 2013). However, this observation does not negate the overall outstanding performance of the oil extract. Observation of inhibition efficiency data in Table 2 shows GFLG adequately inhibited the steel at all concentrates with protection performance values greater than 95% throughout the exposure hours. The inhibition output at 1% and 5% GFLG concentration in both acids at 24 h are 99.35% and 99.51% (H_2SO_4), and 99.60% and 99.60% (HCl). Comparing the data at 24 h to values at 240 h (98.91% and 98.65% in H_2SO_4) and (97.29% and 97.72% in HCl) shows the performance of GFLG as earlier stated is marginally independent of exposure time and concentration as presented in Fig. 1a, b. The trend fluctuation in Fig. 1a, b shows that inhibition output decreases with variation in exposure time. Moreover, the visible fluctuation in Fig. 1a compared to Fig. 1b showed the inhibition output of GFLG in H_2SO_4 is thermodynamically unstable compared to its performance in HCl. The statement has been added to the revised manuscript.

Data on standard deviation (SD), mean, and margin of error for GFLG inhibition values from H_2SO_4 and HCl formulation at set GFLG concentration are depicted in Table 3. The standard deviation data for GFLG in both acids generally vary with concentration. In H_2SO_4 the SD values varies from 0.14 at 4% concentration to 0.25

Table 3 Data on SD, mean, and margin of error for GFLG inhibition efficiency in H_2SO_4 and HCl solution

	H_2SO_4					HCl				
GFLG Conc. (%)	1	2	3	4	5	1	2	3	4	5
SD	0.25	0.20	0.17	0.14	0.25	0.71	0.91	1.01	0.48	0.68
Mean	99.28	99.47	99.76	99.61	99.04	98.84	98.79	98.52	99.36	99.07
	Margin of error	±0%	**Result above 98% inhibition**	100%	**Margin of error**	±0.81%	**Result above 98% inhibition**	88%	**Margin of error**	±0%

Table 4 ANOVA data for statistical influence of GFLG concentration and exposure time on GFLG inhibition performance

H_2SO_4				HCl			
Source of variation	Mean square ratio (F)	Theoretical significance factor	Statistical relevance factor, F (%)	Source of variation	Mean square ratio (F)	Theoretical significance factor	Statistical relevance factor, F (%)
Inhibitor conc.	30.86	2.42	62.38	Inhibitor conc.	15.20	2.42	12.90
Exposure time	4.27	2.1	19.43	Exposure time	41.60	2.1	79.46

at 1% and 5% concentrations. These values show the amount of deviation of GFLG inhibition data from mean value is significantly limited, i.e., it is generally stable analogous to exposure time. However, it tends to be more stable at 4% GFLG concentration. 4% GFLG concentration represents the limit of stability for GFLG performance on LCS in H_2SO_4 solution. Beyond 4% GFLG concentration, the SD values increase to 0.25 due to lateral interaction effect among GFLG ionic species in the electrolyte. Data shows the mean value varies with the 99% range due to steadiness of the inhibitor action upon the steel exterior. The SD data from HCl formulation are generally more than the data obtained in H_2SO_4 solution. The SD values vary from 0.48 at 4% GFLG concentration to 1.01 at 3% concentration. These values show the inhibition effect of GFLG in HCl is less steady akin to exposure time and its performance in H_2SO_4. The inhibition efficiency data in HCl is determined to vary relatively significantly with exposure time signifying increased or decreased performance over time. However, the mean values confirm effective performance. Data on margin of error depicts 100% and 88% of inhibition output in H_2SO_4 and HCl solution are above 98% inhibition performance at margin of error of ±0% and ±0.81%, respectively.

The effect of GFLG concentration and measurement time (sources of variation) on LCS corrosion was assessed with analysis of variance (ANOVA) and the data is laid out in Table 4. The statistical relevance component depicts the percentage degree of influence of the origin of variation on the performance output of

GFLG. Table 4 shows inhibitor concentration dominates the inhibition effect of GFLG on LCS at 62.38% compared to exposure time at 19.43%. The comparable values for HCl reveal exposure time dominates the performance of GFLG in HCl at 79.46% compared to GFLG concentration at 12.9%. The mean square ratio depicts the statistical data which must be more than the theoretical significance factor for the statistical relevance factor to be important. The mean square ratio for the origin of variation in H_2SO_4 and HCl formulation presented in Table 4 is more than the theoretical significance factor. Hence, the statistical relevant factors are valid.

4 Summary

Data assessment of the protection effect of admixed grapefruit and lemongrass oil extract on low carbon steel in dilute acid media was performed. Results showed the distillates adequately inhibited at all concentrations with average inhibition effect of 95%. Inhibition output from H_2SO_4 solution was determined to be significantly subject to concentration compared to the observation from HCl solution which was determined to be dependent on exposure time. The extracts' performance was determined to be more stable in H_2SO_4 solution compared to HCl with respect to exposure time. At 98% inhibition efficiency threshold, 100% of the extracts' inhibition on low carbon steel in H_2SO_4 were above the threshold, while 88% of the values in HCl were above the threshold.

References

Ali, N., & Fulazzaky, M. A. (2020). The empirical prediction of weight change and corrosion rate of low-carbon steel. *Heliyon, 6*(9), e05050. https://doi.org/10.1016/j.heliyon.2020.e05050

Ashassi-Sorkhabi, H., Majidi, M. R., & Seyyedi, K. (2004). Investigation of inhibition effect of some amino acids against steel corrosion in HCl solution. *Applied Surface Science, 225*, 176–185. https://doi.org/10.1016/j.apsusc.2003.10.007

Boumhara, K., Tabyaoui, M., Jama, C., et al. (2015). Artemisia Mesatlantica essential oil as green inhibitor for carbon steel corrosion in 1 M HCl solution: Electrochemical and XPS investigations. *Journal of Industrial and Engineering Chemistry, 29*, 146–155. https://doi.org/10.1016/j.jiec.2015.03.028

Bouoidina, A., Chaouch, M., Abdellaoui, A., et al. (2017). Essential oil of "Foeniculum vulgare": Antioxidant and corrosion inhibitor on mild steel immersed in hydrochloric medium. *Anti-Corrosion Methods and Materials, 64*(5), 563–572. https://doi.org/10.1108/ACMM-10-2016-1716

Branko, N. P. (2015). Chapter 14 – Corrosion inhibitors. In *Corrosion engineering, principles and solved problems* (pp. 581–597). https://doi.org/10.1016/B978-0-444-62722-3.00014-8

Brycki, B. E., Kowalczyk, I. H., Szulc, A., Kaczerewska, O., & Pakiet, M. (2017). *Organic corrosion inhibitors*. IntechOpen. https://doi.org/10.5772/intechopen.72943

Cabrini, M., Fontana, F., & Lorenzi, S. (2015). Effect of organic inhibitors on chloride corrosion of steel rebars in alkaline pore solution. *Journal of Chemistry*, 521507. https://doi.org/10.1155/2015/521507

Cruz, J., Pandiyan, T., & García-Ochoa, E. (2005). A new inhibitor for mild carbon steel: Electrochemical and DFT studies. *Journal of Electroanalytical Chemistry, 583*(1), 8–16. https://doi.org/10.1016/j.jelechem.2005.02.026

Dwivedi, D., Lepkov, K., & Becker, T. (2017). Carbon steel corrosion: A review of key surface properties and characterization methods. *RSC Advances, 7*, 4580–4610. https://doi.org/10.1039/C6RA25094G

El Ouadi, Y., Bouyanzer, A., Majidi, L., Paolini, J., et al. (2014). Salvia officinalis essential oil and the extract as green corrosion inhibitor of mild steel in hydrochloric acid. *Journal of Chemical and Pharmaceutical Research, 6*(7), 1401–1416.

El Ouariachi, E., Bouyanzer, A., Salghi, R., et al. (2015). Inhibition of corrosion of mild steel in 1 M HCl by the essential oil or solvent extracts of Ptychotis verticillata. *Research on Chemical Intermediates, 41*, 935–946. https://doi.org/10.1007/s11164-013-1246-5

Fu, J., Li, S., Cao, L., et al. (2010). L-Tryptophan as green corrosion inhibitor for low carbon steel in hydrochloric acid solution. *Journal of Materials Science, 45*, 979–986. https://doi.org/10.1007/s10853-009-4028-0

Guadalupe, H. J., García-Ochoa, E., Maldonado-Rivas, P. J., et al. (2011). A combined electrochemical and theoretical study of N,N′-bis(benzimidazole-2yl-ethyl)-1,2-diaminoethane as a new corrosion inhibitor for carbon steel surface. *Journal of Electroanalytical Chemistry, 655*(2), 164–172. https://doi.org/10.1016/j.jelechem.2011.01.039

Hamdani, I., El Ouariachi, E., Mokhtari, O., et al. (2015). Chemical constituents and corrosion inhibition of mild steel by the essential oil of Thymus algeriensis in 1.0 M hydrochloric acid solution. *Der Pharma Chemica, 7*(8), 252–264.

Hussin, M. H., Kassim, M. J., Razali, N. N., et al. (2016). The effect of Tinospora crispa extracts as a natural mild steel corrosion inhibitor in 1 M HCl solution. *Arabian Journal of Chemistry, 9*(1), S616–S624. https://doi.org/10.1016/j.arabjc.2011.07.002

Khaksar, L., & Shirokoff, J. (2017). Effect of elemental sulfur and sulfide on the corrosion behavior of Cr-Mo low alloy steel for tubing and tubular components in oil and gas industry. *Materials, 10*, 430. https://doi.org/10.3390/ma10040430

Lahhit, N., Bouyanzer, A., & Desjobert, J. M. (2011). Fennel (Foeniculum Vulgare) essential oil as green corrosion inhibitor of carbon steel in hydrochloric acid solution port. *Electrochimica Acta, 29*(2), 127–138. https://doi.org/10.4152/pea.201102127

Li, Y., Zhao, P., Liang, Q., et al. (2005). Berberine as a natural source inhibitor for mild steel in 1 M H_2SO_4. *Applied Surface Science, 252*, 1245–1253. https://doi.org/10.1016/j.apsusc.2005.02.094

Loto, R. T. (2013). Pitting corrosion evaluation of austenitic stainless steel type 304 in acid chloride media. *Journal of Materials and Environmental Science, 4*(4), 448–459.

Loto, R. T. (2017). Study of the synergistic effect of 2-methoxy-4-formylphenol and sodium molybdenum oxide on the corrosion inhibition of 3CR12 ferritic steel in dilute sulphuric acid. *Results in Physics, 7*, 769–776. https://doi.org/10.1016/j.rinp.2017.01.042

Loto, R. T., & Babalola, P. (2017). Corrosion polarization behaviour and microstructural analysis of AA1070 aluminium silicon carbide matrix composites in acid chloride concentrations. *Cogent Engineering, 4*(1), 1422229. https://doi.org/10.1080/23311916.2017.1422229

Loto, C. A., & Loto, R. T. (2012). Effect of P-phenylenediamine on the corrosion of austenitic stainless steel type 304 in hydrochloric acid. *International Journal of Electrochemical Science, 7*(10), 9423–9440.

Loto, C. A., & Loto, R. T. (2013). Effect of dextrin and thiourea additives on the zinc electroplated mild steel in acid chloride solution. *International Journal of Electrochemical Science, 8*(12), 12434–12450.

Loto, R. T., & Oghenerukewe, E. (2016). Inhibition studies of rosmarinus officinalis on the pitting corrosion resistance 439LL ferritic stainless steel in dilute sulphuric acid. *Oriental Journal of Chemistry, 32*(5), 2813–2832. https://doi.org/10.13005/ojc/320557

Loto, R. T., Leramo, R., & Oyebade, B. (2018). Synergistic combination effect of salvia officinalis and lavandula officinalis on the corrosion inhibition of low-carbon steel in the presence of

$SO_4{}^{2-}$- and Cl^-- containing aqueous environment. *Journal of Failure Analysis and Prevention, 18*(6), 1429–1438. https://doi.org/10.1007/s11668-018-0535-0

Loto, R. T., Okpaleke, P. C., & Ufansi, U. (2020). Inhibition effect of admixed citrus paradise and cymbopogon oil distillates on the corrosion resistance mild steel in dilute acid electrolytes. *Surface Review and Letters*, 2150016. https://doi.org/10.1142/S0218625X21500165

ÖZcan, M. (2008). AC impedance measurements of cysteine adsorption at mild steel/sulphuric acid interface as corrosion inhibitor. *Journal of Solid State Electrochemistry, 12*, 1653–1661. https://doi.org/10.1007/s10008-008-0551-1

Quartarone, G., Ronchin, L., Vavasori, A., et al. (2012). Inhibitive action of gramine towards corrosion of mild steel in deaerated 1.0 M hydrochloric acid solutions. *Corrosion Science, 64*, 82–89. https://doi.org/10.1016/j.corsci.2012.07.008

Singh, W. P., & Bockris, J. O. (1996). *Toxicity issues of organic corrosion inhibitors: Applications of QSAR model*. NACE International.

Turcio-Ortega, D., Pandiyan, T., Cruz, J., et al. (2007). Interaction of imidazoline compounds with Fen (n = 1–4 atoms) as a model for corrosion inhibition: DFT and electrochemical studies. *Journal of Physical Chemistry C, 111*(27), 9853–9866. https://doi.org/10.1021/jp0711038

Ye, Y., Yang, D., Chen, H., et al. (2020a). A high-efficiency corrosion inhibitor of N-doped citric acid-based carbon dots for mild steel in hydrochloric acid environment. *Journal of Hazardous Materials, 381*, 121019. https://doi.org/10.1016/j.jhazmat.2019.121019

Ye, Y., Zhang, D., Zou, Y., et al. (2020b). A feasible method to improve the protection ability of metal by functionalized carbon dots as environment-friendly corrosion inhibitor. *Journal of Cleaner Production, 264*, 121682. https://doi.org/10.1016/j.jclepro.2020.121682

Ye, Y., Chen, H., Zou, Y., et al. (2021). Study on self-healing and corrosion resistance behaviors of functionalized carbon dot-intercalated graphene-based waterborne epoxy coating. *Journal of Materials Science and Technology, 67*, 226–236. https://doi.org/10.1016/j.jmst.2020.06.023

Zarras, P., & Stenger-Smith, J. D. (2017). Corrosion processes and strategies for prevention: An introduction. In *Handbook of smart coatings for materials protection, corrosion effects on steel/alloys*. Elsevier. https://doi.org/10.1016/B978-0-85709-680-7.50023-1

Corrosion Inhibitive Behaviour of *Moringa Oleifera* in Acidic Medium

A. A. Ayoola, S. C. Okwuonu, B. M. Durodola, E. E. Alagbe, O. Oladokun, O. Agboola, and R. Babalola

1 Introduction

Corrosion can be defined as a destructive phenomenon resulting from the electrochemical reaction of a metallic material with the chemical substance(s) in its surrounding leading to a slow and irreversible degradation of the physical and chemical features of such a metallic material. It is also the decay of an engineered metal into its component atoms due to chemical reactions with its environment (Singh et al., 2011). Hence, corrosion interaction of a metal with its environment can be direct, electrochemical, or biochemical in nature (Singh et al., 2016). Varied degrees of losses or destructions are traceable to corrosion of materials used at homes, in industries and on highways. Corrosion of metals can be curtailed by controlling the dissolution (reaction process) of the metal through the use of a suitable corrosion inhibitor. Corrosion inhibitor is a chemical substance added, in minute quantity, to a corrosive environment in order to stop or limit the corrosion reaction of a metal. The effectiveness of an inhibitor depends on its chemical constituents, quantity applied and temperature of the medium (Dass et al., 2015).

A. A. Ayoola (✉) · S. C. Okwuonu · E. E. Alagbe · O. Oladokun · O. Agboola
Chemical Engineering Department, Covenant University, Ota, Ogun State, Nigeria
e-mail: ayodeji.ayoola@covenantuniversity.edu.ng; stephnieokwuonu@stu.cu.edu.ng; edith.alagbe@covenantuniversity.edu.ng; olagoke.oladokun@covenantuniversity.edu.ng; oluranti.agboola@covenantuniversity.edu.ng

B. M. Durodola
Chemistry Department, Covenant University, Ota, Ogun State, Nigeria
e-mail: bamidele.durodola@covenantuniversity.edu.ng

R. Babalola
Chemical/Petrochemical Enginering Department, Akwa Ibom State University, Akpaden, Mpat Enin, Nigeria
e-mail: asheedbabalola@aksu.edu.ng

A. O. Ayeni et al. (eds.), *Advanced Manufacturing in Biological, Petroleum, and Nanotechnology Processing*, Green Energy and Technology,
https://doi.org/10.1007/978-3-030-95820-6_15

The major classes of the corrosion inhibitors are polymeric, organic and inorganic inhibitors. The inorganic inhibitors include calcium nitrite, rare earth metal salts, zinc phosphate, chromates and lanthanide compounds, and most of the inorganic inhibitors are toxic and non-environmentally friendly compounds (Singh et al., 2016). Current research has shown that organic inhibitors are environmentally friendly and they get adsorbed easily on the metal surface. Also the usage of waste organic materials as inhibitors is another attempt in reducing the cost of inhibitor. This research focus on the use of *Moringa oleifera* leaves as an inhibitor for A36 mild steel in 1M HCl environment.

2 Materials and Method

The major materials used in the course of this research work are A36 mild steel and *Moringa oleifera*, while the equipment used include Autolab PGSTAT 101 Metrohm potentiostat/galvanostat.

2.1 Phytochemical Screening of Moringa oleifera Leaves

Moringa leaves were dried and turned to fine powder. The phytochemical screening of the *Moringa oleifera* leaves was performed in order to establish the active compounds responsible for the inhibitory properties of the leaves. The screening procedures were done as reported in previous work of Gupta et al. (2014).

2.2 Preparation of the Mild Steel Test Samples

The mild steel chemical composition determination was carried out. The mild steel bar was cut into pieces, each has a dimension of 2.5 cm × 2.5 cm and the samples were polished using emery papers of different grades, degreased in ethanol, dried in acetone and then stored in desiccators.

2.3 Preparation of Moringa oleifera Extract

Moringa oleifera leaves were properly dried and grinded to powder (particle size of ≤60 μm). The extraction was performed using 100 g of moringa leaves powder and 400 mL of 99% methanol at 60 °C. After the extraction operation, the extract obtained was separated through filtration process. The extract was then subjected to evaporation (at room temperature) to remove any trace of methanol that may be present.

2.4 *Weight Loss Determination*

Each of the weighted test specimens (mild steel samples) was submerged in 1M HCl solution (using 0–4 vol/vol% inhibitor concentration) for weight loss test. At a 3-day interval, the test specimens were recovered, washed with distilled water, rinsed with ethanol to remove loosed deposits from their surfaces and then dried with acetone. The samples were then reweighed (using accurate weighing balance with ±0.0001), and the difference in weight accounted for the weight loss due to corrosion.

2.5 *Potentiodynamic Polarisation Measurement*

This test was performed using Autolab PGSTAT 101 Metrohm potentiostat/galvanostat with NOVA software version 2.1.2 and a three-electrode cell (having mild steel as the working electrode). The electrolyte was 150 mL of 1M HCl with varied concentrations of the *Moringa oleifera* inhibitor (with no inhibitor in some cases). Potentiodynamic polarisation data were obtained from potential of –0.6 to −0.1 V versus open circuit potential at a scan rate of 0.005 V/s. The polarisation potential (E_{corr}) and current density (I_{corr}) data were evaluated from the Tafel plots.

3 Results and Discussion

3.1 *Phytochemical Result and Mild Steel Composition*

The result of the phytochemical analysis showed that the following compounds were present in the *Moringa oleifera* inhibitor used: tannins, saponins, flavonoids and terpenoid. Table 1 revealed the elemental composition of A36 mild steel utilised in the course of this research work. The mild steel is low carbon (0.15%) mild steel, with high iron content 98.70%.

Table 1 Composition of A36 mild steel

Element	Percentage (wt%)
C	0.15
Si	0.22
Mn	0.51
Cr	0.07
Ni	0.09
Cu	0.22
Al	0.03
Fe	98.70

3.2 Mild Steel Composition and Weight Loss

Figures 1, 2 and 3 showed the trend of the weight loss against the exposure time at different inhibitor concentrations and different temperatures. It could be generally observed that the weight loss of the mild steel samples increased with increase in the exposure time during corrosion. Also, at each temperature, the weight loss decreased with increase in the *Moringa oleifera* inhibitor concentration. This implied that the weight loss was least when the *Moringa oleifera* inhibitor concentration of 4 vol/vol% was used. The reduction in weight loss, as the inhibitor concentration increased, could be accredited to the adsorption of the phytochemicals, on the surface of the mild steel. The phytochemicals formed a thin layer barrier on metal surface in the corrosive medium (Olasehinde et al., 2013). Additionally, it was also observed in Figs. 1, 2 and 3 that with an increase in corrosion reaction temperature, there was an increase in weight loss recorded, an indication that corrosion process was favoured by increase in temperature, as commonly reported in the literatures (Ayoola et al., 2020).

Figure 4 showed the corrosion rate obtained at different inhibitor concentrations and different temperatures. The rate of corrosion of the mild steel samples decreased with increase in the inhibitor concentration. Also, the corrosion rate decreased with decrease in temperature. This result corroborated the Arrhenius equation that says that reaction proceeds in the direction of increased temperature (Loto et al., 2011). These results further confirmed the results obtained with the weight loss in Figs. 1, 2 and 3. Similar trend was noticed in the plot of inhibition efficiency against inhibitor concentration (Fig. 5). That is, inhibition efficiency increased with increase in inhibitor concentration, and the efficiency of the inhibitor increased with decrease in temperature.

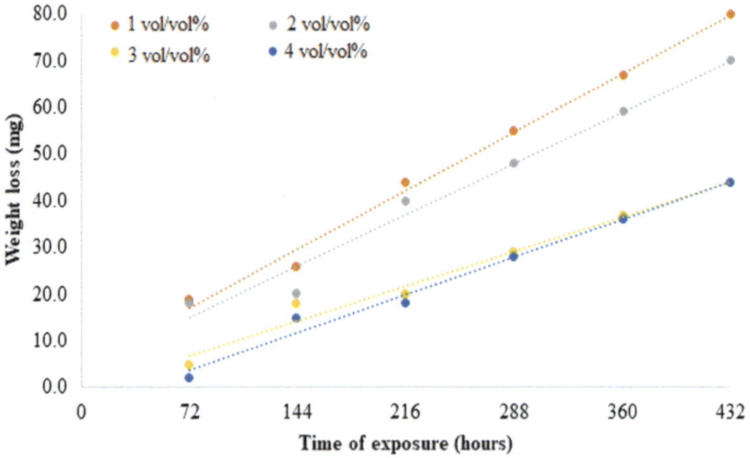

Fig. 1 Weight loss against time of exposure with varied *Moringa oleifera* inhibitor concentration at 5 °C

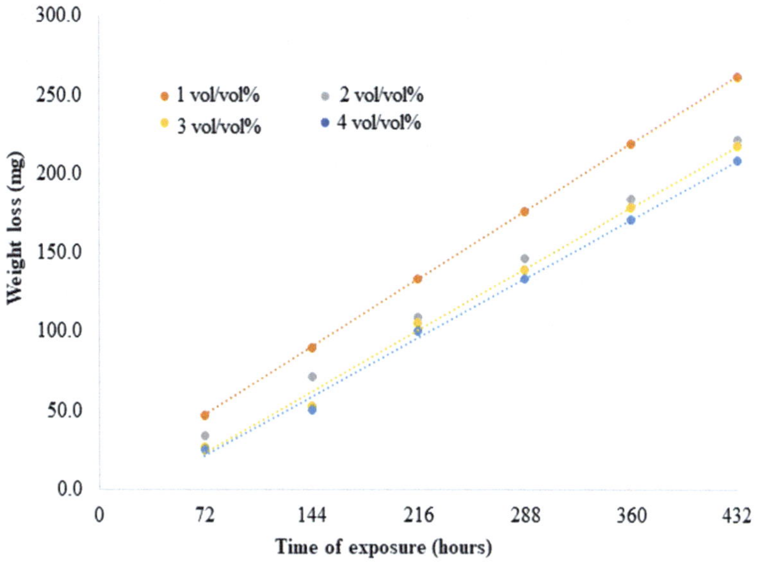

Fig. 2 Weight loss against time of exposure with varied *Moringa oleifera* inhibitor concentration at 31 °C

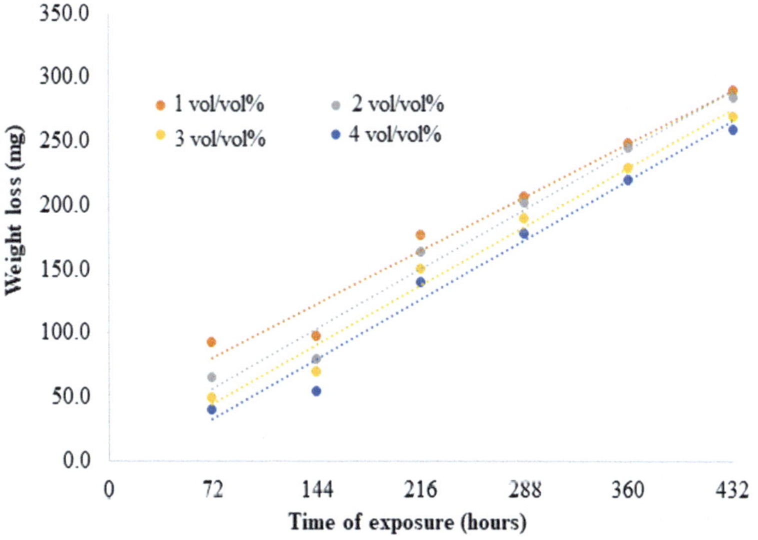

Fig. 3 Weight loss against time of exposure with varied *Moringa oleifera* inhibitor concentration at 50 °C

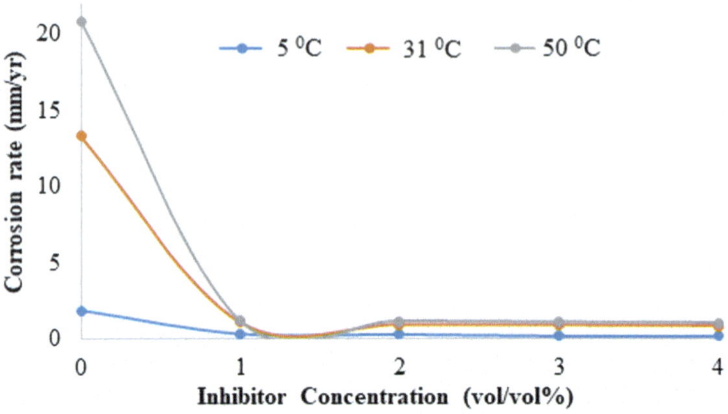

Fig. 4 Corrosion rate obtained at different inhibitor concentrations and different temperatures

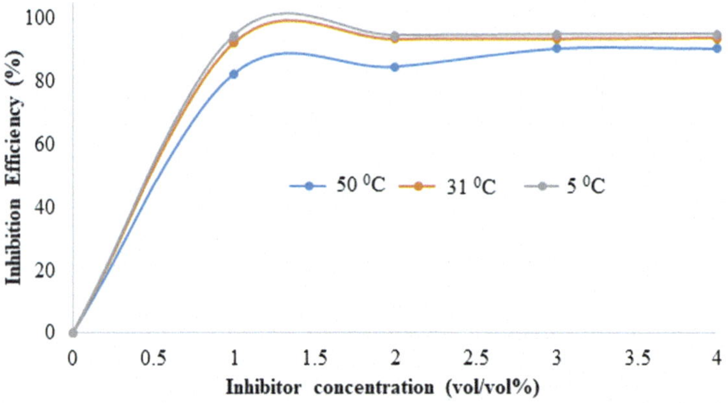

Fig. 5 Inhibition efficiency against inhibitor concentration at three different temperatures

3.3 Potentiodynamic Polarisation Result

Tables 2 and 3 revealed the electrochemical parameters obtained from the potentiodynamic tests at 304.15 K (31 °C) and 323.15 K (50 °C). As the inhibitor concentration increased, the polarisation resistance increased (for each of the temperature considered), thereby causing reduction in the corrosion rate. It could be inferred that the metal sites available for corrosion reaction got increasingly occupied by the adsorbed inhibitor, thereby slowing down the corrosion process. Figures 6 and 7 (Tafel plots) revealed that increase in the inhibitor concentration slowed down the current density recorded, thereby reducing the transfer of electrons responsible for the corrosion process. Also, the nature of the plots showed that the inhibitor behaved as a mixed typed, by reducing both the anodic and cathodic currents (as the inhibitor concentration increased). Comparatively, corrosion of the metal samples was

Table 2 Electrochemical parameters of the inhibitor efficiency at 304.15K (31 °C)

Inhibitor concentration (vol/vol%)	Current density, I_{corr} (μA/cm^2)	Potential applied, E_{corr} (mV)	Polarisation resistance, R_p (Ω)	Corrosion rate, (mm/year)
0	29.964	−451.35	13.15	0.3482
1	16.48	−438.10	25.08	0.1915
2	13.68	−423.52	36.36	0.1590
3	12.83	−413.88	38.95	0.1491
4	9.36	−415.14	42.71	0.1089

Table 3 Electrochemical parameters of the inhibitor efficiency at 323.15K (50 °C)

Inhibitor concentration (vol/vol%)	Current density, I_{corr} (μA/cm^2)	Potential applied, E_{corr} (mV)	Polarisation resistance, R_p (Ω)	Corrosion rate (mm/year)
0	94.22	−325.67	4.9182	1.0948
1	35.96	−454.68	9.5212	0.4541
2	33.07	−440.61	8.0172	0.3843
3	25.02	−447.94	17.026	0.2806
4	23.03	−424.60	16.175	0.2676

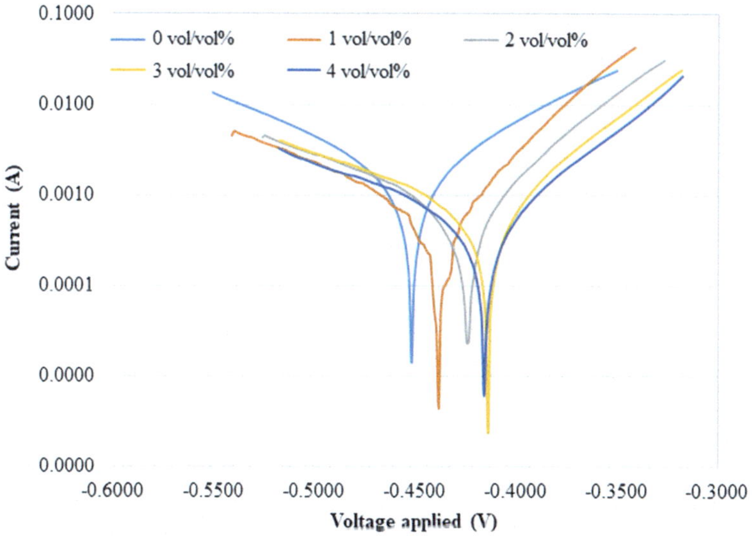

Fig. 6 Tafel plot at 304.15K (31 °C)

favoured by the increase in the corrosion reaction temperature. That is, higher corrosion rate was recorded at 323.15K (50 °C) compared to the results obtained at 304.15 K (31 °C), as shown in Fig. 8.

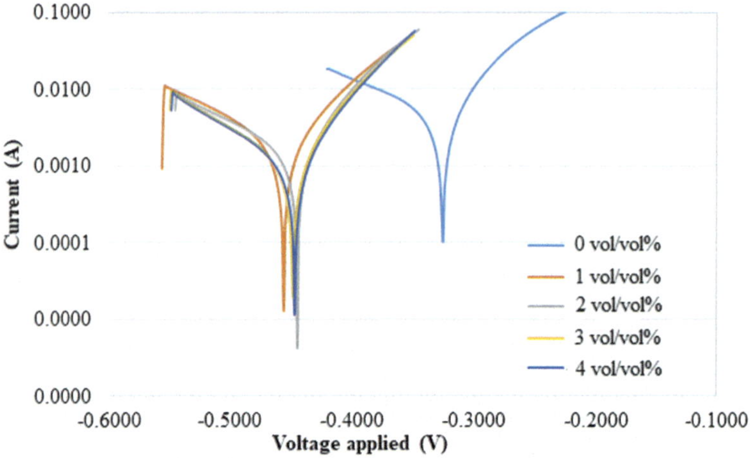

Fig. 7 Tafel plot at 323.15K (50 °C)

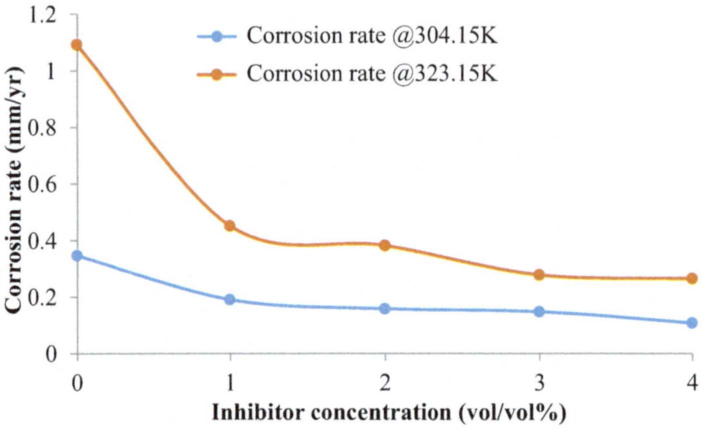

Fig. 8 Corrosion rate against concentration at two different temperatures

3.4 Adsorption of the Inhibitor on A36 Mild Steel

Figure 9 showed the plot of Langmuir isotherm obtained for the adsorbed *Moringa oleifera* inhibitor on A36 mild steel at different temperatures (5, 31 and 50 °C). Langmuir adsorption isotherm was considered in the determination of the adsorption behaviour of the *Moringa oleifera* inhibitor on the surface of the metal samples. The isotherm is expressed as shown in Eq. 1.

$$\frac{C_{\text{inh}}}{\theta} = \frac{1}{K_{\text{ads}}} + C_{\text{inh}} \tag{1}$$

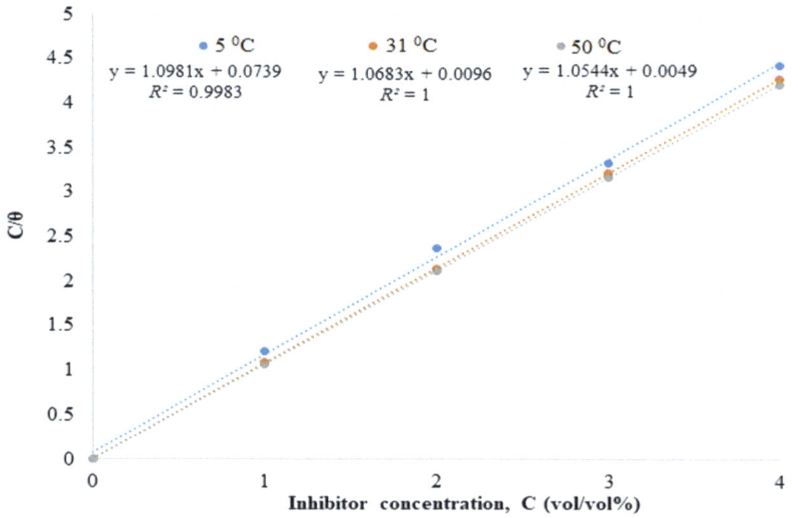

Fig. 9 Langmuir isotherm of the adsorbed *Moringa oleifera* inhibitor at different temperatures

Table 4 Langmuir adsorption parameters for *Moringa oleifera* inhibitor at different temperatures

Temperature (°C)	Equilibrium constant, K_{ads}	Change in Gibbs free energy (ΔG_{ads}, kJ/mol)
5	13.5318	−15.3151
31	104.1667	−21.9085
50	204.0816	−25.0843

where C_{inh} is the inhibitor concentration, θ is the degree of surface coverage and K_{ads} is the adsorption constant.

Using Eq. 2, the values of the change in Gibbs free energy (ΔG_{ads}) during the adsorption process were determined, as shown in Table 4.

$$\Delta G_{ads} = -2.303\text{RT}\log\left(55.5\,K\text{ads}\right) \quad (2)$$

The value of approximately one (1) obtained as the *coefficient of regression* (R^2) from the plots **C/θ** against **C** (at each temperature) indicated that the Langmuir isotherm suitably fit as a model for the prediction of the adsorption of the inhibitor on the metal samples. From Table 4, the equilibrium constant (K_{ads}) increased with an increase in the reaction temperature. But ΔG_{ads} (energy lost to the environment due to work done) decreased as the reaction temperature increased. The trends of the values of these two parameters (K_{ads} and ΔG_{ads}) conform to the Arrhenius equation that states temperature is directly proportional to reaction constant but inversely proportional to the energy loss (Loto et al., 2011).

4 Conclusion

The following conclusions were obtained from the research work:

(i) The corrosion rate of A36 mild steel decreased with an increase in inhibitor concentration (1–4 vol/vol%), but increased with increase in temperature (for both gravimetric and potentiodynamic polarisation tests).
(ii) Inhibitor efficiency of the *Moringa oleifera* inhibitor increased with an increase in inhibitor concentration, but decreased as the reaction temperature increased (for both gravimetric and potentiodynamic polarisation tests).
(iii) Langmuir adsorption isotherm fitted the adsorption behaviour of the *Moringa oleifera* inhibitor on the surface of the A36 mild steel.

Acknowledgement Publication of this conference paper is made possible through the financial commitment of CUCRID Covenant.

References

Ayoola, A. A., Obanla, O. R., Abatan, O. G., Fayomi, O. S. I., Akande, I. G., Agboola, O., Ayeni, O. A., Oyekunle, D., Olawepo, V. A., & Ayo-Aderele, O. O. (2020). Corrosion inhibitive behaviour of the natural honey in acidic medium of A315 mild and 304 austenitic stainless steels. *Analytical and Bioanalytical Electrochemistry, 12*(1), 21–35.

Dass, P., Onen, A., Maitera, O., & Ushahemba, G. (2015). Corrosion inhibition of zinc in acid medium by moringa oleifera and mangifera indica leaves extracts. *International Journal of Sustainable Development, 4*(9), 944–950.

Gupta, J., Gupta, A., & Gupta, K. (2014). Preliminary phytochemical screening of leaves of moringa oleifera lam. *Journal Chemtracks, 16*(1), 285–288.

Loto, A. C., Loto, R. T., & Popoola, A. P. I. (2011). Effect of neem leaf (Azadirachta indica) extract on the corrosion inhibition of mild steel in dilute acids. *International Journal of Physical Sciences, 6*(9), 2249–2257.

Olasehinde, E. F., Olusegun, S. J., Adesina, A. S., Omogbehin, S. A., & Momoh-Yahayah, H. (2013). Inhibitory action of Nicotiana tobacum extracts on the corrosion of mild steel in HCl: Adsorption and thermodynamic study. *Natural Sciences, 11*, 83–90.

Singh, A., Ahmad, I., Yadav, D., Singh, V., & Quraishi, M. (2011). The effect of environmentally benign fruit extract of shajan on the corrosion of mild steel in HCl acid solution. *Chemical Engineering Communications, 199*(1), 6377–6386.

Singh, A. K., Mohapatra, S., & Pani, B. (2016). Corrosion inhibition effect of aloe vera gel: Gravimetric and electrochemical study. *Journal of Industrial and Engineering Chemistry, 33*, 288–297.

Phosphating Technique: A Reliable Approach for Corrosion Resistance of A36 Mild Steel

S. N. Ezekiel, A. A. Ayoola, B. M. Durodola, O. Odunlami, and O. A. Oyeniyi

1 Introduction

Equipment made up of metals are liable to undergo corrosion which could be caused by exposure to water and air. When metals corrode either slightly or totally, their functionality begin to decline (Akpoborie et al., 2021). The equipment can be composed of either ferrous metals or non-ferrous metals. Ferrous metals are metals that contain iron such as the wrought iron, cast iron, carbon steel and mild steel (Mahmoodian, 2018). They possess tensile strength and magnetic properties and are durable, though susceptible to corrosion. Non-ferrous metals are metals that do not contain iron; thus, they do not corrode easily. Examples of non-ferrous metals are aluminium, gold, zinc, brass, nickel, copper and silver. Non-ferrous metals are non-magnetic. Ferrous metals are applied in tall buildings, bridges and automobiles due to their strength, while non-ferrous metals are applied in smaller materials like the roofing sheets, water pipes, wirings or kitchen utensil (Groover, 2017).

Since corrosion on metallic equipment tends to be detrimental with cost, there is the need to prevent or treat corrosion on metals (Ayoola et al., 2020). The phosphating technique is mostly suitable to control metal corrosion either in the form of lubrication, decoration, spraying or painting. Phosphating as a chemical coating process is cost-effective and has a natural bonding effect (Liu et al., 2021). Phosphating also gives high anti-corrosion ability. Generally, phosphating has been

S. N. Ezekiel (✉) · A. A. Ayoola · O. Odunlami · O. A. Oyeniyi
Chemical Engineering Department, Covenant University, Ota, Nigeria
e-mail: neza.ezekielpgs@stu.cu.edu.ng; ayodeji.ayoola@covenantuniversity.edu.ng; olayemi.odunlami@covenantuniversity.edu.ng; esther.oyeniyi@covenantuniversity.edu.ng

B. M. Durodola
Chemistry Department, Covenant University, Ota, Nigeria
e-mail: bamidele.durodola@covenantuniversity.edu.ng

A. O. Ayeni et al. (eds.), *Advanced Manufacturing in Biological, Petroleum, and Nanotechnology Processing*, Green Energy and Technology,
https://doi.org/10.1007/978-3-030-95820-6_16

used in various industries such as the automobiles, electronic/electrical industries and cold processing (Jiang et al., 2020).

Controlling or preventing metal corrosion through phosphating involves the following steps: the removal of impurities from the surface, water-rinsing of the metal either with deionized or distilled water, the phosphating process, second stage water-rinsing, chromate coating (treatment using dilute chromic acid) and drying.

The scanning electron microscopy with energy-dispersive X-ray analysis (SEM-EDX) is carried out to provide high magnification of constituents present on the metal surface, since constituents are usually of micro sizes that cannot be seen with the naked eyes. The analysis is carried out to compare the structure, defects, thickness measurements and mechanisms on the metal surface before and after coating. It also analyses metals before and after corrosion. This analysis provides accuracy and precision in results after testing (Larizza, 2015).

The XRF (X-ray fluorescence) spectroscopy examines and gives data on the chemical compositions in percentage contained in particles.

Phosphate coatings, as an advantage, possess good anti-corrosion and anti-friction properties through the use of either paint or oil. It offers these properties for protection on metals a minimal cost (Pastorek et al., 2016). A disadvantage is that the wastes from phosphating cause negative effects to the environment as the process emits greenhouse gases (e.g. carbon dioxide, CO_2) and hazardous materials from the sludge and heavy metals (Tamilselvi et al., 2015).

To tackle the negative effects on human health, toxic releases need to be reduced and appropriate metals for phosphating need to be selected. Metals like cadmium should be avoided for phosphating; thus, iron and zinc are mostly utilized. Zinc and iron undergo more phosphate coatings since they give acceptable coating characteristics compared to heavy metals like cadmium (Narayanan, 2005).

Heavy metals that are subjected to phosphating take more processing time and highly expensive to handle – since it operates at higher temperature (60–100 °C). Since a higher temperature is involved, more phosphoric acid is needed which increases cost. Reducing to a room temperature or acceptable temperature could be difficult (Duszczyk et al., 2018).

Since coating process has the ability to withhold oils and soaps, when used as lubricants, phosphating is used to decrease the wear in machines and locomotives (Qian et al., 2015).

Phosphating can be applied to reduce the chemical oxidation of pyrite in acid carrier, thereby giving away iron in preference (Narayanan, 2005, p. 165).

This work intends to provide the coating condition on the A36 mild steel so as to provide the steel the best corrosion resistance through the phosphating process in focus.

2 Materials and Methods

2.1 Materials, Reagents and Equipment

The work involved the use of materials such as A36 mild steel, silicon carbide waterproof abrasive papers (P150, P320, P600 and P800) and periwinkle particles. Reagents used for this work included 85% phosphoric acid, H_3PO_4, 50% hydrogen peroxide, H_2O_2, 96% extra pure zinc nitrate hexahydrate, zinc oxide, sodium nitrate ($NaNO_3$), NaOH pellets and calcium oxide (from periwinkle particles). Equipment used were the muffle furnace and the SEM (scanning electron microscope – JOEL-JSM 7600F).

2.2 Methods

2.2.1 Sample Preparation

The A36 mild steels were cut in dimensions 2.8 cm by 2.8 cm and divided into 288 chips. The surface of each metal sample was smoothened with silicon carbide waterproof abrasive papers of different grades – P150, P320, P600 and P800. The order of the silicon carbide waterproof abrasive paper is from the roughest to the smoothest. The weight of the samples was taken with the samples orderly labelled.

The smoothening was done until white and clear parts were visible. The silicon papers were applied in order to remove rust particles, roughages or dirt so the chemicals used for the pre-treatment and phosphate process would have better effect in the treatment and coating process, respectively.

2.2.2 Pre-treatment Process

The samples were passed through treatment process where the A36 mild steels were suspended in NaOH for 1 min as a degreasing step. The NaOH solution removed grease, oils and dirt from the sample surface. The samples were then water-rinsed to remove the effect of the NaOH so as not to contaminate further stages. After water-rinsing the samples, they were suspended in HCl solution (de-rusting stage) for another minute to remove the rust properties. The samples were water-rinsed for a minute again and suspended in H_2O_2 solution (activation stage) to sterilize the samples in greater activity. The H_2O_2 solution also provided the sample surface with stronger bonding for crystal formation. With the samples gone through activation stage, they were then water-rinsed for another minute and finally air-dried.

Table 1 Concentration of reagents for the phosphating bath

S/no	Reagents	Concentration
1	Zinc oxide, ZnO	5 g/L
2	Zinc nitrate, $Zn(NO_3)$	0.2 g/L
3	Sodium nitrate, $NaNO_3$	0.1 g/L
4	Sodium saccharin	(0.1 – 0.2) g/L
5	Calcium oxide, CaO from periwinkle shell particles	1.0, 1.5, 2.0 and 2.5 g/L
6	Phosphoric acid, H_3PO_4	20 mL/L

2.2.3 Phosphating Process

The phosphating bath solution was prepared with the contents showed in Table 1 in no order; phosphoric acid was added lastly to dissolve other reagents completely.

Weights of 5 g of zinc oxide, 0.2 g of zinc nitrate, 0.1 g of sodium nitrate, 0.1–0.2 g of sodium saccharin and 1.0–2.5 g of calcium oxide in sequence of experiments were added, with 20 mL of phosphoric acid to a 1 litre graduated glass beaker. The solution was then topped with distilled water to 1 litre in the beaker.

The treated samples were then immersed into the prepared phosphating bath for the specified time and temperature with the concentration of calcium varied accordingly. The temperature considered were 40 and 60 °C for 30, 40, 50 and 60 min.

2.2.4 X-Ray Fluorescence (XRF) on Periwinkle Particles

The calcined particles (periwinkle) made up of CaO were analysed via XRF to determine chemical composition of the elements via the XRF spectrometer by absorbing X-rays. The calcined periwinkle particles were exposed to the entire spectrum of photons consisting of primary radiations emitted from a standard X-ray tube. These then bright forms of the particles made the elements in it to release secondary fluorescence with their characteristic X-ray line spectra.

The energies and intensities of the emitted lines were determined by the detection system which consisted of two parts: the primary channel simultaneous wavelength dispersive spectrometer and the personal computer for control and data processing. The rapid detection system is used in analysing crystal around the calcined periwinkles. These then gave dispersed wavelength of the secondary radiation.

The magnitude of the individual wavelength was measured in a mass gas flow detector. The mass gas flow detector allowed simultaneous measurements of up to ten elements at peak and background positions. The output signals from the detector were fed into the analyser, where the photon counts were stored in the computer memories. The count rate was calibrated for each element by comparing it to the count rate from a standard of accurately pre-determined composition.

The spectral line energies of wavelengths of the emitted lines were used in the quantitative analysis of the element in the periwinkle particles. The magnitudes of the emitted line were related to their concentration for quantitative analysis.

3 Discussion of Results

3.1 Metal Compositions

The metal, A36 mild steel, contained chemical compositions as shown in Table 2. The chemical compositions were obtained with the aid of the equipment, X-ray fluorescence (XRF) spectroscopy.

From Table 2, iron was dominant in the mild steel sample since the composition of Fe (98.00%) is more than the composition of Cu, Al, Mo, Ni, Cr, Mn, Si and C.

3.2 XRF (X-Ray Fluorescence) Analysis

Table 3 showed the chemical compositions of the particles from periwinkle shells as sources of calcium. From Table 3, periwinkle shell particles showed the very high amounts of calcium oxide with 81.60% from which calcium can be obtained.

Table 2 Compositions of elements in A36 mild steel

S/N	Elements	Compositions (%)
1	Copper (Cu)	0.22
2	Aluminium (Al)	0.03
3	Molybdenum (Mo)	0.62
4	Nickel (Ni)	0.09
5	Chromium (Cr)	0.07
6	Manganese (Mn)	0.51
7	Silicon (Si)	0.22
8	Carbon (C)	0.24
9	Iron (Fe)	98.00

Table 3 XRF analysis of periwinkle shell particles

Compounds/chemicals	Compositions (%)
Silicon (IV) oxide (SiO_2)	6.95
Aluminium oxide (Al_2O_3)	2.59
Iron (III) oxide (Fe_2O_3)	2.40
Manganese (II) oxide (MnO)	0.42
Calcium oxide (CaO)	81.60
Phosphorus (V) oxide (P_2O_5)	0.55
Potassium oxide (K_2O)	0.30
Titanium (IV) oxide (TiO_2)	0.15
Sulphur (VI) oxide SO_3	1.20
Sodium oxide (Na_2O)	0.002
Magnesium oxide (MgO)	3.07

3.3 *Coated Mild Steels*

Table 4 shows the visual inspections and physical appearance of the mild steel coated using periwinkle shell particles as source of calcium in the phosphating process. The metals were coated at various times and concentrations of calcium and temperatures at 40 °C and 60 °C.

3.3.1 Coated Mild Steels with Periwinkle Shell Particles at Coating Temperature of 40 °C

From Table 4 mild steels coated at 40 °C with phosphating process, parameter Pt_1C_340 gave mild steel coating of grey appearance and merely homogenous with rough surface texture. This observation is found also in Pt_2C_340, Pt_3C_340 and Pt_4C_340. Mild steel coated at condition Pt_2C_440 at 40 °C produced better results compared to Pt_1C_340, Pt_2C_340, Pt_3C_340 and Pt_4C_340 since the coating on the mild steel surface was rough, crystalline and homogenous in appearance. At condition Pt_1C_440, the coating appearance on the mild steel was more crystalline and homogenous with rougher surface texture. However, conditions Pt_3C_440 and Pt_4C_440 produced better coating nature than Pt_1C_440 where the coating with Pt_3C_440 and Pt_4C_440 produced a more crystalline and rougher texture in homogeneity. Hence, the better condition for coating mild steel with phosphating process is either Pt_3C_440 or Pt_4C_440. However, coating at condition Pt_4C_440 was observed to give more crystalline appearance than at Pt_3C_440. The results revealed that better result in coating could be obtained in the phosphating bath solution at 40 °C with a calcium concentration of 2.5 g/L. Coatings though become better as temperature increases to a certain level, but at different concentration.

3.3.2 Coated Mild Steels with Periwinkle Shell Particles at Coating Temperature of 60 °C

From Table 4 (mild steels coated at 60 °C with phosphating process), parameter Pt_1C_360 gave mild steel coating of dark crystalline appearance and merely homogenous with smooth surface texture. Pt_2C_360 gave a grey appearance with rough surface texture in mere homogeneity but lacked the crystalline appearance; hence, mild steel at Pt_1C_360 was more coated than the ones at Pt_2C_360 and Pt_3C_360. Coated mild steel at condition Pt_3C_360 resulted in dark crystalline appearance and merely homogenous with rough surface texture. Coating condition Pt_4C_360 was better than Pt_1C_360, Pt_2C_360 and Pt_3C_360 due to its rougher texture. Mild steel coated at condition Pt_1C_460 produced coating nature of dark appearance and homogenous with smooth surface texture. With this observation, coatings at Pt_2C_460, Pt_3C_460 and Pt_4C_460 have better appearance on the mild steel surface due to their rough,

Table 4 Physical appearance of mild steel coated with periwinkle shell particles as additive at 40 and 60 °C

S/n	Phosphating parameter	Physical appearance	Visual observation
1	$Pt_1C_3$40		Grey appearance and merely homogenous with smooth surface texture
2	$Pt_2C_3$40		Grey appearance and merely homogenous with smooth surface texture
3	$Pt_3C_3$40		Grey appearance and merely homogenous with smooth surface texture
4	$Pt_4C_3$40		Grey appearance and merely homogenous with smooth surface texture
5	$Pt_1C_4$40		Grey, more crystalline appearance and homogenous with rougher surface texture
6	$Pt_2C_4$40		Grey crystalline appearance and homogenous with rough surface texture
7	$Pt_3C_4$40	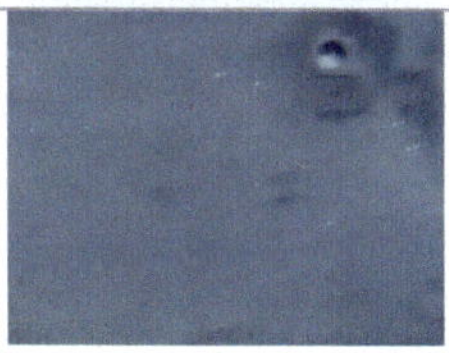	Grey crystalline appearance and more homogenous with rougher surface texture

(continued)

Table 4 (continued)

S/n	Phosphating parameter	Physical appearance	Visual observation
8	Pt_4C_440	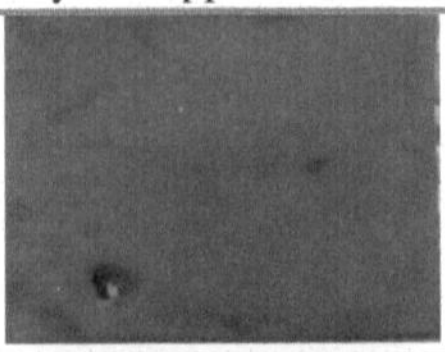	Grey, more crystalline appearance and more homogenous with rougher surface texture
9	Pt_1C_360		Dark crystalline appearance and merely homogenous with rough surface texture
10	Pt_2C_360	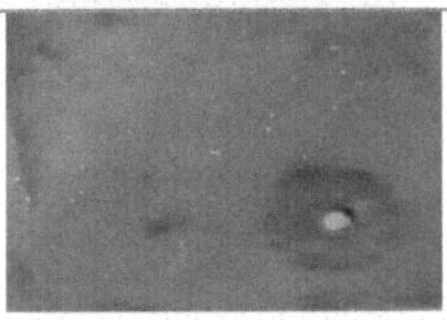	Grey, crystalline appearance and merely homogenous with rough surface texture
11	Pt_3C_360		Dark crystalline appearance and merely homogenous with rough surface texture
12	Pt_4C_360		Dark crystalline appearance and merely homogenous with rougher surface texture
13	Pt_1C_460		Dark appearance and homogenous with smooth surface texture
14	Pt_2C_460		Dark crystalline appearance and homogenous with rough surface texture
15	Pt_3C_460		Dark homogenous appearance, crystalline with rough surface texture

(continued)

Table 4 (continued)

S/n	Phosphating parameter	Physical appearance	Visual observation
16	Pt_4C_460		Dark homogenous appearance, crystalline with rough surface texture

Key: P = CaO obtained from periwinkle, $t_1 = 30$ min, $t_2 = 40$ min, $t_3 = 50$ min, $t_4 = 60$ min, $C_3 = 2.0$ g/L, $C_4 = 2.5$ g/L

crystalline and homogenous appearance. Hence, a better coating condition of mild steel at 60 °C is at Pt_2C_460, Pt_3C_460 and Pt_4C_460 due to homogeneity in texture.

Comparing Pt_3C_440 and Pt_4C_440 with Pt_2C_460, Pt_3C_460 and Pt_4C_460, it was observed that Pt_2C_460, Pt_3C_460 and Pt_4C_460 were rougher, more homogenous and crystalline in texture than mild steels coated at 40 °C in conditions Pt_3C_440 and Pt_4C_440. In summary, to obtain optimum coating results, the mild steel should be coated with phosphating process at 60 °C for conditions Pt_2C_460, Pt_3C_460 and Pt_4C_460. This means that phosphating bath solution should be heated at 60 °C with calcium concentration of 2.5 g/L. The mild steel should be suspended in the bath for 40, 50 and 60 min to get the optimum coating.

3.4 SEM Images

The coating formations deposited on the surface of the metal depended on the mild steel. Figure 1 shows different structures and crystal forms. Figure 1 also shows the morphology on the mild steel surface due to coatings at 60 °C with the mild steel immersed at various times t_3 and t_4. Figure 1a showed coating formation of Zn-Ca phosphate layer and the distribution of grain structures (of coating layers on the mild steel surface). This confirmed that layers on the surface were homogenous with no gaps. Gaps show vulnerability to corrosion according to the work of Aref (2017).

It could be observed from Figure 1b that the coating layers were distributed across the surface of the mild steel but few gaps are seen. The grain structures were of diameters 10μm. This meant that with coating time of 60 min, gaps were formed in the layer that left room for the phosphoric acid to act on the crystal structures to damage the Zn-Ca layer (Asadi et al., 2015). As coating time changed, the structures on the A36 mild steel changed in various forms, which was in accordance with Figure 1b which produced needle-like structures (Chantorn et al., 2018) with larger gaps. The crystal structures were not broken enough to cover the gaps in the phosphate layer as this was due to lesser concentration of calcium ions (Ca^{+2}) (Jiang et al., 2020).

Figure 1c–d showed sparse and plate-like crystals, respectively, formed on its surface with no gaps observed. With no gaps observed meant coating layers were

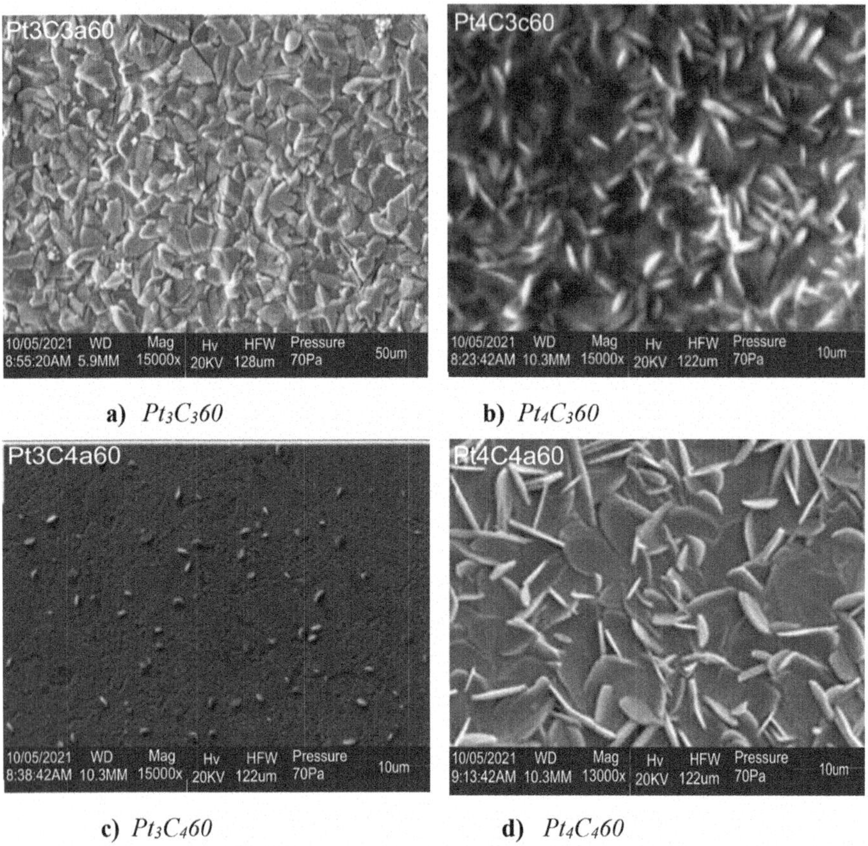

Fig. 1 The surface morphology for A36 mild steel coated through phosphating process at coating temperature of 60 °C. The coating was done using periwinkle shell particles as source of calcium. The concentration of calcium at C_3 and C_4, at coating time of t_3 and t_4
(a) *A36 mild steel coatings at coating time* $t_3 = 50$ *min in presence of periwinkle particles with concentration* $C_3 = 2.0$ *g/L*
(b) *A36 mild steel coatings at coating time* $t_4 = 60$ *min in presence of periwinkle particles with concentration* $C_3 = 2.0$ *g/L*
(c) *A36 mild steel coatings at coating time* $t_3 = 50$ *min in presence of periwinkle particles with concentration* $C_4 = 2.5$ *g/L*
(d) *A36 mild steel coatings at coating time* $t_4 = 50$ *min in presence of periwinkle particles with concentration* $C_4 = 2.5$ *g/L*

fully distributed over the surface; hence, there is strong corrosion resistance as mentioned by Aref (2017). From Fig. 1c, at concentration 2.5 g/L, the crystal structures were gaps in phosphate layer of Zn-Ca which meant there was a chance for corrosion to occur. An increase in the concentration resulted in breakdown of the crystals in the Zn-Ca layer, hence covering a wider surface area and tinier gaps. Figure 1c showed no gaps and the coating formation was homogenous. Increased calcium concentration (in the form of Ca^{2+}) on the mild steel coatings led to more compact phosphate layer. As coating time increased from 50 min to 60 min at 2.5 g/L, the morphology changed into plate-like structures as revealed in Fig. 1d. However, Fig. 1c showed finer form than Fig. 1d. In summary, appropriate coating condition from the SEM morphology was at Pt_3C_4. This was depicted to have uniform coating formation in fine crystal structures. Coating on the mild steels should be done with calcium concentration of 2.5 g/L for 50 min through the phosphating process. Mild steels coated in condition Pt_4C_3 showed weaker corrosion resistance due to the gaps found within the morphology on the steel surface.

4 Conclusion

From this research work, the following conclusions were made:

(i) Coatings on the A36 mild steel bodies using calcium reagents in phosphating provide long-lasting corrosion resistance.
(ii) Quality coatings on A36 mild steel are known when the mild steel surfaces are of rough, crystalline, dark or grey in homogenous nature.
(iii) Periwinkle shell is a good source of a locally produced calcium that produced fine coating results on the mild steel via phosphating technique.
(iv) From the visual observation, the best coating conditions on the A36 mild steel to mitigate corrosion is $Pt_3C_4$60 using periwinkle as source of calcium to accelerate the coating formation. This condition showed a very good corrosion resistance.
(v) From the SEM analysis, mild steels need to be coated at coating conditions of $Pt_3C_4$60 since it showed strong corrosion resistance.

Acknowledgements Covenant University's CUCRID has made a significant financial contribution to the publication of this article.

References

Akpoborie, J., et al. (2021). Electrochemical corrosion phenomenon and prospect of materials selection in curtailing the challenges. In *IOP conference series: Materials science and engineering* (Vol. 1107, p. 12072). IOP Publishing.

Aref, M. (2017). Modified zinc phosphate coatings: A promising approach to enhance the anti-corrosion properties of reinforcing steel. *MOJ Civil Engineering, 3*(5), 370–374.

Asadi, V., et al. (2015). The effect of immersion time and immersion temperature on the corrosion behavior of zinc phosphate conversion coatings on the corrosion behavior of zinc phosphate conversion coatings on carbon steel. *The Journal of Materials Research, 18*(4), 706–713.

Ayoola, A., et al. (2020). Inhibitive corrosion performance of the eco-friendly aloe vera in acidic media of mild and stainless steels. *Journal of Bio- and Tribo-Corrosion, 6*(3), 1–13. https://doi.org/10.1007/s40735-020-00361-y

Chantorn, R., et al. (2018). Phosphating of Mn modified Zn for low alloy steels SA335P22. *Materials Today: Proceedings, 5*(3), 9635–9641. https://doi.org/10.1016/j.matpr.2017.10.151

Duszczyk, J., et al. (2018). Manganese phosphatizing coatings: The effects of preparation conditions on surface properties. *Materials, 11*(12), 2585. https://doi.org/10.3390/ma11122585

Groover, M. (2017). *Fundamentals of modern manufacturing* (2nd ed., pp. 1–180). Wiley.

Jiang, C., et al. (2020). Electrochemistry communications phosphate conversion coatings on 35CrMnSi steels subjected to different heat treatments. *Electrochemistry Communications, 110*, 106636. https://doi.org/10.1016/j.elecom.2019.106636

Larizza, P. (2015). Measurement, testing, and diagnosis for micro-manufacturing systems. *Micromanufacturing Engineering and Technology, 2*, 675–704. https://doi.org/10.1016/B978-0-323-31149-6.00029-3

Liu, B., et al. (2021). Influence of surface post-processing on crystal refinement and characteristics of hopeite coating by phosphating. *Coatings, 11*(5), 1–8. https://doi.org/10.3390/coatings11050541

Mahmoodian, M. (2018). *Reliability and maintainability of in-service pipelines* (pp. 1–43). Gulf Professional Publishing. https://doi.org/10.1016/b978-0-12-813578-5.00001-9

Narayanan, S. (2005). Surface pretreatment by phosphate conversion coatings – A review. *Reviews on Advanced Materials Science, 9*, 130–177.

Pastorek, F., et al. (2016). Effect of surface pretreatment on quality and electrochemical corrosion properties of manganese phosphate on S355J2 HSLA steel. *Coatings, 6*(4), 46. https://doi.org/10.3390/coatings6040046

Qian, Y., et al. (2015). The application of anti-corrosion coating for preserving the value of equipment asset in chloride-laden environments: A review. *International Journal of Electrochemical Science, 10*, 10756–10780.

Tamilselvi, M., et al. (2015). Progress in zinc phosphate conversion coatings: A review. *International Journal of Advanced Chemical Science and Applications (IJACSA), 3*(1), 25–41.

Adoption of Environmental Engineering Strategies for Public Health and Sustainable Development

O. J. Oyebode

1 Introduction

Environmental process engineering needs to be applied for public health and sustainable development. Process engineering is the understanding and application of the fundamental principles and laws of nature that allow humans to transform raw material and energy into products that are useful to society, at an industrial level. By taking advantage of the driving forces of nature such as pressure, temperature and concentration gradients, as well as the law of conservation of mass, process engineers can develop methods to synthesize and purify large quantities of desired chemical products. Process engineering focuses on the design, operation, control, optimization and intensification of chemical, physical and biological processes. Process engineering encompasses a vast range of industries, such as agriculture, automotive, biotechnical, chemical, food, material, development, mining, nuclear, petrochemical and pharmaceutical and software development. The application of systematic computer-based methods to process engineering is process systems engineering. Environmental engineers play important role in sustainable development by planning and building projects that preserve natural resources that are cost-efficient and support human and natural environments. A closed-loop human ecosystem can be used to illustrate the many activities of engineers that support sustainable development.

Public health is defined as the art and science of preventing disease, prolonging life and promoting health through the organized efforts of society (Jackson, 2018). A process engineer is responsible for the equipment, system and methods used in a manufacturing facility to transform raw ingredients to their final product form.

O. J. Oyebode (✉)
Civil and Environmental Engineering Department, Afe Babalola University, Ado-Ekiti (ABUAD), Ado-Ekiti, Ekiti State, Nigeria

A. O. Ayeni et al. (eds.), *Advanced Manufacturing in Biological, Petroleum, and Nanotechnology Processing*, Green Energy and Technology,
https://doi.org/10.1007/978-3-030-95820-6_17

Process engineering is the understanding and application of the fundamental principles and laws of nature that allow humans to transform raw material and energy into products that are useful to society, at an industrial level.

Environmental pollution has been in existence since man began to live in settlements. In the earlier days of nomadic hunting communities, the tribal group moved on when food in their current location became depleted and the area around their camp became polluted or soiled. In Nigeria, efforts at bringing about a cleaner environment have relied on a philosophy of pollution control. This has in some cases involved costly measures and controversial political decisions (Oyebode, 2018). Environmental management plan, contingency plans and equipment that are to be used to carry out a specific work on all construction contracts should be specified, in order to have a friendly environment (Oyebode, 2019). Engineering infrastructures are required in all aspects of human endeavour, they are found at various stages of human economic and social economic life, and buildings are common one around. Researchers all over the world are progressively aiming to develop effective aids to minimize and eradicate the spread of the virus. Strategies for alleviating flood and erosion should be implemented supported by the latest technology for sustainable infrastructural development (Oyebode, 2021).

Process engineering involves the utilization of multiple tools and methods. Depending on the exact nature of the system, processes need to be simulated and modelled using mathematics and computer science. Processes where phase change and phase equilibria are relevant require analysis using the principles and laws of thermodynamics to quantify changes in energy and efficiency. Disciplines within the field of mechanics need to be applied in the presence of fluids or porous and dispersed media. Materials engineering principles also need to be applied, when relevant. In engineering, a process is a series of interrelated tasks that, together, transform inputs into a given output. These tasks may be carried out by people, nature or machines using various resources; an engineering process must be considered in the context of the agents carrying out the tasks and the resource attributes involved.

Sustainability addresses the ability of societies to maintain and improve quality of life while preserving both the quality and availability of its natural resources. Engneers play important role in sustainable development by planning and building projects that preserve natural resources that are cost-efficient and support human and natural environments. An environmental engineer is a public health scientist who studies nature and communities to determine future outcomes and impacts on one to the other. This job includes monitoring changes over time and keeping track of chemicals that are naturally or artificially present in the water, air and soil. While sustainability professionals technically fall under the umbrella of environmental engineers, they do tend to focus more on a product's entire life cycle with regard to its impacts on the environment. Figure 1 indicates case studies for chemical and environmental applications, while Fig. 2 presented the essence of governance in technology, society, economy and ecology.

Fig. 1 Case studies for chemical and environmental applications

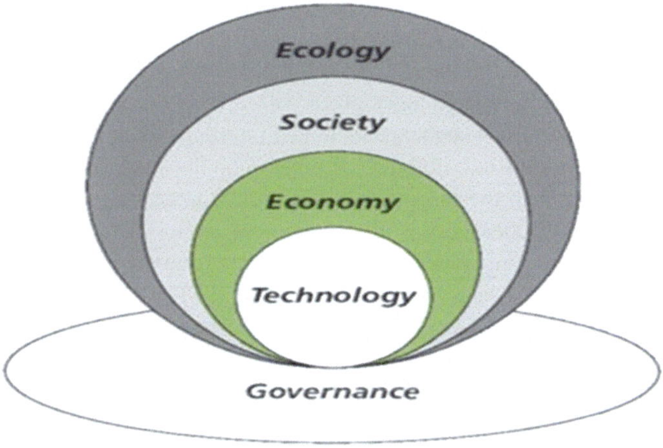

Fig. 2 Essence of governance in technology, society, economy and ecology

1.1 Environmental Process Engineering

The built environment serves as a dynamic interface through which the human society and the ecosystem interact and influence each other. Understanding this interdependence is a major key to understanding sustainability in environmental engineering as it applies to civil engineering construction works. Engineering design and construction have traditionally been dominated by a narrow, one-dimensional view of technological efficiency with the implicit assumption that nature is an infinite supplier of resources, perpetually regenerative, with an indefinite capacity to absorb all waste. It was only in the latter half of the twentieth century, particularly during the energy crisis of the 1970s, that the negative impacts of overreliance on technological advancement surfaced as a problem to the economic world and the essential interconnection of society, economics, technology and environment came under scrutiny. Responsible manipulation of material and energy flows requires an understanding of the interactions in and between the natural, industrial and social subsystems and their impacts.

A safe, environmentally sound and economically viable energy pathway that will sustain human progress into the distant future is clearly imperative.

The issue is not just numbers of people, but how those numbers relate to available resources. Urgent steps are needed to limit extreme rates of population growth.

Drought affects food supply and increases fire risks in urban areas; flooding also affects food supply and the economy; businesses cannot open; and local and national governments have a large bill for a clean-up operation, as well as putting an enormous strain on public services.

Environmental engineers already play a huge part in managing water supply – in droughts and in floods and outside of disaster areas, identifying where there are problem areas not getting enough water – irrigation or inadequate facilities.

Many disciplines have adopted a systems approach for studying these interactions, focusing on the processes that take place within certain boundaries, where a system "is any portion of the universe which is set aside to study" and a process is "the method by which a change takes place within the system". More generally, a process can be defined as "a sequence of linked interdependent activities that taken together, transform inputs into outputs". Emphasis on the study of industrial chemical processes began with the inception of chemical engineering as an independent discipline, combining chemistry with mechanical engineering. Recognition that "any chemical process, on whatever scale conducted, may be resolved into a coordinate series of what may be termed Unit Operations, as pulverizing, drying, roasting, crystallizing, filtering, evaporation, electrolyzing and so on", where "the number of these basic unit operations is not large and relatively few of them are involved in any particular process" (Arthur D. Little 1915, in Yoo et al., 2021), opened up a large reservoir of knowledge from many industrial branches and allowed rapid advancement in industrial manufacturing. Similarly, the focus of EPE (environmental process engineer) on the principles of the processes and interactions in and between the technical and natural subsystems and the conditions necessary for change allows the modification and/or application of these processes with any media, so that the traditional medial boundaries (air/water/solids) can be transcended (Fig. 3).

2 Environment Protection and Sustainability

Environment protection and sustainability are harmonious, and sustainability can be achieved by protecting our natural resources. The recent changes in global climate are believed to be the result of growing anthropogenic greenhouse gas (GHG)

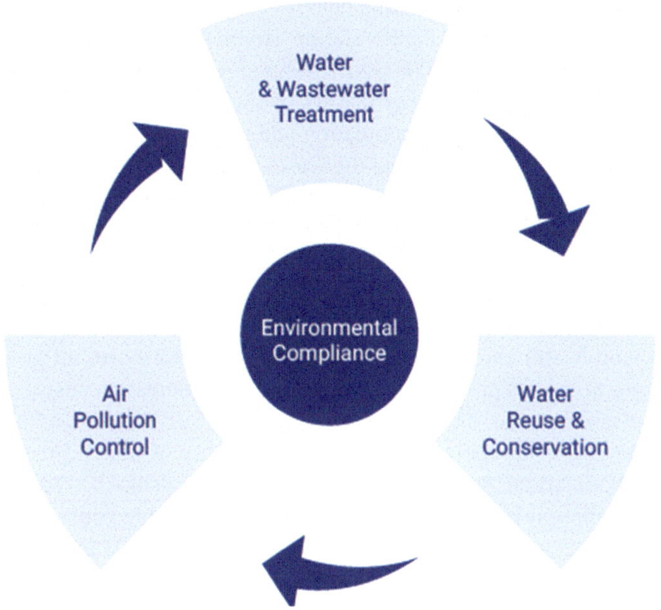

Fig. 3 Environmental compliance wheel

emissions, mainly carbon dioxide and methane, resulting from the increased industrial activities over the years. One of the main emission sources that add to the anthropogenic greenhouse gas concentrations in the atmosphere are derived from the processes of solid waste disposal (Malik & Grohmann, 2012). Solutions to global environmental and development problems require engineers to design and construct ecologically and socially just systems, within the carrying capacity of nature and society, without compromising the welfare of future generations. Engineers must also have international perspectives if the discipline is to take a leadership role in improving the lives of the global community (Mihelcic et al., 2006).

Current processes of economic development, while alleviating many social and health problems, are increasingly linked to environmental health threats, ranging from air pollution and physical inactivity to global climate change. Sustainable development practices attempt to reduce environmental impacts and should, in theory, reduce adverse environmental health consequences compared to traditional development (Furie & Balbus, 2012).

A world of healthy people living in healthy ecosystems has proven to be an elusive goal of the sustainable development agenda. Disease prevention and control interventions that link health to environmental problems and to local development tend to be the exception more than the rule. Degraded ecosystems affect everyone, but the poor suffer the consequences in a disproportionate manner. Poor health and degraded ecosystems represent losses in both natural and social capital. Not surprisingly, the prevailing consensus is that sustainable development depends on reducing poverty while protecting and promoting health. A common challenge is to conciliate the objectives of development, health and environmental protection with those of social equity (Boischio et al., 2009; UNEP, 2008; PAHO, 2007).

Health and well-being are the key drivers of social and economic development, as well as a major priority for the population and communities, as the deterioration in health generates the loss of life and the waste of resources in all sectors, besides suffering to people. Factors contributing to the prosperity also contribute to an increase in the health of the population, since fair access to education, decent living conditions and decent incomes contribute to maintaining health.

Occupational accidents represent an area of contemporary public health with major consequences in the economic, social, legal or political sector. Concerns about improving health and safety at work must be a priority for all governments, given that these accidents generate both societal and organization costs but also suffering and disability that can compromise the life of the victim and family. Sustainable development is closely linked to the quality of life and the life expectancy of the population. At the same time, the relationship between economic growth and population health is evident, because sustainable economic growth is based on a healthy workforce and, in turn, economic well-being contributes to maintaining health. Therefore, public health can be considered as one of the major factors of socio-economic development (Marcuta et al., 2018).

Another aspect to be pursued in public health policy-making is the value of GDP, based on several assumptions, namely, the role that GDP has in developing to improve health; health investment, although indispensable for healthcare, has a subsidiary, additive or multiplier impact on public health; a national economic and social development programme should cover the requirements of health insurance and be subordinated to the need to increase the quality of life (Enachescu & Marcu, 1998).

There are four processes which affect the amount of oxygen in the water: reaeration, photosynthesis, respiration and the oxidation of wastes. There are many diseases which are associated with the contamination of water supplies by animal or human wastes. They include cholera, typhoid fever, paratyphoid fever, dysentery, tularaemia and infectious hepatitis. It has only been in the last century that these diseases have come under control in the developed world, largely by means of some relatively simple sanitation measures. A simple way to categorize water-associated health hazards is according to whether they are communicable. Of the communicable water-associated diseases, those of most interest here are caused by the ingestion of biological agents. There are, however, other modes of transmission including simple water contact, as in the case of schistosomiasis, or insect vectors, as in the case of malaria, which are also important (Masters, 1997).

Waste and pollution transported by stormwater poses quantity and quality problems, affectingpublic health and the quality of the environment. Best management practices should be seen as an opportunity for development and improvement of social, educational and environmental conditions in urbanized and surrounding areas. Therefore, they require an ample perspective and the participation of different stakeholders. High-quality decision needs time and a fair overview of the problem:

the purpose of this document is to contribute to sustainable stormwater management, informing on the most relevant factors that should be assessed and their interaction (Barbosa et al., 2012).

Traditionally, EPE (environmental process engineers) have dealt with treatment processes for gaseous, liquid or solid waste streams. However, EPE cannot be viewed as just the sum of the individual environmental unit operations. The study of pollutant pathways and interactions between the technical, natural and social systems as well as management of their distribution and impact must be included.

3 Methodology

Methodology adopted includes literature survey and other secondary data for process engineering. Environmental pollution is not a new phenomenon, yet it remains the world's greatest problem facing humanity and the leading environmental causes of morbidity and mortality. Man's activities through urbanization, industrialization, mining and exploration are at the forefront of global environmental pollution. Both developed and developing nations share this burden together, though awareness and stricter laws in developed countries have contributed to a larger extent in protecting their environment. Despite the global attention towards pollution, the impact is still being felt due to its severe long-term consequences. In the world today, there are so many things one can talk about concerning environmental pollution. Almost everyone is involved in it; we pollute our environment in so many ways that the scientists have predicted the issue of global warming. The wastes from our houses, public buildings and industries that are not disposed of carefully cause soil pollution and air pollution. The chemical wastes thrown into the rivers, the ponds or the ocean are killing the organisms in the water, and it's also spoiling our water system; there's also air pollution. This chapter talks about the various ways/factors in which our environment can or has been polluted.

3.1 Poor Governance

According to economists like Nicholas Stern, the climate crisis is a result of multiple market failures. Economists and environmentalists have urged policy-makers for years to increase the price of activities that emit greenhouse gases (one of our biggest environmental problems), the lack of which constitutes the largest market failure, for example, through carbon taxes, which will stimulate innovations in low-carbon technologies. To cut emissions quickly and effectively enough, governments must not only massively increase funding for green innovation to bring down the costs of low-carbon energy sources, but they also need to adopt a range of other policies that address each of the other market failures. Further, organizations such as the United Nations are not fit to deal with the climate crisis: it was assembled to prevent another world war and is not fit for purpose. Anyway, members of the UN are not mandated to comply with any suggestions or recommendations made by the organization. Further, the issue of equity remains a contentious issue whereby developing countries are allowed to emit more in order to develop to the point where they can develop technologies to emit less, and it allows some countries, such as China, to exploit this.

3.2 Food Waste

A third of the food intended for human consumption – around 1.3 billion tons – is wasted or lost. This is enough to feed 3 billion people. Food waste and loss accounts for 4.4 gigatons of greenhouse gas emissions annually; if it was a country, food waste would be the third highest emitter of greenhouse gases, behind China and the USA. Food waste and loss occurs at different stages in developing and developed countries; in developing countries, 40% of food waste occurs at the post-harvest and processing levels, while in developed countries, 40% of food waste occurs at the retail and consumer levels. At the retail level, a shocking amount of food is wasted because of aesthetic reasons; in fact, in the USA, more than 50% of all produce thrown away is done so because it is deemed to be "too ugly" to be sold to consumers – this amounts to about 60 million tons of fruits and vegetables. This leads to food insecurity, another one of the biggest environmental problems on the list.

4 Results and Discussions

This focuses on the solutions for the environmental pollutions listed or explained in the previous chapter. The role of an environmental process engineer and an environmental engineer is to design facility for repair and enhancement of our environment. Figure 4 gave the relationship on resource availability, economic capacity and environmental sustainability.

Figure 5 presented food waste collection for circular economy. Figure 6 indicates various aspects of sustainability perspectives.

Sustainable development is an increasingly controversial concept of the global ecological crisis of 1929–1930, nowadays representing a new path of humanity and being integrated with the economic, social and human spheres. Through sustainable development, we aim to ensure the best quality of life for all the inhabitants of the planet, both for the present generation and for future generations, putting man and his needs in a central place alongside the natural environment and protecting and preserving (Tita & Oprean, 2015). Figure 7 gave the three pillars of sustainability. Figure 8 presented the fact that environmental engineering design has high impact on environment.

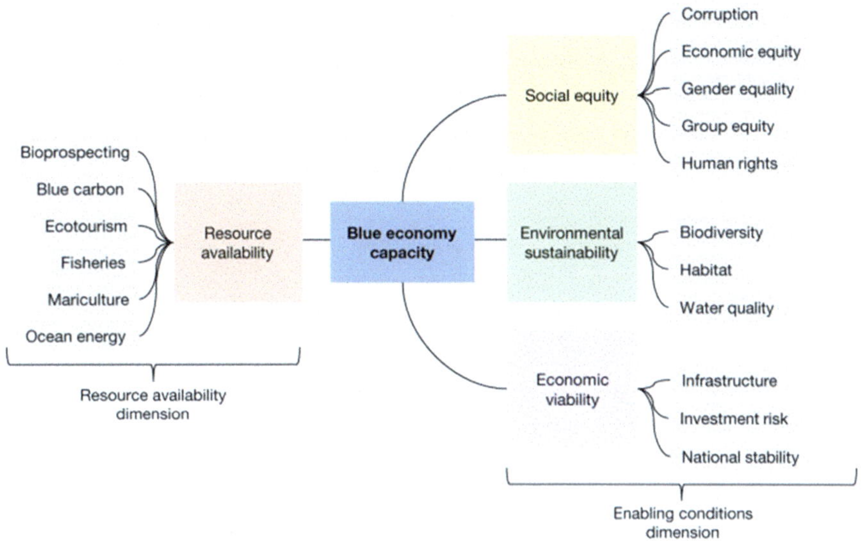

Fig. 4 Information on resource availability, economic capacity and environmental sustainability

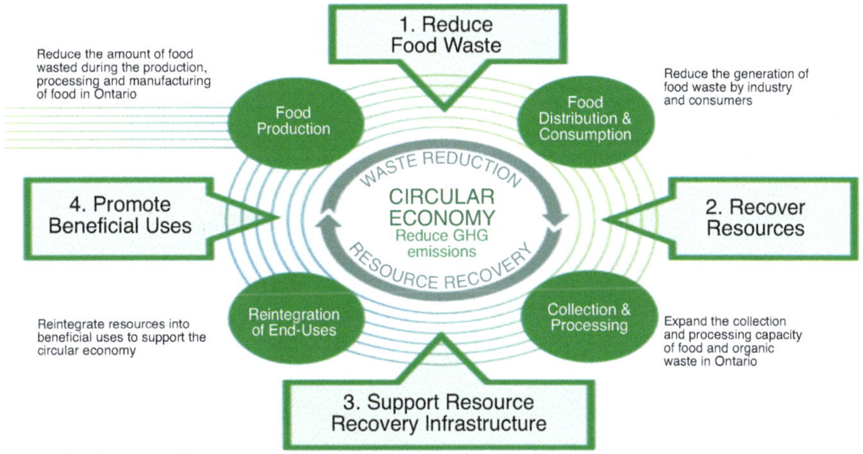

Fig. 5 Food waste collection for circular economy

4.1 What Is Pollution Prevention?

Pollution prevention is any practice that reduces, eliminates or prevents pollution at its source. P2, also known as "source reduction", is the ounce-of-prevention approach to waste management. Reducing the amount of pollution produced means less waste to control, treat or dispose of. Less pollution means less hazards posed to public health and the environment.

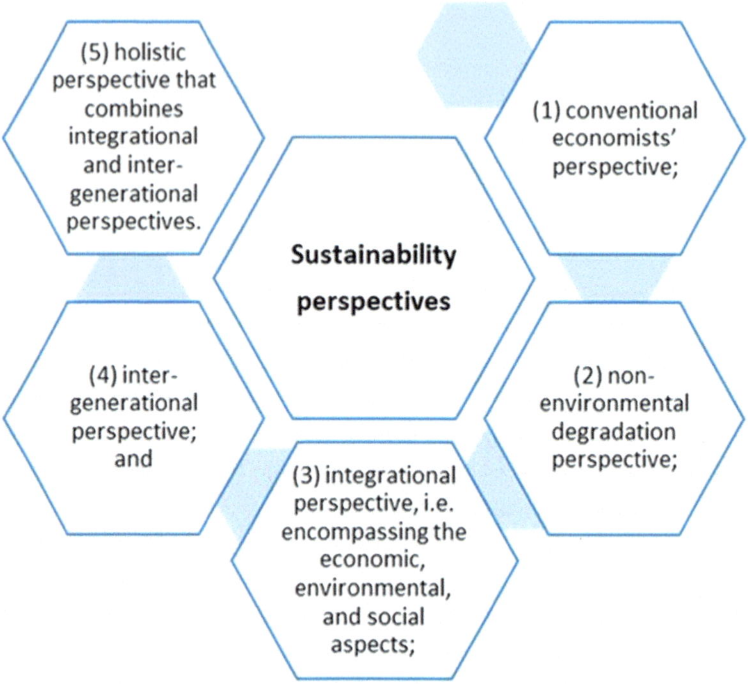

Fig. 6 Aspects of sustainability perspectives

Fig. 7 The three pillars of sustainability

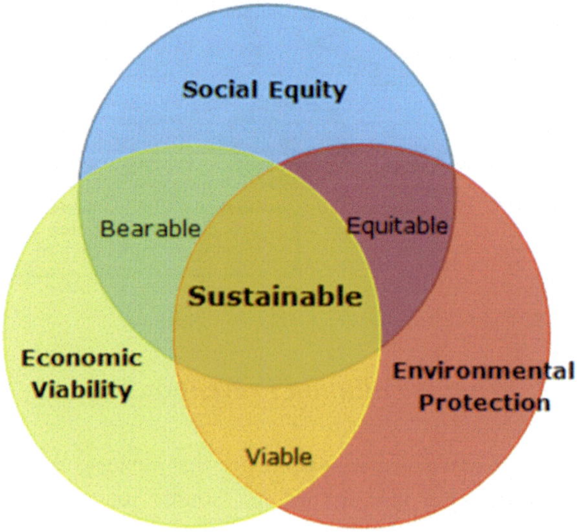

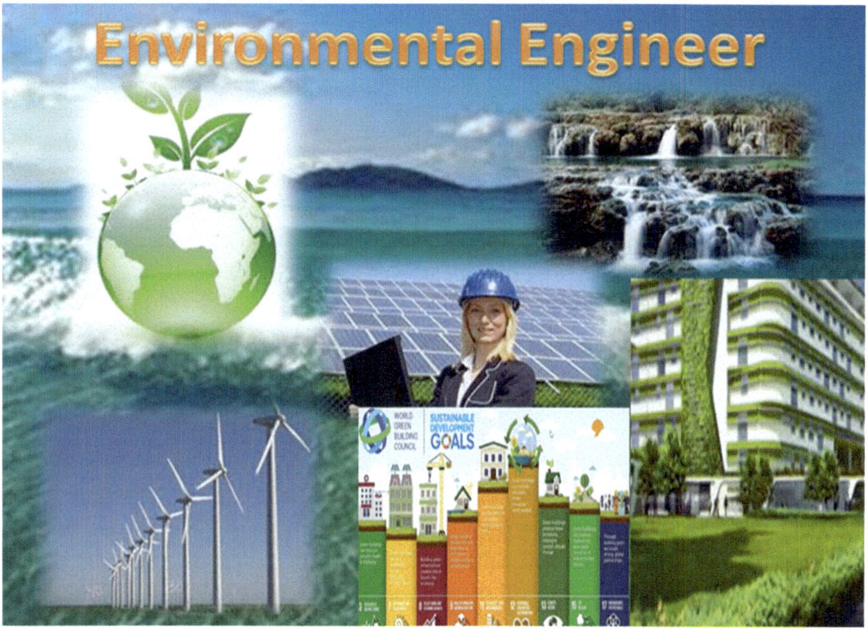

Fig. 8 Impact of environmental engineering design on environment

4.2 *Why Is Pollution Prevention Important?*

Pollution prevention reduces both financial costs (waste management and clean-up) and environmental costs (health problems and environmental damage). Pollution prevention protects the environment by conserving and protecting natural resources while strengthening economic growth through more efficient production in industry and less need for households, businesses and communities to handle waste.

4.3 *Specific Pollution Prevention Approaches*

Pollution prevention approaches can be applied to all potential and actual pollution-generating activities, including those found in the energy, agriculture, federal, consumer and industrial sectors. Prevention practices are essential for preserving wetlands, groundwater sources and other critical ecosystems – areas in which we especially want to stop pollution before it begins.

In the energy sector, pollution prevention can reduce environmental damages from extraction, processing, transport and combustion of fuels. Pollution prevention approaches include:

(i) Increasing efficiency in energy use and use of environmentally sustainable fuel sources.
(ii) In the agricultural sector, pollution prevention approaches include reducing the use of water and chemical inputs.
(iii) Adoption of less environmentally harmful pesticides or cultivation of crop strains with natural resistance to pests; and protection of sensitive areas.
(iv) In the industrial sector, examples of reducing pollution practices include modifying a production process to produce less waste.
(v) Using non-toxic or less toxic chemicals as cleaners, degreasers and other maintenance chemicals.

5 Conclusion

Protection of environment against environmental hazards, climate change issues, ecological problems, issues from poor waste management and unhygienic behaviours are very crucial for public health, energy, wealth and environmental sustainability. Creating a sustainable world that provides a safe, secure, healthy, productive and sustainable life for all peoples should be a priority for the engineering profession. Engineering input through various applications within the economic development, industrialization and sustainable development of Nigeria cannot be undermined. Engineers have an obligation to meet the basic needs of all humans for water, sanitation, food, health and energy, as well as to protect cultural and natural diversity. Improving the lives of the five billion people whose main concern is staying alive each day is no longer an option; it is an obligation. The environmental process engineer has the task of being involved with the society or the environment; a good environment can bring forth good health same way a bad environment can be bad for the body. The environmental engineer helps in keeping our environment clean and also to create new technological tool or chemicals that can help in this aspect; what good would it be if science and technology wasn't involved when we know that they are there to make man's life easier. Human beings are more than a billion in this world, and the number of people that would be environmental process engineers in the future will be more than enough to enforce the health rules in our society for our benefits; science can be applied in so many ways for the benefits of environmental engineers, and so it can be made sure that the problem of pollution would be eradicated with the proper approach using viable strategies and science and technology.

6 Recommendations

The following recommendations will yield tremendous benefits:

(i) More effort should be put in place to improve recycling, waste disposal, public health and water and air pollution control.
(ii) Engineers need to design more systems for pollution control.

(iii) Protection of public health should be given adequate priority.
(iv) Promoting alternative design and engineering approach incorporating the use of natural products in the polymers for PPEs.
(v) Adoption of environmental impact assessment of major projects.
(vi) New policies or frameworks that can assist environmental engineer need to be developed.
(vii) Proper funding and adequate remuneration of medical, engineering and environmental health-related research.
(viii) Systematic management and implementation of the circular economy strategies.
(ix) There is a need for the implementation of water and energy conservation practices.

References

Barbosa, A. E., Fernandes, J. N., & David, L. M. (2012). Key issues for sustainable urban stormwater management. *Water Research, 46*(20), 6787–6798.

Boischio, A., Sánchez, A., Orosz, Z., & Charron, D. (2009). Health and sustainable development: Challenges and opportunities of ecosystem approaches in the prevention and control of dengue and Chagas disease. *Cadernos de saúde pública, 25*, S149–S154.

Enachescu, D., & Marcu, M. G. (1998). *Public health and health management*. All Publishing House.

Furie, G. L., & Balbus, J. (2012). Global environmental health and sustainable development: The role at Rio+ 20. *Ciencia and saude coletiva, 17*, 1427–1432.

Jackson, A. (2018). *Considerations for the health system for addressing the double burden of malnutrition* (No. IAEA-CN--268).

Malik, A., & Grohmann, E. (Eds.). (2012). *Environmental protection strategies for sustainable development* (Vol. 520). Springer.

Marcuta, A., Simionescu, A., Tindeche, C., & Marcuta, L. (2018). Relationship between sustainable development and public health. Case study Romania. *Scientific Papers: Management, Economic Engineering in Agriculture and Rural Development, 18*(3).

Masters, G. M. (1997). *Introduction to environmental science and engineering*. Prentice-Hall.

Mihelcic, J. R., Phillips, L. D., & Watkins, D. W., Jr. (2006). Integrating a global perspective into education and research: Engineering international sustainable development. *Environmental Engineering Science, 23*(3), 426–438.

Oyebode, O. J. (2018). Impact of environmental laws and regulations on Nigerian environment. *World Journal of Research and Review, 7*(3), 262587.

Oyebode, O. J. (2019). Impacts of civil engineering infrastructures in the sustainability of the environment. In *Paper published in the conference proceedings of 17th international conference and AGM tagged "PEACE 2019" of Nigerian Institution of Civil Engineers (NICE).*

Oyebode, O. J. (2021, March). Hazard assessment and implementation strategies for mitigating of flood and erosion in Nigeria. *IOP Conference Series: Materials Science and Engineering, 1036*(1), 012066.

Pan American Health Organization. (2007). *Health in the Americas*. Pan American Health Organization.

Tita, O., & Oprean, C. (2015). *Current perspectives on sustainable development*. Pro Universitaria Publishing House.

United Nations Environment Programme. *Global environment outlook 4*. http://www.unep.org/geo/geo4/. Accessed 28 Apr 2008.

Yoo, H., Kim, B., Kim, J. W., & Lee, J. H. (2021). Reinforcement learning based optimal control of batch processes using Monte-Carlo deep deterministic policy gradient with phase segmentation. *Computers & Chemical Engineering, 144*, 107133.

Adsorptive Performance of Immobilized Activated Carbon for Effective Removal of Dibenzothiophene from a Synthetic Petroleum Distillate in a Packed-Bed Column

O. O. Sadare and M. O. Daramola

1 Introduction

The release of sulfur oxides during the direct combustion of diesel has necessitated urgent research in this area. Emission of sulfur oxides results into acid rain, which destroys plants, building, and animals (Jiang et al., 2003). The Environmental Protection Agency has passed a stringent law to refineries around the world to reduce the emission of sulfur-containing compounds into the environment. For this standard to be met, an efficient, less expensive technique is required. Researchers all over the world had employed different desulfurization methods such as oxidation, reduction, biodesulfurization, and adsorption (Gawande & Kaware, 2018). Hydrodesulfurization is the most commonly used method in the refineries nowadays. Although it has been successful in removing sulfides and disulfides from petroleum distillates, however, thiophenes and its derivatives have been difficult to remove using this method (Isam et al., 2013). In addition, HDS is operated at high temperature and pressure, which makes the process energy-intensive. The use of hydrogen without possibility of regeneration also makes the process expensive. Therefore, more efficient, cheap, and less energy-intensive process is urgently required. Adsorption technique has been discovered to be a promising alternative method to HDS, due to its ability to be operated at low temperature and pressure with possibility of regeneration of adsorbent (Gawande & Kaware, 2018). The process is also cheap and less energy-intensive.

Activated carbon is the most commonly used adsorbent, owing to its large surface area and porous structure. Different researchers have investigated adsorption of

O. O. Sadare · M. O. Daramola (✉)
Department of Chemical Engineering, Faculty of Engineering, Built Environment and Information Technology, University of Pretoria, Pretoria, South Africa
e-mail: michael.daramola@up.ac.za

A. O. Ayeni et al. (eds.), *Advanced Manufacturing in Biological, Petroleum, and Nanotechnology Processing*, Green Energy and Technology,
https://doi.org/10.1007/978-3-030-95820-6_18

activated carbon in batch adsorption mode. Although batch adsorption mode is needed for collecting basic data, continuous adsorption mode is preferable in commercial application (Muftah et al., 2017). Furthermore, continuous adsorption process is feasible, is cheap, and requires less amount of adsorbent and can be easily controlled. High hydrostatic pressure causes disintegration of the adsorbent in the bed column. This challenge can be overcome by immobilizing the adsorbent (Jawad et al., 2018; Sadare, 2018). Furthermore, immobilization of adsorbent could overcome these challenges of column clogging and regeneration in a fixed-bed column. The demands for sustainable techniques of entrapping adsorbent have increased in the recent times. Contribution of immobilization technology has attracted the attention of researchers in this direction. Immobilization technology also offers better reusability, high adsorbent loading, and minimal clogging in continuous mode (Jawad et al., 2018). Activated carbon has been widely reported in literature for column adsorption of sulfur compound from petroleum distillates. However, only few studies have been conducted on the removal of DBT from diesel using immobilized adsorbent in a fixed-bed adsorption column. Therefore, activated carbon was immobilized into pellets for use in a continuous packed-bed adsorption column.

2 Experimental

2.1 Materials

Activated carbon was purchase from Merck (Pty) Ltd, South Africa; acetonitrile (99% purity), n-hexane (98% purity), and dibenzothiophene (DBT) were purchased from Sigma-Aldrich (Pty) Ltd, South Africa. All other chemicals were of analytical grade, commercially available and used without further purification.

2.2 Preparation of Immobilized Activated Carbon

About 5 g of activated carbon was mixed with 250 mL of deionized water and allowed to hydrate for 10 min. Then, 3% (w/v) sodium alginate was added to the solution. 0.2 M calcium chloride was prepared, and the sodium alginate-adsorbent mixture was added dropwisely into the calcium chloride solution via a tube connected to a peristaltic pump to get even-sized pellets. The droplets of sodium alginate-adsorbent form pellets immediately they were in contact with calcium chloride solution, entrapping the adsorbent in it. The pellets were allowed to solidify for about 30 min and washed with 0.9% sodium chloride in order to remove excess calcium ion. The generated activated pellets had a diameter ranging from 3 mm to 4 mm (Seepe, 2015).

2.3 Characterization of Immobilized Adsorbent

Fourier transform infrared analyzer Bruker Tensor 27 at a spectra range of 500 cm^{-1} to 5000 cm^{-1} was used to determine the attachment of surface functionalities and the type of functional groups on the surfaces of the adsorbent samples. Prior to the commencement of the scanning, the 2 mg of the oven-dried powdered samples were mixed with 200 mg of potassium bromide (KBr) and pressed for 3 min at 20 Mpa.

N_2 physisorption experiments at 77 K were performed on the adsorbents for textural properties of the adsorbents. Static volumetric analysis using a Micrometrics TriStar 3000 unit was used to measure the surface area, cumulative pore volume, and pore diameters of the adsorbents. This information was obtained from liquid nitrogen (@ −196 °C) equilibrium isotherms.

Scanning electron microscopy (SEM) was used to check the morphology of the adsorbents. The samples were coated with 60% palladium and 40% gold (Pd/Au) prior to SEM analysis to prevent charge up. Carl Zeiss sigma field electronic scanning electron microscope (FESEM) was used to observe the surface morphology of the immobilized activated carbon at different magnifications.

2.4 Desulfurization of Model Diesel

Desulfurization of model was carried out in a lab-scale designed adsorption bed column. The column was made with a glass, with an internal diameter of 2.4 cm and a bed depth of 20 cm. The performance study of the column was conducted at different DBT concentrations (100–500 mg/L), bed height (50–150 mm), and flow rate (0.5–1.5 mL). Adsorption reaction time was 3 h. The immobilized activated carbon (pellets) was packed in the column with a layer of glass filter at the bottom. The DBT solutions were pumped in a downward direction using HLPC pump (HPLC Consta Metric 3500 MS. RIA VIS). Samples were taken at 10-min intervals using the experimental set-up in Fig. 1 and the experimental samples were analyzed.

The desulfurized model diesel samples were analyzed using a pre-calibrated Shimadzu (Japan) gas chromatography GC-MS unit equipped with fused silica column RXi-5MX. The dimension of the column was 0.25 μm thickness, 0.25 mm diameter, and 30.0 mm length. The column temperature was set to hold at 90 °C for 2 min and then increase to 300 °C at the rate of 10 °C/min and kept at 300 °C for 30 min. Helium was used as the carrier gas. MS ion source temperature was 200 °C. The pressure was set at 100 kpa with column flow of 1.39 mL/min, linear velocity of 43.7 cm/s, purge flow of 3 mL/min, and stop time of 10 min.

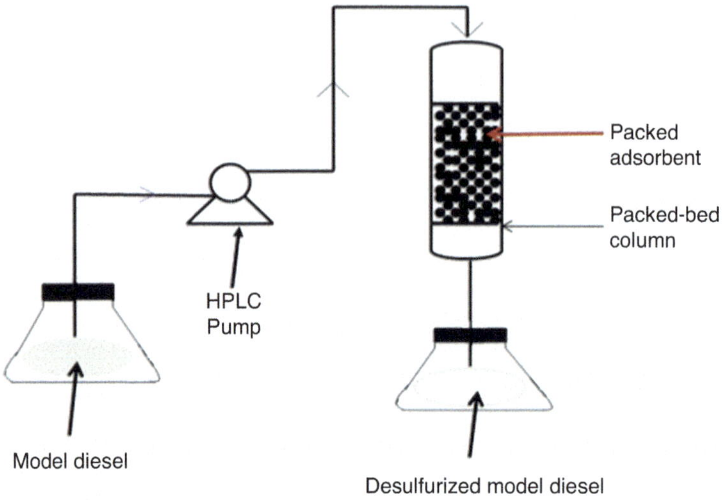

Fig. 1 Experimental set-up for continuous adsorption in a packed-bed column

Table 1 Surface area, pore volume, and pore size determination of the adsorbent by the Brunauer-Emmett-Teller (BET) analysis

Adsorbents	Surface area (m^2/g)	Pore volumes (cm^3/g)	Pore sizes (nm)
Commercial AC	1154.87	1.02	3.45

3 Results and Discussion

3.1 Physicochemical Characterization of Adsorbent

Table 1 shows the textural properties of the adsorbents; surface area, pore volume, and pore size of the commercial activated carbon measured to be 1154.87 m^2/g, 1.02 cm^3/g, and 3.45 nm, respectively. AC has high surface area and pore column and pore size. Adsorption performances of the adsorbents are dependent on their textural properties. Therefore, high textural properties could be instrumental to the adsorption performance of the adsorbent in this study (Sadare & Daramola, 2019).

Figure 2 depicts the surface morphology of (a) fresh NaAlg-AC and (b) NaAlg-AC after use. From Fig. 2a, it can be observed that the surface of the fresh NaAlg-AC was rough with cracks on the surface of the adsorbent. However, after adsorption experiment, the cracks on the surface of the adsorbent have been covered up. This could be due to the adsorption of DBT onto the surface of the adsorbent as seen in Fig. 2b.

Figure 3 shows the FTIR spectra of sodium alginate (NaAlg) and AC trapped in sodium alginate (NaAlg-AC). The results show that there was a stretching vibration of O-H bands that appeared at band 3264 cm^{-1}. The stretching vibration of aliphatic C-H was also observed at 2992 cm^{-1} for both raw alginate and alginate-AC. The

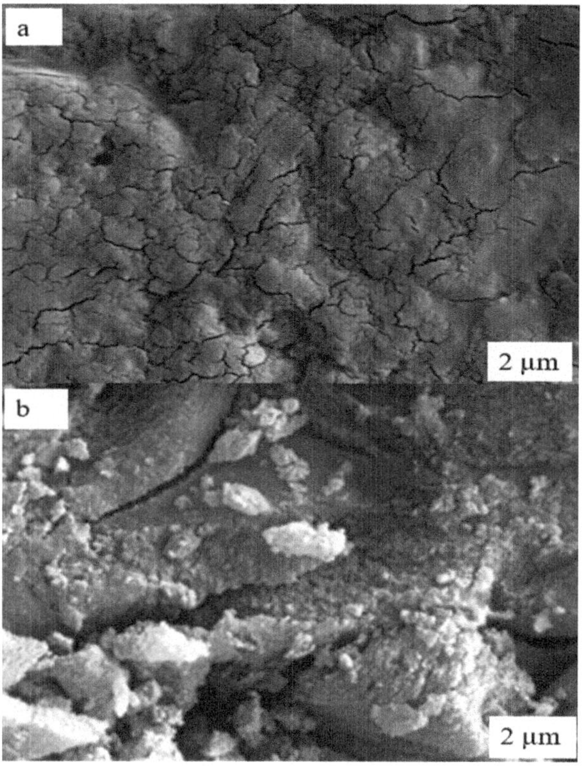

Fig. 2 SEM images of (**a**) fresh alginate-AC and (**b**) alginate-AC after use

observed band in 1596 cm^{-1} and 1416 cm^{-1} could be attributed to C=O symmetric stretching vibration and C-H in-plane deformation. The peaks at 1033 cm^{-1} could be attributed to the C-O stretching vibration and C-O stretching with contribution from C-C-H and C-OH deformation (Daemi & Barikani, 2012). The FTIR spectrum showed a shift in the band around 1176 cm^{-1} and 1249 cm^{-1} in alginate-AC which is absent in FTIR spectrum of raw alginate. This could be attributed to the C-C and C-O stretch because of the AC entrapped in it. It could also be observed that there was weakening of the band –OH group on the FTIR spectrum of alginate-AC. This could be due to the utilization of some of –OH group on the NaAlg during the formation with the AC (Kummara et al., 2013).

3.2 *Performance Evaluation of Adsorbent in the Packed-Bed Column*

To understand the performance of immobilized AC adsorbents during the desulfurization of DBT in a packed-bed column, the effect of operating parameters such as bed depth, flow rate, and initial DBT concentration on column performance for the

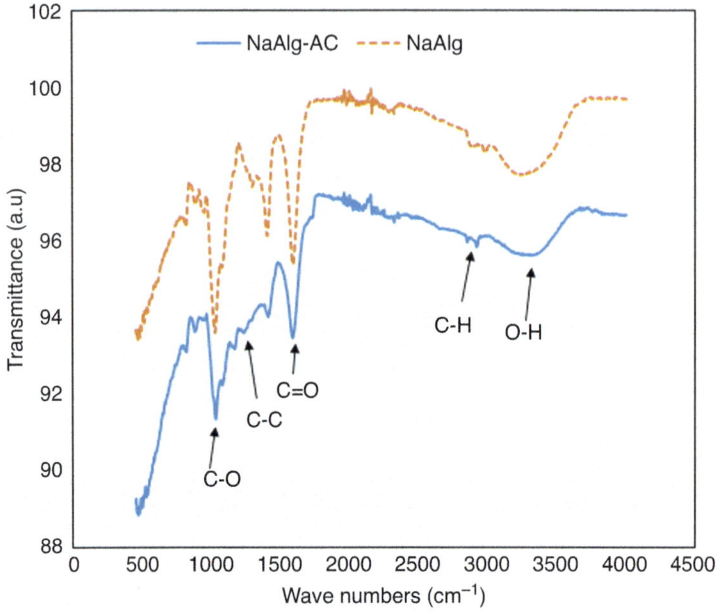

Fig. 3 FTIR spectra of alginate and alginate-AC

removal of DBT from the model diesel and typical real diesel was investigated. DBT adsorption in a packed-bed column depends on the amount of adsorbent in the column. Column breakthrough experiments were conducted at a constant flow rate of 0.5 mL/min, constant DBT concentration of 100 mg/L, and varying bed heights of 5 cm, 10 cm, and 15 cm. The breakthrough curve was obtained by plotting the ratio of DBT outlet concentration C to initial DBT concentration. Figure 4 depicts the effect of bed height on the adsorption of DBT onto immobilized AC in a continuous packed-bed column. It could be observed from the results that breakthrough time increased with increasing bed height. This may be due to the increase in the number of active binding sites on the surface of the adsorbent in the column, which broadens the adsorption transfer zone. The DBT adsorption capacity increased from 0.066 to 1.53 mg/g as the bed height increased from 5 cm to 15 cm. The adsorption capacity for DBT adsorption in this column study was found to be maximum at 15 cm bed height. Therefore, this height was used in subsequent column experiments.

In order to evaluate the efficiency of adsorbent for continuous packed-bed column studies on pilot or industrial scale, flow rate is an essential parameter to consider. Column experiments were conducted in this study at constant bed height of 15 cm and constant initial DBT concentrations of 100 mg/L by varying the flow rate from 0.5 mL/min to 1.5 mL/min. Effect of flow rate is depicted in Fig. 5. The superficial velocity increased as the flow rate increased. The superficial velocities were calculated to be 0.031, 0.063, and 0.094 m/min at 0.5, 1.0, and 1.5 mL/min, respectively. Breakthrough curve is depicted in Fig. 5. It was observed that breakthrough

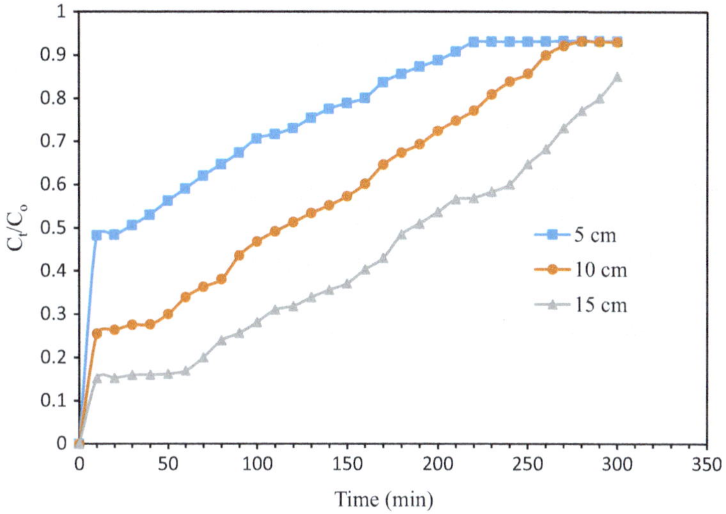

Fig. 4 Effect of bed height on the adsorption of DBT in model oil in a packed-bed column. *Experimental conditions*: initial DBT concentration, 100 mg/L; flow rate, 0.5 mL/min

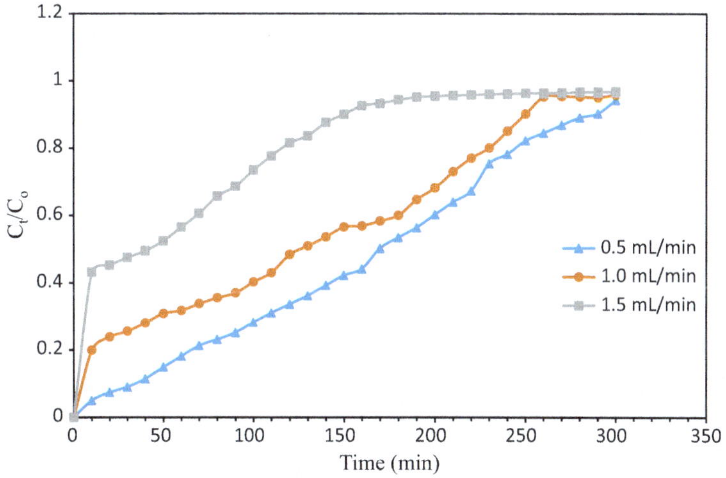

Fig. 5 Effect of flow rate on adsorption of DBT in a packed-bed column. *Experimental conditions*: bed height, 15 cm; initial DBT concentration, 100 mg/L

time decreased from 60 min to 30 min with increasing flow rate from 0.5 mL/min to 1.5 mL/min. This could be due to the increase in occupancy of the adsorption zone as the influent velocity increased, which invariably resulted in decrease in time required to attain the specific breakthrough concentration. Flow rate also influenced the adsorption of DBT onto immobilized activated carbon. The adsorption capacities of activated carbon at different flow rates were 0.066, 0.050, and 0.033, indicating that the adsorption capacity decreased as flow rate increased from 0.5, 1.0, and 1.5 mL/min, respectively. This could be because of insufficient time for adsorption and diffusion limitation of DBT molecules on the surface of the adsorbent in the column at higher flow rate. Therefore, breakthrough time was reached faster at higher superficial velocity. The results obtained in this study are in agreement with Muzic (Muzic et al., 2011). Result also showed that there was higher removal of DBT at lower superficial velocity, because there was enough contact time for DBT with the immobilized activated carbon. This result is similar to what was obtained by other researchers with different sorbent-sorbate studies (Kiran & Kaushik, 2008; Auta, 2012).

Figure 6 depicts the effect of initial concentration of DBT on its adsorption onto an immobilized AC in a continuous packed-bed column. Since the concentration of DBT varies in real diesel, it is important to understand its effect on its adsorption in a column study. The initial concentration of effluent DBT solution was varied from 100 to 500 mg/L. Other operating conditions such as bed height (15 cm) and flow rate (0.5 mL/min) remained constant. The results show that breakthrough time decreased from 70 min to 40 min with increasing initial DBT concentration from 100 to 500 mg/L. This could be attributed to increase in DBT uptake rate, which results in the decrease in adsorption zone length. In addition, these results explained that change of concentration gradient affects the saturation rate and breakthrough time. The resident time experienced at higher initial concentration affected its

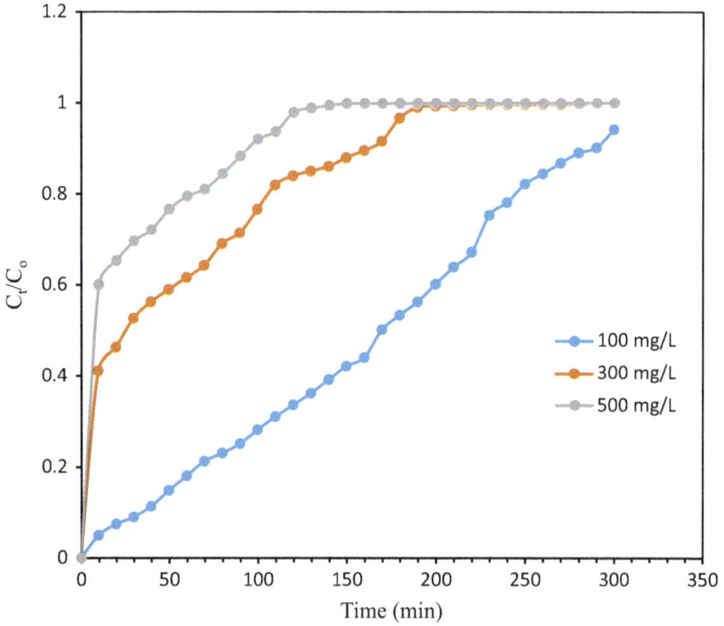

Fig. 6 Effect of initial concentration on adsorption of DBT from a model oil in a packed-bed column. *Experimental conditions*: bed height, 15 cm; flow rate, 0.5 mL/min

solid-phase concentration, resulting in the reduction in adsorption capacity even as the concentration increased. The same trend was observed in studies reported by Muzic et al. (2011) and Han et al. (2009).

The results obtained in this study are comparable with literature. Muzic et al. (2010) studied the adsorption of DBT in diesel in a continuous fixed-bed column with initial DBT concentration of 27 mg/L, in a bed column of height 28.4 cm and internal diameter 2.2 cm. The breakthrough time was reported to be 11.8 h when the feed flow rate was 1.0 mL/min at reaction temperature of 50 °C. The final concentration achieved at the various operating conditions was 0.7 mg/L. Comparing this result with what was obtained in this study (bed height 15 cm, at room temperature, initial DBT concentration 100 mg/L, flow rate 0.5 mL/min), the results show that the breakthrough time was about 60 min. It could be observed that the breakthrough time obtained in this study is lower than what was obtained in Muzic et al. (2010). This could be attributed to lower initial sulfur concentration in the synthetic diesel used in their study. In addition, the bed height used by Muzic et al. (2010) was almost twice what was used in this study which could also favor the adsorption breakthrough time in their study. Meanwhile, a reasonable effluent concentration was achieved in this study accounting for about 90% adsorption performance of the bed column.

4 Conclusion

As demonstrated in this study, activated carbon was successfully immobilized in sodium alginate and evaluated for adsorption of DBT from model diesel in a continuous packed-bed column. The following conclusions were drawn from the study:

- Results obtained from FTIR analysis showed that AC adsorbent was successfully immobilized in the NaAlg.
- Results from the adsorption experiment indicate that adsorption of DBT in the packed-bed column is dependent on superficial velocity of the sorbate through the adsorption column. The best result was obtained at the lowest superficial velocity of 0.031 m/min (0.5 mL/min) at 15 cm bed height and lowest initial DBT concentration of 100 mg/L.
- Breakthrough time increased with decreasing DBT initial concentration, with decreasing flow rate and with increasing bed height. Increasing the superficial velocity of the DBT solution decreased the breakthrough time needed for the contact of adsorbate with the adsorbent in the bed column resulting in decreased amount of DBT adsorbed by the immobilized adsorbent. This means that a longer breakthrough time is needed for higher performance of the bed column, which will successively result in higher adsorption capacity.
- Immobilization technology could therefore be a promising approach of entrapping adsorbent for maximum desulfurization performance in a packed-bed column.

Recommendation Though the adsorption capacity of the adsorbent is still very low, this could be attributed to the minimum amount of AC that the sodium alginate could accommodate. Only a small amount of adsorbent was immobilized on the alginate. Therefore, further study should consider increasing the capacity of the alginate. In addition, influence of alginate ratio on the performance of the immobilized activated carbon should be considered. Finally, kinetics and thermodynamic study should be attempted.

Acknowledgments The authors are thankful to L'Oréal-UNESCO For Women in Science, sub-Saharan African fellowship for the financial support for OOS toward her degree program.

References

Auta, M. (2012). Fixed bed adsorption studies of Rhodamine B dye using oil palm empty fruits bunch activated carbon. *Journal of Engineering Research and Studies, 3*(3), 03–06.

Daemi, H., & Barikani, M. (2012). Synthesis and characterization of calcium alginate nanoparticles, sodium homopolymannuronate salt and its calcium nanoparticles. *Scientia Iranica, 19*(6), 2023–2026.

Gawande, P. R., & Kaware, J. P. (2018). Isotherm and kinetics of desulphurization of diesel by batch adsorption studies. *International Journal of Chemical and Process Engineering Research, 10*(1)), 1–16.

Han, R., Yu, W., Xin, Z., Yuanfecy, W., Fulng, X., Cheng, J., & Tang, M. (2009). Adsorption of methyl blue by phoenix tree leaf power in a fixed bed column: Experiments and prediction of breakthrough curves. *Desalination, 245*(1–3), 284–297.

Isam, A., Zubaidy, H., Fatma, B., Noora Naif, D., Balsam, S., Sana, A., Aysha, A., & Lamis, A. (2013). Adsorption process of sulphur removal from diesel oil using sorbent materials. *Journal of Clean Energy Technologies, 1*(1), 66–68.

Jawad, R. J., Ismail, M. H. S., & Siajam, S. I. (2018). Adsorption of heavy metals and residual oil from palm oil mill effluent using a novel adsorbent of alginate and mangrove composite beads coated with chitosan in a packed bed column. *The IIUM Engineering Journal, 19*(1), 1–14.

Jiang, Z., Liu, Y., Sun, X., Tian, F., Sun, F., Liang, C., You, W., Han, C., & Li, C. (2003). Activated carbons chemically modified by concentrated H2SO4 for the adsorption of the pollutants from wastewater and the dibenzothiophene from fuel oils. *Langmuir, 19*, 731–736.

Kiran, B., & Kaushik, A. (2008). Cyanobacteria biosorption of Cr (VI): Application of two parameter and Bohart Adams models for batch and column studies. *Chemical Engineering Journal, 144*, 391–399.

Kummara, M. R., Basetty, M. K., Rao, K. S. V., & Subha, M. C. S. (2013). Synthesis and characterization of pH sensitive poly-(hydroxyl ethyl methacrylate and acrylamidoglycolic acid) based hydrogels of controlled release studies of 5-fluorouracil. *Journal of Applied Pharmaceutical Science, 3*(06), 061–069.

Muftah, H. E., Manal, A. A., & Sulaiman, A. (2017). Evaluation of an activated carbon packed bed for the adsorption of phenols from petroleum refinery wastewater. *Environmental Science and Pollution Research, 24*, 7511–7520. https://doi.org/10.1007/s11356-017-8469-8

Muzic, M., Gomzi, Z., & Sertic-Bionda, K. (2010). Modeling of the adsorptive desulfurization of diesel fuel in a fixed-bed column. *Chemical Engineering & Technology, 33*(7), 1137–1114.

Muzic, M., Sertic-Bionda, K., & Adzamic, T. (2011). Desulfurization of diesel fuel in a fixed bed adsorption column: Experimental Study and simulation. *Petroleum Science and Technology, 29*(22), 2361–2371. https://doi.org/10.1080/10916461003716657

Sadare. (2018). *Development and evaluation of adsorption coupling Biodesulphurization (AD/BDS) process for the desulphurization of South African petroleum distillates*. A PhD thesis submitted to the Department of Chemical Engineering, University of Witwatersrand South Africa. Available at: http://wiredspace.wits.ac.za/bitstream/handle/10539/26589/PhD%20thesis%20-SADARE-OO-1592992.pdf?sequence=2&isAllowed=y. Accessed 29 June 2021.

Sadare, O. O., & Daramola, M. O. (2019). Adsorptive removal of dibenzothiophene from petroleum distillates using pomegranate leaf (*Punica granatum*) powder as a greener adsorbent. *Chemical Engineering Communications, 206*(3), 333–345. https://doi.org/10.1080/00986445.2018.1488691

Seepe, L. (2015). *The use of cassava waste in the removal of cobalt, chromium and Vanadium ions from synthetic effluents*. A dissertation submitted to the University of the Witwatersrand, South Africa.

Overview of Nano-agro-composite Additives for Wastewater and Effluent Treatment

S. A. Afolalu, O. Okwilagwe, O. O. Yusuf, O. R. Oloyede, S. O. Banjo, and F. Ademuyiwa

1 Introduction

An equivalent upsurge in waste products usually accompanies the upsurge in the standard of living and consumption of goods, which means that the supply of freshwater and the treatment of effluent water become a top priority for different industries. In an attempt to meet the need for freshwater supply, standard practices employed by a bulk of these industries involve the use of wastewater treatment plants (Bv, 2017). Water pollution due to increased industrial, agricultural, and municipal activities, unplanned urbanization, population growth, and the unskilled use of natural water resources has severely deteriorated global water quality. Consequently, it has become a significant challenge to the health of humans, animals, plants, and the environment (Bhatnagar et al., 2015). Since the last decade, many water treatment techniques like chemical precipitation, filtration membrane, microbial degradation, and adsorption have been made available with varying degrees of success to reduce or control water pollution. Nevertheless, the limitations of most of these methods are high operation and maintenance (O&M) costs, complicated treatment procedures, and generation of toxic sludge

S. A. Afolalu (✉) · O. Okwilagwe · S. O. Banjo · F. Ademuyiwa
Department of Mechanical Engineering, Covenant University, Ota, Nigeria
e-mail: sunday.afolalu@covenantuniversity.edu.ng; okwilagwe.osise@covenantuniversity.edu.ng; solomon.banjo@covenantuniversity.edu.ng; faith.ademuyiwa@covenantuniversity.edu.ng

O. O. Yusuf
Department of Microbiology, Obafemi Awolowo University, Ile-Ife, Nigeria
e-mail: onolola.olabisi@covenantuniversity.edu.ng

O. R. Oloyede
Department of Mechanical Engineering, Afe Babalola University, Ado-Ekiti, Nigeria
e-mail: olamilekun.oloyede@abuad.edu.ng

A. O. Ayeni et al. (eds.), *Advanced Manufacturing in Biological, Petroleum, and Nanotechnology Processing*, Green Energy and Technology,
https://doi.org/10.1007/978-3-030-95820-6_19

(Ali & Hassan, 2017). Also, as a means of controlling the level of pollution, the World Health Organization (WHO), in conjunction with the Environmental Protection Agency (EPA), has stipulated an acceptable maximum discharge level (Ibisi & Asoluka, 2018). Adsorption technology, which involves the rapid separation of adsorbate from liquid to a surface, is the most preferred method for contaminant removal from effluent water due to its sustainability, simplicity of design, low treatment cost, and ease of operation (Afolalu et al., 2015a, b). Generally, an absorbent is considered "cost-effective" if it is abundant in nature, is a by-product/residue gotten from industries, or needs little to no processing (Bhatnagar et al., 2015).

In recent times, there have been many studies surrounding the development and application of a variety of novel technologies for the treatment of organic and inorganic contaminants in water or wastewater. Among these technologies, the use of agro-industrial wastes (AIW) is an evolving method identified by investigators as an attractive alternative approach to the conventional treatment of water and wastewaters (Mo et al., 2018). These agro-wastes are not utilized in their original state but are modified in a diversified number of ways to reinforce the adsorption surface area and porosity of the given material in the adsorption of organic (dyes) and inorganic (heavy metal) pollutants in water. Common modification technologies of agro-waste absorbents include carbonization, nano-structuring, grafting, and activation (Abbasi et al., 2013; Idan 2017). Furthermore, the shift to agro-industrial waste adsorbents is because they are eco-friendly and economical owing to their unique chemical properties, chemical stability, renewable nature, peculiar structure, and abundant availability processes (Mo et al., 2018; Bhatnagar et al., 2015). This paper reports a broad concept of the development and application of nano-agro-composite additives for wastewater and effluent treatment.

2 Additive-Based Treatment Methods

Additives are widely used in the scope of wastewater treatment. The addition of several chemicals, like oxidation agents, coagulants, and so on, has been developed as a promising way to treat highly contaminated and complex effluents (Kayiwa et al., 2019; Mohan et al., 2013). They are used to regulate pH conditions, improve settling, balance influent C/N/P ratio, and introduce specific exogenous bacteria/enzymes or to supply micro-elements into influent (Moletha et al., 2007). The additives could be biological, chemical, physical, or physiochemical. The additive-based method involves the addition of a single chemical compound or multiple chemical compounds for effluent water treatment.

2.1 *Agro-waste Adsorbents*

Agro-waste refers to the residue obtained from agricultural and forestry production and processing operations. The significant types of biomass gotten from these operations are husks, straws, grouts, crumbs, and trash. Agro-wastes are grouped generally grouped into four categories, namely, agricultural industry wastes (AIWs), fruit and vegetable wastes (FVWs), crop residues, and livestock waste. An adsorbent is a sold substance used in the removal of pollutants from liquid or gas that can harm the environment (Yunus et al., 2019). The search for low-cost adsorbents, unlike their commercial counterpart, is due to the need of adsorbents that are readily available in nature and require less processing. Some properties of agricultural waste are to be used as potential bio-sorbents in the treatment of wastewater include their large specific surface area and high affinity which enables easy physical adsorption of heavy metals like lead, mercury, nickel, and copper due to the existence of abundant functional groups in the cellulose, hemicellulose, and lignin of the plant (Afolalu et al., 2018a, b, c, d). The binding groups present such as carboxyl, phenolic, alcohols, acetamido, amino, amido, and ester donate the electron pair that forms complexes with metal ions or replaces the hydrogen ions with metal ions in the solution, therefore removing the heavy metals (Sudaryanto et al., 2006). There are an enormous number of bio-sorbents, which have been investigated and reported by researchers to remove heavy metals in effluent waters. Some of these bio-sorbents are shown in Table 1.

Besides the existence of heavy metals, dyes are highly toxic compounds present in wastewater, which can cause mutagenic and carcinogenic effects, which can lead to allergic dermatitis, permanent damage to the eyes, skin irritation, mutation, and cancer of humans and animals. These dyes, along with pigments, are extensively used in textile, food, plastic, paper, leather, cosmetic, and printing industries to add

Table 1 Different types of agro-waste used in the removal of adsorbates (heavy metals)

Agro-waste	Heavy metal	References
Rice husks	Cu(II)	Pellera et al. (2012)
Rice straw	Pb(II)	Jiang et al. (2012)
Oakwood and oak bark	Cd(II) and Pb(II)	Liang et al. (2013)
Coconut shell char	Ni(II)	Abesekara et al. (2020)
Palm kernel shell	Zn(II), Fe(II), Pb(II), and Ni(II)	Afolalu et al. (2018b)
Lemon peels	Co(II)	Bhatnagar et al. (2010)
Garlic peels	Pb (II), Cu(II), and Ni(II)	Liang et al. (2013)
Sulfuric acid-treated jackfruit peels	Cd(II)	Inbaraj and Sulochana (2004)
Nitric acid- and thioglycolic acid-treated cassava peels	Pb(II), Cu(II), and Cd(II)	Suharso and Buharu (2011)
$ZnCl_2$-treated grapefruit peel	Pb(II)	Afolalu et al. (2018c)

Table 2 Different types of agro-waste used in the removal of absorbate (dyes)

Agro-waste	Dyes removed	References
Coconut shell fibers	Methylene blue (MB), methyl orange (MO)	Alfa et al. (2012)
Pomelo peels	Congo red (CR)	Zheng et al. (2020)
Kenaf	Reactive red (RR)	Palma et al. (2016)
Grapefruit peels	Crystal violet (CV)	Saeed et al. (2010)
Garlic peels	Methylene blue (MB)	Hameed and Ahmad (2009)
Jackfruit peel	Rhodamine	Javaragan et al. (2011)
Lemon peels	Methyl orange (MO), Congo red (CR)	Bhatnagar et al. (2009)
Acorn	Brilliant green (BG)	Ghaedt et al. (2011)
Soybean hull	Safranin	Afolalu et al. (2018d)
Avocado peels	Reactive Black, Naphthol Blue Black, Basic Blue 41	Palma et al. (2016)

color to the end product, for analytical purposes in chemical laboratories, and as a biological stain in many biomedical and biological laboratories (Ghaedi et al., 2011). Dyes are either natural or synthetic (primarily used in industries) in nature. Their discharge into water bodies hinders light penetration, resulting in the delay of biological processes, which causes damage to aquatic life. Hence, it creates an urgent need for the removal and decoloration of dyes present in wastewater. Some bio-sorbents investigated for the removal dyes are presented in Table 2.

3 Agro-waste Peel-Based Adsorbents

Fruit and vegetable wastes (FVWs) consist of different types of garbage and by-products, including the peels/skins of unprocessed fruits and vegetables, such as orange, mangoes, bananas, pineapples, tomatoes, jackfruits, garlic, pomegranate, and many more, generated during industrial sorting, selecting, and boiling processes (Bhatnagar et al., 2015). These peels are usually discarded, fed to livestock, or used as fertilizers and constitute the highest percentage of wastes found in most kitchen garbage bins. VFWs generated from fruit and vegetable industries are enormous such that the inappropriate disposal of these organic wastes poses serious environmental pollution concerns due to their highly perishable nature and need to be managed and utilized (Pattanaik et al., 2019). Over the past decade, efforts have been made toward the improvement and reuse of fruit and vegetable wastes. Among these, the valorization of bio-agro-wastes from food processing industries is one of the most important efforts. In recent years, wastes generated from fruits and vegetables have been developed to produce a diversified range of value-added products like edible oils, essential oils, food additives, enzymes, pigments, biodegradable plastics, bioethanol, polyphenolic and anti-carcinogenic compounds, dietary fibers, and other miscellaneous products. Agricultural waste skins are economic, environmentally benign, and natural sources of adsorbents, which can be utilized in removing a wide range of aquatic contaminants as well as in contaminant reduction. These waste skins are an attractive resource for water and wastewater treatment due to the characteristics of hemicellulose, cellulose, pectin, and lignin, which provide metal binding sites to treat metal pollution, particularly in the aspect of environmental technology which is extensively studied due to the growing interest of the scientific community.

3.1 Citrus Waste Peels

In recent decades, there has been an exponential rise in the worldwide production and consumption of citrus fruits. In 2010, the Food and Agriculture Organization of the United Nations (FAOSTAT) records revealed that over 100 million tons of citrus fruits were produced globally and estimated to even more increase in years to come. The high rate of consumption is responsible for the generation of a large number of wastes, which may be as high as 50% of the fruit weight. These wastes are majorly disposed of by unplanned landfilling, burning, or dumping or used in the manufacture of livestock feed or as raw materials for the extraction of active substances, which create different environmental pollution challenges and resource wastage. Orange, pomelo, grapefruit, and lemon peels are the commonly available citrus waste peels, but for the course of this paper, only two of these citrus fruits will be discussed.

3.1.1 Orange Peels

In the soft drink and orange juice industries, vast amounts of orange peels are generated and discarded as waste worldwide. The accumulation of waste orange peels in these industries and their dumping have resulted in two critical problems which are contamination with phenolic compounds and land space occupancy. Recent research has shown that orange peels can have potential applications in the development of high-value products like methane. These discarded orange peels can be utilized as adsorbents for the removal of pollutants, such as dyes, harmful metal ions, etc., from wastewater (Bhatnagar et al., 2015). Orange peels primarily contain chlorophyll pigments, cellulose, hemicellulose, pectin, lignin, and hydrocarbons of low molecular weight, including limestone, which comprise several hydroxyl functional groups, making them a prospective adsorbent for many contaminants. Orange peels can control the following parameters, such as pH, chlorine, COD (chemical oxygen demand), TDS (total dissolved solids), TSS (total suspended solids), total phosphorus, total nitrogen, volatile acidity, nitrate, and nitrite in wastewater. Their abundance in nature, biodegradability, and nontoxicity are also reasons. The general preparation of orange peels as adsorbents involves the removal of dust particles or soluble impurities from the obtained waste peels (from houses or local market areas, etc.) by thoroughly cleaning with distilled water, after which they are cut into smaller pieces and then dried in sunlight for a few days until the peels are dried and crisp. The dried orange peels are then grounded into uniform-size particles using a mortar (Abdurrahman et al., 2013). The ability of orange peels to remove inorganic pollutants like zinc, copper, lead, nickel, and chromium from aqueous solution by adsorption has been studied. Studies have also highlighted the modification of orange peels with a range of chemical reagents as bio-sorbents, in the removal of cadmium ions. The compositional chemistry of raw orange peels shows that it contains 51.4% oxygen, 42.2% carbon, 5.4% hydrogen, and 1.0% nitrogen. Furthermore, the outcomes of several chemical modifications, including varying acids and alkalis saponification change after saponification with NaOH on the absorbent properties of orange peels, have been investigated.

Abdurrahman et al. (2013) researched on the removal of dyes from textile effluent water using a cost-effective adsorbent. The adsorbent was formulated from orange skins in which the trial was carried out at the laboratory scale. The research was done to discover the outcome of adsorbent dosage, optimum pH, and retention time. It was deduced that the removal percentage was between 60% and 70% at 7pH, with a retention time of 120 min and an adsorbent dosage of 1.5 g/2.5 mL. Mahir et al. (2015) studied the treatment of dairy wastewater using cost-effective adsorbents. The skins of orange and fish scales were employed as adsorbents by dehydration and carbonization methods. Their study evaluated the outcome of pH, contact time, adsorbent particle size, and dosage in the removal of contaminants from dairy wastewater. It was observed that the carbonization method was more efficient that the dehydration method for both orange skins and fish scales. It was noted that the orange skins were more active in the removal of pollutants in both modes with a percentage removal of 50.1% and 14.3%, respectively, compared to the percentage removal of 31.25% and 8.2%, respectively, of fish scales. Orange skins are a new natural coagulant alternative in limiting the use of chemical coagulants in wastewater treatment. Dollah et al. (2014) experimented with different dosages, to find out the optimum dosage of orange skins to be utilized as a natural coagulant, the optimum pH and percentage of turbidity removal in the synthetic water sample. Their results revealed that the percentage removal of turbidity in the water sample was affected by an increase in the amount of dosage, and it was concluded that 5.0 is the optimum pH and 60 mg/l the optimum dosage, where the peak removal of turbidity reached a percentage of 88.40%. Their result proved the ability of orange skins as a novel composite coagulant in the treatment of water and how the utilization of this natural coagulant can limit the health risk associated with the prolonged use of chemical coagulant together with the production of reduced chemical sludge to the environment.

3.1.2 Pomelo Peels

The pomelo fruit is regarded as one of the natural fruits of China, which is consumed in large quantities yearly. This fruit has more peels and segment membrane compared to other citrus fruits, causing a significant amount of pomelo waste peels. The pomelo peels (PPs) are rich in plant fiber. They contain insoluble polysaccharides (such as pectin, cellulose, and hemicellulose) as well as numerous functional groups like amidogen, hydroxyl, and carboxyl, making it an attractive adsorbent (Tan et al., 2019; Zhu et al., 2017). These peels are the primary residue, corresponding to 44–54% (w/w), which are inedible to humans and release CO_2 and noxious gases when discarded in landfills, causing environmental pollution (Afolalu et al., 2020; Ai et al., 2011). As a cost-effective natural adsorbent, literature has revealed PP as having elevated oil sorption performance because of its elevated pore structure, especially macropores (2–20 μm), which presents the potential to adsorb oil and organic pollutants.

Mohapatra et al. (2010) in their experimental employed the use of eco-friendly hydrophobic aerogels with low density and porosity that had been synthesized from waste pomelo peels as precursors to separate a variety of organic pollutant or oils from oily industrial wastewater. The result of their study disclosed that PP-based aerogels exhibited exceptional adsorption capacity for a broad group of organic solvent or oil products (chloroform 71.3 gg^{-1}, soya bean oil 62.33 gg^{-1}, crude oil 49.8 gg^{-1}, etc.), compared to a chain of sponge aerogels (HPSA-0, HPSA-1, HPSA-2) simply prepared through high-speed dispersion, freeze-drying, and salinization with methyltrimethoxysilane. The further recorded a slight decrease in the absorption capacity from 94.66% to 83.82% after ten consecutive cyclic tests, which designate high recyclability. The removal of Congo red (CR) dye from effluent waters was studied by Silva et al. (2013), Liu et al. (2012), and Anwar et al. (2010) using nano-porous adsorbents of pomelo fruit peels. The effect of adsorption isotherm was investigated at different adsorbent dosages (1.0, 2.0, 3.0 gL^{-1}), pH (5.99, 6.72, 8.73), and temperatures (30, 40, and 60 °C), respectively. It was found that the absorption performance of the dye could be expressed rationally by Langmuir and Freundlich models. It was observed that the monolayer adsorption performance was between 1.08 and 0.75 mg g^{-1}. It was concluded that pomelo peel nano-adsorbents are an attractive alternative in removing dye from wastewater.

3.2 *Banana Peels*

Bananas, botanically berries (*Musa* spp.), are among the most popular tropical fruits consumed globally and cultivated in over 130 countries. They are the second-largest fruit produced after citrus, adding roughly 16% of the global fruit production (Akpomie & Conradie, 2020). India, the largest producer of banana, provides 27% of the world's banana production. Several tons of banana skins are produced daily in market areas and household garbage, creating disposal problems and

Table 3 Percentage chemical composition of banana peels

Hemicellulose	Cellulose	Lignin	Ash	Proteins	Lipids	Others	References
6.4–9.4	7.6–9.6	6–12	–	6–9	1.70	24.26	Mohaptra et al. (2010)
–	14	9.6	–	8.7	–	62.0	Akpomie and Conradie (2020)

Table 4 Percentage chemical composition of cassava peelings

Hemicellulose	Cellulose	Lignin	Ash	Proteins	Lipids	Others	References
32.36	9.71	16.89	11.38	3.70	1.70	24.26	Kouteu et al. (2016)
23.40	14.17	10.88	3.70	5.29	–	29.84	Pooja and Padmaja (2015)

environmental menace (Afolalu et al., 2019a, b). The skins obtained from the banana plant account for 30–40% of the total fruit weight, which is primarily used in manufacturing animal feed, compositing, and production of enzymes, ethanol, methane, proteins, and pectin. Owing to the presence of carboxyl and hydroxyl groups of pectin, the banana peels/skins present a high adsorption capacity for inorganic and organic compounds. Banana peels are rich in starch, crude protein, polyunsaturated fatty acids, total dietary fiber, essential amino acids, pectin, and micronutrients (Kouteu et al., 2016; Pooja & Padmaja, 2015). They are also good sources of cellulose, hemicellulose, lignin, and galacturonic acid. Possible chemical compositions of banana peels influenced by their type, variety, altitude, and climate are presented in Table 3. The feasibility of utilizing banana peels as an adsorbent in removing atrazine and ametrine from waters was studied at laboratory scale by Omorogie et al. (2014). The conditions of their study for the removal of these pesticides include 50 mL sample volume, 3.0 g of banana mass, 40 min stirring time, and no necessary pH adjustments. In conclusion, they recorded a linear analytical curve reaching 10 $\mu g\ L^{-1}$, good recoveries between 82.9% and 106.6%, precise results higher than 4.5%, and removal efficiency of about 90% that were attained for both pesticides. The removal of lead (II) and cadmium (II) from wastewater by adsorption on banana peels was studied in batch mode by Anwar et al. (2010). Also, the adsorption of cobalt (II) and nickel (II) using banana peels by flame absorption spectroscopy has been investigated (Udo et al., 2018; Ren et al., 2018; Owa, 2013; Bhatnagar & Sillanpaa, 2010).

3.3 *Cassava Peels*

Cassava or its scientific name *Manihot esculenta* Crantz is among the most important crops cultivated in many countries like Nigeria, Mozambique, Indonesia, India, etc., especially in the tropics and sub-tropic regions of the world. It is the third-largest source of carbohydrate after rice and corn and is ranked the fifth most important food crop globally (Udo et al., 2018). Cassava is a plant that serves different purposes of use. In general, cassava serves as the raw material used in the production of various traditional foods, cassava starches, cakes, animal feed supplements, bio-composites, bio fibers, and biopolymers (Ren et al., 2018). Its leaves can be used as natural medicine or vegetables since they are rich in protein and other bioactive compounds, and its wood is frequently utilized as firewood for cooking. Cassava peels from cassava starch processing have been termed an ecological menace to society. This is because of the large number of solid wastes produced causing environmental problems. Cassava peels account for 10–20% of the total wet weight of fresh cassava. Table 4 shows the compositional analysis of cassava peel. The difference in these composition results, according to Afolalu et al. (2019a, b), is probably a result of the different climatic conditions, soil type and fertility, variety of species used, and harvesting period.

Literatures reveal that cassava peels like many lignocellulose biomass-rich materials are a precursor to produce cost-effective, renewable, and sustainable adsorbents for the treatment of wastewater. For instance, the manufacture and ability of activated carbon synthesized from cassava peels for the adsorption of active pharmaceutical ingredients (APIs) from effluent water were highlighted and discussed in a review paper by Chandane and Singh (2016). Activated carbon can be synthesized from cassava peels either physically or chemically. Physical activation involves the production of activated carbon via continuous processes of carbonization and activation, utilizing heat energy in an inert atmosphere, while chemical activation is the process of heating the dried peels with an activating reagent for 7–14 h at a temperature ranging from 200 to 1050, depending on the activating agent (Pei & Liu, 2011).

4 Preparation and Modification of Agro-waste Adsorbents

Different agro-wastes require different methods of preparation before their application as agro-waste-derived absorbents. A general preparative approach involves obtaining the desired agro-waste(s) from local farms, markets, industrial plants, household garbage, or dumping sites to undergo specific modification processes, which could be physical or chemical (Afolalu et al., 2018b). These processes include thorough washing with distilled water, drying in sunlight for a stipulated number of days, milling, sieving grafting copolymerization, and derivation, such that the creation of smaller particle sizes and higher surface areas result in improved performance as well as outstanding adsorption capacity for many pollutants. The direct use of nanoparticles in treating water and effluent waters is expensive. However, the modification of agro-wastes with nanoparticles to produce a new class of cost-effective and easy-to-prepare hybrid materials rich in functionalities and with uncompromised efficiency for the removal of both metal and organic pollutants from aqueous solutions is an area that requires further investigation (Odisu et al., 2019). For instance, the modification of *Nauclea diderrichii* (ND) seed waste biomass with graphene oxide (GO) and mesoporous silica (MS) nanoparticles, in making two novel hybrid materials, namely, MND and GND adsorbents, for agricultural waste modified with MS and GO + MS nanoparticles, respectively, for the adsorption of chromium and lead ions was studied by Omorogie et al. (2014). It was observed from the analysis of experimental data that the modified ND agro-wastes exhibited improved specific surface area and pore structure (pore size and pore volume) and were subsequently responsible for the increased rate of adsorption and diffusion coefficients for the adsorption of chromium and lead ions over their unmodified ND counterparts. It was also discovered that GND absorbent was the most rapid of all adsorption processes (Anastopoulos et al., 2017; Afolalu et al., 2015a, b; Gupta et al., 2009).

5 Justification

The degradation and pollution of the environment due to human, industrial, and agricultural practices over the last few decades have had a tremendously adverse effect on water bodies, creating the need for comprehensive water remediation and management techniques (Owa, 2013). The utilization of agro-industrial wastes (AIWs) in treating heavy metals and organic pollutants in water and wastewater is still an emerging field, thus making their recycle and reuse as a resource a very critical process (Mo et al., 2018). Furthermore, the utilization of agro-waste materials as a resource in the development of low-cost, biocompatible, and nontoxic adsorbents is desirable, due to their contributions in terms of waste management, reduced costs of waste disposal, and environmental pollution load, which in the long-run contributes to environmental protection (Bhatnagar & Sillanpaa, 2010). Comprehensively, agro-waste-based adsorbents offer noteworthy advantages ranging from their outstanding adsorption capacity to their high selectivity, which has achieved satisfactory results in the removal of both organic and inorganic pollutants.

6 Conclusion

The development and application of agro-waste peel adsorbents, like orange, banana, cassava, etc., and their effectiveness in removing a diversified range of pollutants such as heavy metals, dyes, active pharmaceutical ingredients, and oils from flowing waters were reviewed in this report. The use of these agro-waste materials instead of commercial chemicals, as revealed by several bodies of literature, is an effective and efficient resource in overcoming water pollution, thus contributing to environmental protection.

Acknowledgments We acknowledge the financial support offered by Covenant University in the actualization of this research work for publication.

References

Abbasi, Z., Alikarami, M., Nezhad, E. R., Moradi, F., & Moradi, V. (2013). Adsorptive removal of Co2+ and Ni2+ by peels of banana from aqueous solution. *Universal Journal of Chemistry, 1*(3), 90–95.

Abdurrahman, F., Bin, M., Akter, M. Z., & Abedin, M. Z. (2013). Dyes removal from textile wastewater using orange peels. *International Journal of Scientific & Technology Research, 2*(9), 47–50.

Abesekara, M. S., Kosvinna, K. N., & Amarasinghe, B. M. W. P. K. (2020). Adsorption and desorption studies of Ni2+ ions on to coconut shell char. *IOP Conference Series: Earth and Environmental Science, 427*(1), 012005.

Afolalu, S. A., Adejuyigbe, S. B., Adetunji, O. R., & Olusola, O. O. (2015a). Effects of additives on ASTM A53 mild steel during machining. In *IOP conference series.*

Afolalu, S. A., Adejuyigbe, S. B., Adetunji, O. R., & Olusola, O. I. (2015b). Production of carbon additives cutting tools from recycled steel with palm kernel shell as carbon additives. *International Journal of Advanced Research, 3*(5), 1–7.

S.A. Afolalu, A.A. Abioye, M.O. Udo, O.R. Adetunji, O.M. Ikumapayi, & S.B Adejuyigbe, (2018a). Data showing the effects of temperature and time variances on adsorption of heavy metals from cement wastewater using activated carbon from palm.

Afolalu, S. A., Asonaminasom, E. H., Ongbali, S. O., Abioye, A. A., Udo, M. O., & Salawu, E. Y. (2018b). Dataset on experimental investigation of optimum carburizing temperature and holding time of Bi-Nano additives treatment of AISI 5130 steel. *Data in Brief, 19*(1), 2279–2283.

Afolalu, S. A., Orenuga, O., Samson, O., Ongbali, O., Abiodun, A., Abioye, A., Okokpujie, I. P., & Oloyede, O. R. (2018c). The study of the impact of nano carbon nano-additives treatment of mild steel during machining. *Data in Brief, 19*(2018), 456–461.

Afolalu, S. A., Ongbali, S. O., Abioye, A. A., Udo, M. O., & Akintayo, T. C. (2018d). Experimental investigation of the effects of Bi-Nano additives on the mechanical properties of AISI 5130 mild steel during machining. *International Journal of Mechanical Engineering and Technology, 9*(12), 264–274.

Afolalu, S. A., Efekodha, G. E., Ongbali, S. O., Abioye, A. A., Salawu, E. Y., Ajayi, O. O., & Oluwabunmi, A. P. (2019a). Experimental analysis of the effect of Tri-Nano additives on wear rate of mild steel during machining. *Procedia Manufacturing, 35*, 395–400.

Afolalu, S. A., Oladipupo, S., Bose, M. E., Abioye, A. A., Adejuyigbe, S. B., Ajayi, O. O., & Ongbali, S. O. (2019b). Agro waste a sustainable source for steel reinforcement-review. In *Journal of Physics: Conference Series* (Vol. 1378, No. 3, p. 032032). IOP Publishing.

Afolalu, S. A., Samuel, O. D., & Ikumapayi, O. M. (2020). Development and characterization of nano-flux welding powder from calcined coconut shell ash admixture with FeO particles. *Journal of Materials Research and Technology, 9*(4), 9232–9241.

Ai, Y., Yu, Z., Chen, Y., Zhu, X., Ai, Z., Liu, S., … Zhang, Y. (2011). Development of catalystlized substance for imprint. *Environmental Sciences, 6*(3), 271–275.

Akpomie, K. G., & Conradie, J. (2020). Banana peel as a biosorbent for the decontamination of water pollutants. A review. *Environmental Chemistry Letters, 18*, 1–28.

Alfa, H., et al. (2012). Agricultural waste materials as potential adsorbent. *International Journal of Chemistry Research, 2*(April), 48–54.

Ali, H. R., & Hassaan, M. A. (2017). Applications of bio-waste materials as the green synthesis of nanoparticles and water purification. *Advances in Materials, 6*(5), 85.

Anastopoulos, I., Bhatnagar, A., Hameed, B. H., Ok, Y. S., & Omirou, M. (2017). A review of waste-derived adsorbents from the sugar industry for pollutant removal in water and wastewater. *Journal of Molecular Liquids, 240*, 179–188.

Anwar, J., Shafique, U., Salman, M., Dar, A., & Anwar, S. (2010). Removal of Pb (II) and Cd (II) from water by adsorption on peels of banana. *Bioresource Technology, 101*(6), 1752–1755.

Bhatnagar, A., & Sillanpää, M. (2010). Utilization of agro-industrial and municipal waste materials as potential adsorbents for water treatment – A review. *Chemical Engineering Journal, 157*(2), 277–296.

Bhatnagar, A., Kumar, E., Minocha, A. K., Jeon, B. H., Song, H., & Seo, Y. C. (2009). Removal of anionic dyes from water using citrus limonum (lemon) peel: Equilibrium studies and kinetic modeling. *Separation Science and Technology, 44*(2), 316–334.

Bhatnagar, A., Minocha, A. K., & Sillanpää, M. (2010). Adsorptive removal of cobalt from aqueous solution by utilizing lemon peel as biosorbent. *Biochemical Engineering Journal, 48*(2), 181–186.

Bhatnagar, A., Sillanpää, M., & Witek-Krowiak, A. (2015). Agricultural waste peels as versatile biomass for water purification – a review. *Chemical Engineering Journal, 270*, 244–271.

Bv, C. (2017). *Wastewater treatment additives chemicals B.V. 2*. Retrieved from carburization on mechanical properties of recycled steel with perm kernel shell.

Chandane, V., & Singh, V. K. (2016). Adsorption of safranin dye from aqueous solutions using a low-cost agro-waste material soybean hull. *Desalination and Water Treatment, 57*(9), 4122–4134.

Dollah, Z., Hamzah, N., Hasan, D., Rosseli, S. R., & Ash, S. B. (2014). Potential of orange peels as natural coagulant in water treatment potential of orange peels as natural coagulant in water treatment. *Environmental Management, 23*(5), 967.

Ghaedi, M., Hossainian, H., Montazerozohori, M., Shokrollahi, A., Shojaipour, F., Soylak, M., & Purkait, M. K. (2011). A novel acorn based adsorbent for the removal of brilliant green. *Desalination, 281*, 226–233.

Gupta, V. K., Carrott, P. J. M., Ribeiro Carrott, M. M. L., & Suhas. (2009). Low-cost adsorbents: A viable approach to wastewater treatment: A review. *Critical Reviews in Environmental Science and Technology, 39*(10), 783–842.

Hameed, B. H., & Ahmad, A. A. (2009). Batch adsorption of methylene blue from aqueous solution by garlic peel, agricultural waste biomass. *Journal of Hazardous Materials, 164*(2), 870–875.

Ibisi, N. E., & Asoluka, C. A. (2018). Use of agro-waste (Musa paradisiaca peels) as a sustainable biosorbent for toxic metal ions removal from contaminated water. *Chemistry International, 4*(1), 52.

Idan, I. J. (2017). Adsorption of anionic dye using cationic surfactant-modified kenaf core fibers. *OALib, 04*(07), 1–18.

Inbaraj, B. S., & Sulochana, N. (2004). Carbonized jackfruit peel as an adsorbent for the removal of Cd(II) from aqueous solution. *Bioresource Technology, 94*(1), 49–52.

Jayarajan, M., Arunachala, R., & Gurusamy, A. (2011). Agricultural wastes of jackfruit peel nano-porous adsorbent for removal of rhodamine dye. *Asian Journal of Applied Sciences, 4*, 263–270.

Jiang, T., Jiang, J., Xu, R., & Li, Z. (2012). Chemosphere adsorption of Pb (II) on variable charge soils amended with rice-straw derived biochar. *Chemosphere, 89*(3), 249–256.

Kayiwa, R., Olupot, P. W., & Lubwama, M. (2019). Production and potential of activated carbon from cassava peels for remediation of active pharmaceutical ingredients from wastewater: A review. *Journal of Materials Science Research and Reviews, 4*(4), 1–24.

Kouteu Nanssou, P. A., Jiokap Nono, Y., & Kapseu, C. (2016). Pretreatment of cassava stems and peelings by thermohydrolysis to enhance the hydrolysis yield of cellulose in the bioethanol production process. *Renewable Energy, 97*, 252–265.

Liang, S., Guo, X., & Tian, Q. (2013). Adsorption of Pb2+, Cu2+, and Ni2+ from aqueous solutions by novel garlic peel adsorbent. *Desalination and Water Treatment, 51*(37–39), 7166–7171.

Liu, C., Ngo, H. H., Guo, W., & Tung, K. L. (2012). Optimal conditions for preparation of banana peels, sugarcane bagasse, and watermelon rind in removing copper from water. *Bioresource Technology, 119*, 349–354.

Mahir, T., Khusaibi, A., Dumaran, J. J., Devi, M. G., & Rao, L. N. (2015). Treatment of dairy wastewater using orange and banana peels treatment of dairy wastewater using orange and banana peels. In *Materials science and engineering* (Vol. 413, No. 2, p. 012028). IOP Publishing.

Mo, J., Yang, Q., Zhang, N., Zhang, W., Zheng, Y., & Zhang, Z. (2018). A review on agro-industrial waste (AIW) derived adsorbents for water and wastewater treatment. *Journal of Environmental Management, 227*, 395–405.

Mohan, D., Kumar, H., Sarswat, A., et al. (2013). Cadmium and lead remediation using magnetic oak wood and oak bark fast pyrolysis bio-chars. *Chemical Engineering Journal, 236*, 513–528.

Mohapatra, D., Mishra, S., & Sutar, N. (2010). Banana and its by-product utilization: An overview. *Journal of Scientific and Industrial Research, 69*(5), 323–329.

Moletta, R., Rambaud, A., Maunoir, S., & Philip, H. (2007). Biologic additives for the wastewater treatment: Application to anaerobic digestion and septic tanks. In *Techniques – Sciences – Methods* (pp. 31–53).

Odisu, T., Edomwonyi-Out, L. C., & Anih, E. C. (2019). Comparative studies of adsorption of heavy metals from cement waste water using activated carbon from palm kernel husk, coconut and groundnut shells. *International Journal of Innovation and Applied Studies, 12*(1), 110.

Omorogie, M. O., Babalola, J. O., Unuabonah, E. I., & Gong, J. R. (2014). Hybrid materials from agro-waste and nanoparticles: Implications on the kinetics of the adsorption of inorganic pollutants. *Environmental Technology, 35*(5), 611–619.

Owa, F. D. (2013). Water pollution: Sources, effects, control, and management. *Mediterranean Journal of Social Sciences, 4*(8), 65–68.

Palma, C., Lloret, L., Puen, A., Tobar, M., & Contreras, E. (2016). Production of carbonaceous material from avocado peel for its application as an alternative adsorbent for dyes removal. *Chinese Journal of Chemical Engineering, 24*(4), 521–528.

Pattanaik, L., Pattnaik, F., Saxena, D. K., & Naik, S. N. (2019). Biofuels from agricultural wastes. In *In the second and third generation of feedstocks: The evolution of biofuels*. Elsevier.

Pei, Y., & Liu, J. (2011). Adsorption of Pb^{2+} in wastewater using adsorbent derived from grapefruit peel. *Advanced Materials Research, 391–392*, 968–972.

Pellera, F., Giannis, A., Kalderis, D., Anastasiadou, K., Stegmann, R., Wang, J., & Gidarakos, E. (2012). Adsorption of Cu (II) ions from aqueous solutions on biochars prepared from agricultural by-products. *Journal of Environmental Management, 96*(1), 35–42.

Pooja, N. S., & Padmaja, G. (2015). Enhancing the enzymatic saccharification of agricultural and processing residues of cassava through pretreatment techniques. *Waste and Biomass Valorization, 6*(3), 303–315.

Ren, K., Zhang, W., Cao, S., Wang, G., & Zhou, Z. (2018). Carbon-based fe3o4 nanocomposites derived from waste pomelo peels for magnetic solid-phase extraction of 11 triazole fungicides in fruit samples. *Nanomaterials, 8*(5), 302.

Saeed, A., Sharif, M., & Iqbal, M. (2010). Application potential of grapefruit peel as dye sorbent: Kinetics, equilibrium, and mechanism of crystal violet adsorption. *Journal of Hazardous Materials, 179*(1), 564–572.

Silva, C. R., Gomes, T. F., Andrade, G. C. R. M., Monteiro, S. H., Dias, A. C. R., Zagatto, E. A. G., & Tornisielo, V. L. (2013). Banana peel as an adsorbent for removing atrazine and ametryne from waters. *Journal of Agricultural and Food Chemistry, 61*(10), 2358–2363.

Sudaryanto, Y., Hartono, S. B., Irawaty, W., Hindarso, H., & Ismadji, S. (2006). The high surface area activated carbon prepared from cassava peel by chemical activation. *Bioresource Technology, 97*(5), 734–739.

Suharso, & Buhani. (2011). Biosorption of Pb(II), Cu(II) and Cd(II) from aqueous solution using cassava peel waste biomass. *Asian Journal of Chemistry, 23*(3), 1112–1116.

Tan, F., Shi, G., Qian, Y., Tan, F., & Cai, W. (2019). Controllable synthesis of pomelo peel-based aerogel and its application in adsorption of oil/organic pollutants. *Royal Society Open Science, 6*(2), 181823.

Udo, M., Esezobor, D., Afolalu, A., Onovo, H., Ongbali, S., & Okokpuji, I. (2018). Investigation of balling characteristics of mixture of iron oxide bearing wastes and iron ore concentrates. In *IOP conference series materials science and engineering* (Vol. 413, No. 2, p. 012042). IOP Publishing.

Yunus, Z. M., Tun, U., & Onn, H. (2019). Chemistry towards wastewater treatment and remediation (series 1) chapter 1 utilisation of agro-waste as potential adsorbent in wastewater purification. In *Green chemistry towards wastewater treatment and remediation – Series 1* (pp. 1–13). Universiti Tun Hussein Onn.

Zheng, H., Sun, Q., Li, Y., & Du, Q. (2020). Biosorbents prepared from pomelo peel by hydrothermal technique and its adsorption properties for congo red. *Materials Research Express, 7*(4), 045505.

Zhu, L., Wang, Y., Wang, Y., You, L., Shen, X., & Li, S. (2017). Environmentally friendly carbon aerogels derived from waste pomelo peels for the removal of organic pollutants/oils. *Microporous and Mesoporous Materials, 241*, 285–292.

Sustainable Waste Management Towards Circular Economy in Nigerian Context: Challenges, Prospects and Way Forward

O. J. Oyebode

1 Introduction

Sustainable waste management is a method of preserving the environment and eliminating wastes without negative impact on the environment. Wastes can be reduce, reuse and recycle by engineering technology for the benefit of populace. A circular economy is an economic system geared towards the removal of wastes and the incessant utilization of resources. Weed means plants in the wrong place to botanists, in the same way wastes means resources in the wrong location to environmental expects.

Reuse is the utilization of a product without any reprocessing, whether or not for its initial purpose or to satisfy a dissimilar perform. This conserves energy, time and different resources. It assists in job creation and entrepreneurship and contributes to the economy. This is totally different from usage; this is often the breaking down of used things to make raw materials for the manufacture of latest merchandise. Recycling will forestall pollution reduction, materials reduction and resource recovery. Circular systems employ reclaim, regenerative approach, allotment, revamp, restoration and re-manufacturing in the environment. Figure 1 indicates the difference between linear and circular waste disposal system.

Circular economy is more environmentally friendly than linear economy approach. Reduction of solid wastes and manufacturing of recycling systems will conserve natural resources and render great assistance in the reduction of environmental pollution. There are cases which will need totally different or further ways, like buying new, additional energy-efficient instrumentation.

O. J. Oyebode (✉)
Civil and Environmental Engineering Department, Afe Babalola University,
Ado-Ekiti, Nigeria

A. O. Ayeni et al. (eds.), *Advanced Manufacturing in Biological, Petroleum, and Nanotechnology Processing*, Green Energy and Technology,
https://doi.org/10.1007/978-3-030-95820-6_20

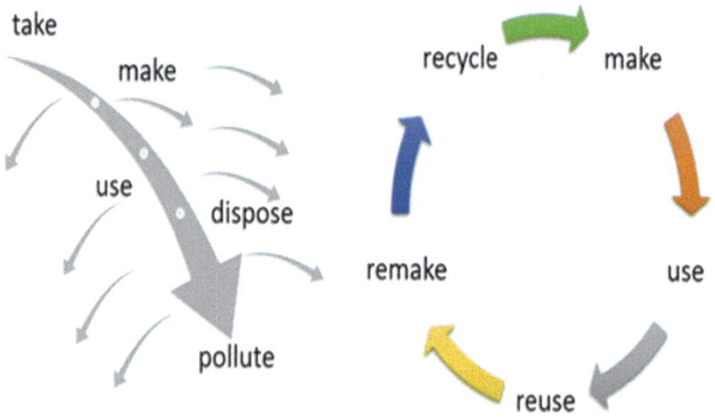

Fig. 1 Difference between linear and circular waste disposal system. (Source: Genovese et al., 2017)

2 Solid Waste Management Towards Circular Economy

Waste management through circular economy gives better opportunities for greener environment, wealth of people, energy generation and effective policies in developed and developing countries. There is deficit of information, and scarcities of data have slowed down its implementation in many countries (Ezeudu & Ezendu, 2019). Management of wastes in the cities and communities of Nigeria is mainly handled by state government agencies. This arrangement needs to be strengthened because it has limited capabilities to handle a lot of waste generated on a daily basis. The circular economy (CE) is rising in greater pace; it is a space of study that concentrates on regenerative approach to natural resources management, as a critical linear technique that's massively unsustainable as a result of the finite accessibility of raw resources for production and also the associated environmental degradation. Many nations of the world have enraptured from the implementation of ancient waste management practices and policies and have absolutely adopted circular economy principles (Ellen et al., 2019; Mathews & Tan, 2011; Lehmann et al., 2014).

Three ways of aggregation municipal solid waste (MSW) indicated in two are indicated in Fig. 2.

The support for circular economy approach has to be deployed to waste management in developing countries for health, wealth and greener environment. The safety problems are available in form of creating certain that informal staff have healthy operating conditions, safety kits and pension edges. Considering the extent of monetary condition and state within the country, there are nice prospects for attracting additional informal staff once the circular economy is formally declared through applicable policies and incentives. These measures are repeatedly stressed in previous analysis efforts (Scarlet et al., 2015; Ferronato et al., 2019; Nzeadibe, 2009). Most methods for handling wastes in Nigeria are terribly restricted technical aspects and monetary capacities, with none funding from abroad (Lamond et al., 2012; Wilson et al., 2013) (Fig. 3).

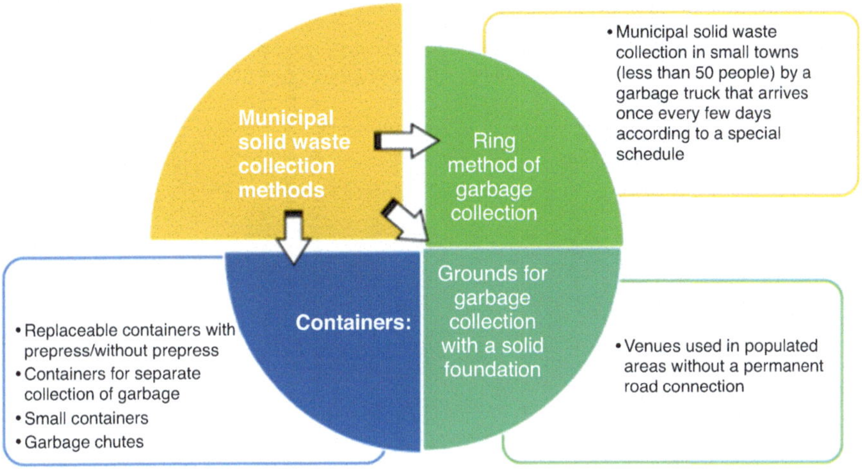

Fig. 2 Main methods of municipal solid waste collection from the population and infrastructure. (Source: Scarlat et al., 2015)

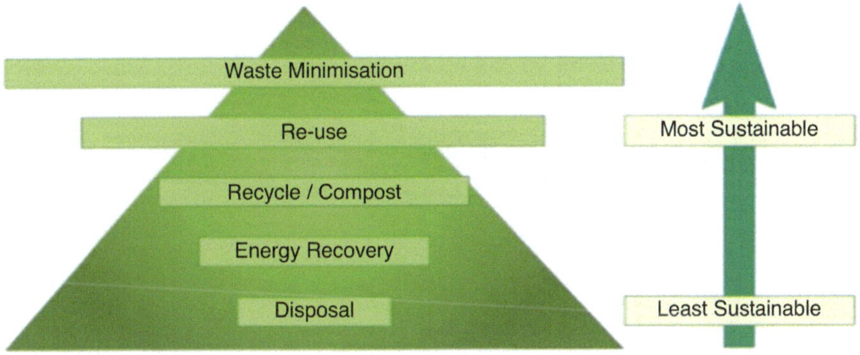

Fig. 3 Waste minimization is the most sustainable method. (Source: Jones et al., 2013)

This is to say that alternative physical elements, like environmentally sound treatment and disposal and resource recovery from waste, need to be compelled to be thought-about from the showtime therefore on guarantee and capitalize on the synergism of actions towards strategic long-run goals of the system. In alternative words, it is necessary and helpful to verify coherence of policy goals and cohesion among the instruments applied to achieve them (Howlett & Rayner, 2007; Nilsson et al., 2012). Yet on the increase is the demand for good waste management service for public health and environmental protection (Oyebode, 2018a). Table 1 presented the relationship between Sustainable Development Goals and solid waste management.

Solid waste management (SWM) can be defined as the systematic interaction between various activities of waste generation, storage, collection, transfer and

Table 1 Relationship between Sustainable Development Goals and solid waste management

Driver		Sustainable Development Coal (SDG)[I]	Sustainable development: Specific target	Solid waste management (SWM): SWM related 'virtual SDG'[II]
Protection of public health		SDG 11: Sustainable cities	11.1 Ensure access for all to adequate, safe, and affordable basic services; upgrading slums	→ **Goal 1. Ensure access for all to adequate, safe, and affordable solid waste collection services.**
		SDG 3: Good health and well-being	3.2 End preventable deaths of children under 5 years	*Uncollected waste is often dumped in waterways or burned in the open air, thus directly causing pollution and contamination. Waste also clogs the drains, which exacerbates floods, keeping stagnant water and contributing to water-borne diseases and malaria. Children are among the most vulnerable, so they are affected the most.*
			3.3 End malaria and combat water-borne diseases	
			3.9 Reduce illnesses from hazardous chemicals and air, water and soil pollution, and contamination	
		SDG 11: Sustainable cities	11.6 Reduce the adverse environmental impact of cities; special attention to waste management	
Protection of the environment	**Local**	SDG 12: Responsible consumption and production	12.4 Environmentally sound management of chemicals and all wastes in order to minimize their adverse impacts on human health and the environment	→ **Goal 2. Eliminate uncontrolled dumping and open burning**, *as the first stepping-stone to achieving environmentally sound SWM practices.*
		SDG 6: Clean water and sanitation	6.3 Improve water quality by reducing pollution, eliminating dumping and minimizing release of hazardous materials	→ **Goal 3. Achieve environmentally sound management of all wastes, particularly hazardous wastes** (*either chemical or biological hazardous wastes*).
		SDG 15: Life on land	15.1 Ensure the conservation of terrestrial and inland freshwater ecosystems and their services	
	Global	SDG 7: Affordable and clean energy	7.2 Increase the share or renewable energy in the global energy mix	→ **Goal 3.** *SWM technologies can derive renewable energy from (organic) waste.*
		SDG 13: Climate action	SDG 13: Take urgent action to combat climate change and its impacts	→ **Goal 3.** *Adequate SWM practices can prevent emissions of large amounts of greenhouse gases.*[III]
		SDG 14: Life below water	14.1 Prevent marine pollution of all kinds, in particular from land-based activities, including marine debris	→ **Goal 1** and **Goal 2**. *Extending waste collection to all and eliminating uncontrolled dumping will prevent waste (particularly plastics) ending up in the oceans.*

transport, intermediate treatment and final disposal (Oyebode, 2018b). It is sure that goal wastewater management will further improve water supply and sanitation in most cities and solid waste management (Oyebode, 2021). Cleaner environment and pollution control mechanisms will assist the health and wealth of people living in the country. This has in some cases involved costly measures and controversial political decisions (Oyebode, 2018b). Most problems affecting waste management include insufficient adequate drainage facilities, enforcement-related issues, corrupt practices and systems' failure, pitiable supervision and poor culture for maintenance of infrastructures (Oyebode, 2021).

In Nigeria, an average citizen produces about 0.43 kg solid waste per head per day. Organic matter constitutes about 60–80% of the total waste stream along with plastics, nylon and scrap metal as major recyclable constituents.

3 Disposal Site Inspections and Waste Management Methods

The methodology adopted includes inspection of disposal sites, surveys and data from literatures. There is an urgent need for effective waste management in majority of rural and urban centres in Nigeria. Challenges, prospects and way forward were identified for sustainable waste management in the country. Figure 4 indicates how waste is managed in some residential areas in Nigeria. Figure 5 presented the negative and positive sides of circular economy. Figure 6 presented the different ethnic groups and waste management cultures in Nigeria. Figure 7 gave the current waste management framework in Nigeria.

Figure 8 presented the comparative view of linear economy and circular economy, while Fig. 9 gave the seven crucial material efficiency strategies for the reduction of emissions. Public health is a fundamental requirement to human existence.

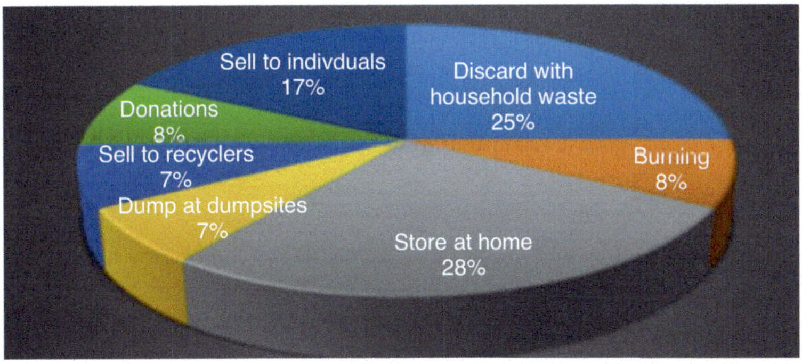

Fig. 4 How waste is managed in some residential areas in Nigeria. (Source: Ezeudu et al., 2021)

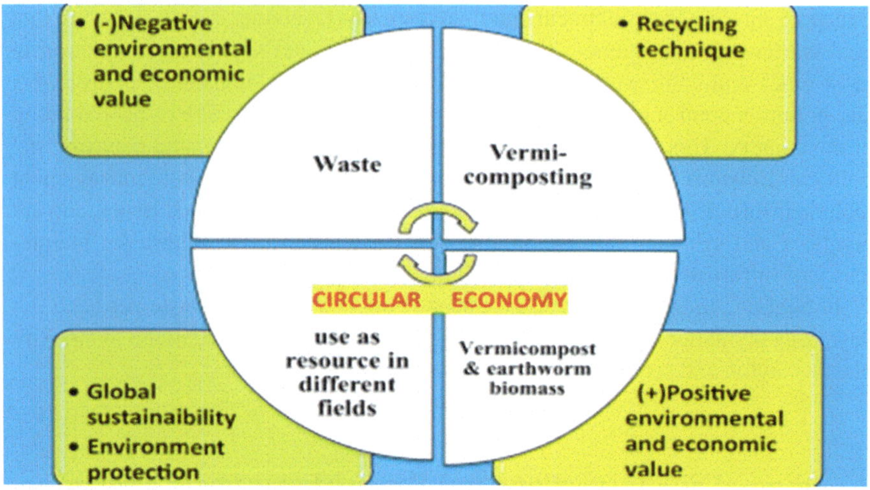

Fig. 5 Circular economy negative and positive sides. (Source: Muranko et al., 2019)

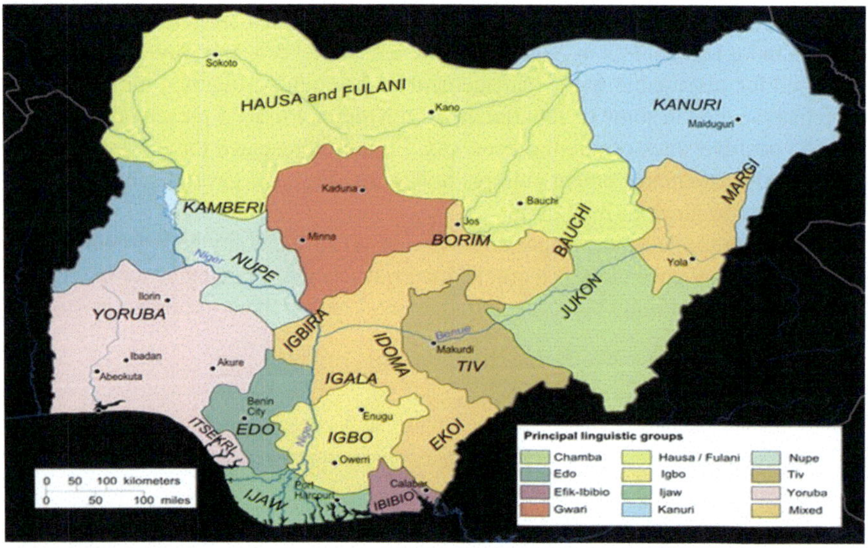

Fig. 6 Different ethnic groups and waste management cultures in Nigeria. (Source: Nas & Jaffe, 2004)

Waste is directly linked to human development, both technologically and socially; some components of waste have economical value and can be recycled once correctly recovered (Oyebode, 2013). Figure 10 presented the business opportunities for material-efficient housing. Figure 11 indicates the policy, material efficiency and greenhouse impacts, while Fig. 12 presented the circular strategies, innovations, life cycle stages and enablers.

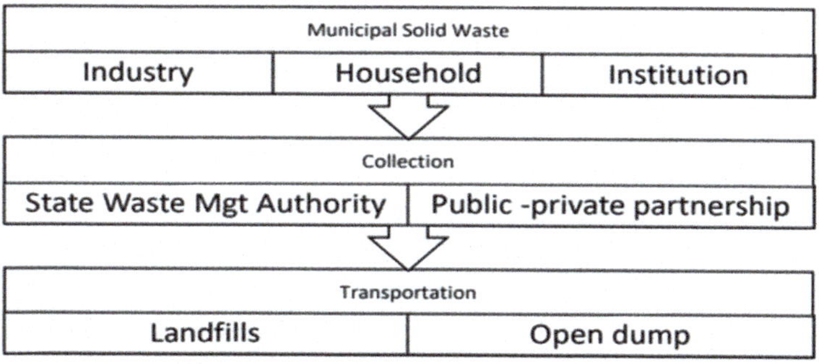

Fig. 7 Current waste management framework in Nigeria. (Ike et al., 2018)

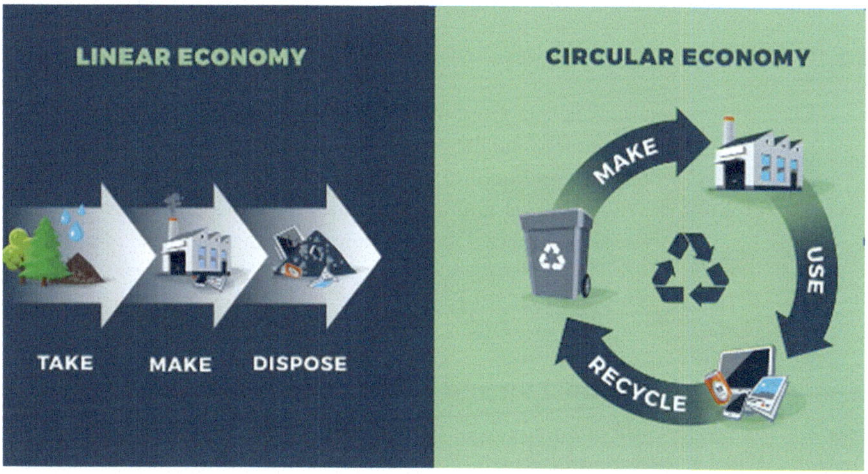

Fig. 8 Comparative view of linear economy and circular economy

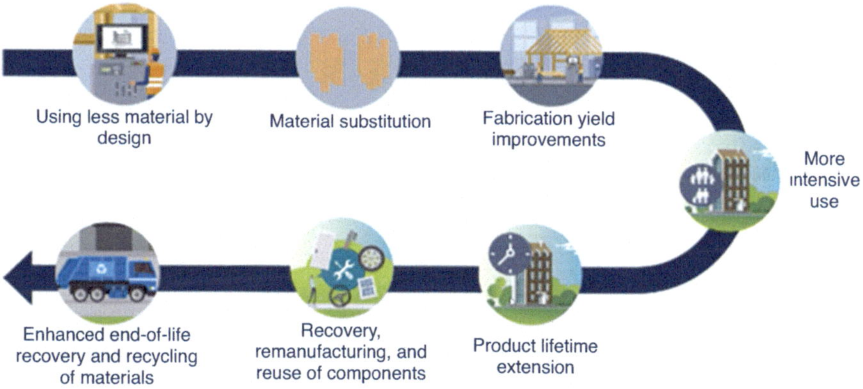

Fig. 9 Seven crucial material efficiency strategies for the reduction of emissions

Fig. 10 Business opportunities for material-efficient housing

Fig. 11 Policy, material efficiency and greenhouse impacts

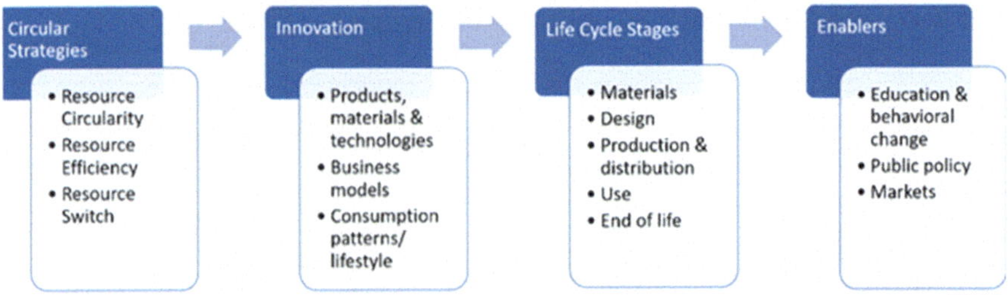

Fig. 12 Circular strategies, innovations, life cycle stages and enablers

3.1 Challenges

There are a lot of challenges facing the management of wastes in Nigeria. There are issues with the collection, inadequate treatment and disposal and recycling systems. There are problems of air and groundwater contamination. There are problems of flooding traceable to dumping of refuse at canals and drainages in Nigeria. There are insufficient formidable legal framework and implementation strategies for sustainable management and circular economy. Other challenges are traceable to inadequate funding, capacity building and entrepreneurship in waste recycling.

3.2 Prospects from Sustainable Waste Management Towards Circular Economy

The major prospect is wealth and hygienic environment for all and sundry. When waste is properly managed, it will be harmless to human and animal life – the ecology and environment generally. Public health will be preserved. Waste to wealth, health and energy will enhance environmental sustainability. Resource conservation will also be another prospect of sustainable waste management.

4 Way Forward

There is a need to attract private investment into waste management. Job creation and entrepreneurship programme on waste management should be given priority. All materials can be subjected to efficiency and resource recovery. Recycling of water, use of recycled products and reduction in the extraction of natural resources can also be achieved.

Research and innovation into waste management should be supported by government and all stakeholders. There is a need to create robust and workable sustainable waste management systems. The idea of circular economy should be given adequate awareness, planning and execution by institutions and all tiers of the government. Manpower development, capacity building, stakeholder collaboration and consumer education campaign should be embraced and supported. Nigeria needs to attract recycling infrastructural investors for effective waste management. Policy and regulatory advocacy will yield tremendous benefits. Relevant attention should be given to all aspect of waste management and circular economy. Treatment, disposal, recycling systems and other facilities should be designed and put in place. Stakeholders' responsibility and legal requirement are very useful in circular economy.

5 Conclusion

Waste management requires adequate attention by all and sundry for the achievement of cleaner environment, circular economy and effective waste management. Involvement of government agencies, laws and guidelines for waste minimization, industries, institutions, building owners, appropriate capacity building and all stakeholders will yield tremendous rewards. This requires adequate attention in Nigeria for health, wealth and energy. The quantity and the rate of solid waste generation in Nigeria have outgrown the capacity of nature to naturally absorb them. Energy generation, wealth and cleaner environment can be achieved through this waste management approach. To achieve long-term and holistic solution, this approach is very central and critical for sustainable development. It follows the concept of reducing, reuse and better ways of handling consumption and production processes. Waste management facilities need proper design and improvement. There are issues linked to planning, maintenance, recycling culture among Nigerians, inadequate storage and disposal facilities.

6 Recommendations

The following recommendations will assist circular economy in Nigerian environment:

(i) There should be absolute reduction of carbon footprint in Nigerian environment.
(ii) Research related to solid waste management should be adequately funded.
(iii) Robust systems and strategies should be deployed towards the collection, transportation, sorting, disposal and recycling of various wastes.
(iv) Seminars, public awareness, workshops, conferences and other capacity building activities should be periodically organized for the attainment of circular economy.
(v) The government should reinforce waste collection and disposal systems in every state while strengthening and enforcing the appropriate laws. To prevent serious environmental disaster in Nigeria, priority should be given to waste management.
(vi) Electronic wastes should be disposed carefully in urban and rural areas.
(vii) Adoption of engineering and technological-driven initiatives.
(viii) There is a need for movement towards cleaner fuel and low-carbon economy.
(ix) Solid waste management policies and enforcement of sanitation laws in various Nigerian states should be energized, and various environmental organizations and societies to do more until the dreamed clean environment in Nigeria becomes a reality.
(x) There is a need for disclosure of emission-related information to consumer for better-informed decision.

(xi) Increasing disclosures on product-specific carbon footprint is very important.
(xii) Policy intervention programmes towards circular economy by the Nigerian government will enhance Nigerian environment.
(xiii) Strategic policies towards environmental sustainability will be of great benefit.

References

Ellen MacArthur Foundation. (2019). *Concept: What is circular economy? A framework for an economy that is restorative and regenerative by design*. Available online https://www.ellenmacarthurfoundation.org/circular-economy/concept. Accessed 9 Sept 2019.

Ezeudu, O. B., & Ezeudu, T. S. (2019). Implementation of circular economy principles in industrial solid waste management: Case studies from a developing economy (Nigeria). *Recycling, 4*(4), 42.

Ezeudu, O. B., Oraelosi, T. C., Agunwamba, J. C., & Ugochukwu, U. C. (2021). Co-production in solid waste management: analyses of emerging cases and implications for circular economy in Nigeria. *Environmental Science and Pollution Research, 28*(37), 52392–52404.

Ferronato, N., Rada, E. C., Portillo, M. A. G., Cioca, L. I., Ragazzi, M., & Torrenta, V. (2019). Introduction of the circular economy within developing regions: A comparative analysis of advantages and opportunities for waste valorization. *Journal of Environmental Management, 230*, 366–378.

Genovese, A., Acquaye, A. A., Figueroa, A., & Koh, S. L. (2017). Sustainable supply chain management and the transition towards a circular economy: Evidence and some applications. *Omega, 66,* 344–357.

Howlett, M., & Rayner, J. (2007). Design principles for policy mixes: Cohesion and coherence in 'New Governance Arrangements'. *Policy and Society, 26*, 1–18.

Ike, C. C., Ezeibe, C. C., Anijiofor, S. C., & Daud, N. N. (2018). Solid waste management in Nigeria: Problems, prospects, and policies. *The Journal of Solid Waste Technology and Management, 44*(2), 163–172.

Jones, P. T., Geysen, D., Tielemans, Y., Van Passel, S., Pontikes, Y., Blanpain, B., ... & Hoekstra, N. (2013). Enhanced Landfill Mining in view of multiple resource recovery: a critical review. *Journal of Cleaner Production, 55*, 45–55.

Lamond, J., Bhattacharya, N., & Bloch, R. (2012). The role of solid waste management as a response to urban flood risk in developing countries, a case study analysis. In D. Proverbs, S. Mambretti, & C. A. Brebbia (Eds.), *Flood recovery, innovation and response III* (pp. 193–205). WIT Press. Available online http://eprints.uwe.ac.uk/21458. Accessed 1 Dec 2016.

Lehmann, M., Leeuw, B. D., Fehr, E., & Wong, A. (2014). Circular economy. In *Improving the management of natural resources*. World Resource Forum.

Mathews, J. A., & Tan, H. (2011). Progress toward a circular economy in China. *Journal of Industrial Ecology, 15*, 435–475.

Muranko, Z., Andrews, D., Chaer, I., & Newton, E. J. (2019). Circular economy and behaviour change: Using persuasive communication to encourage pro-circular behaviours towards the purchase of remanufactured refrigeration equipment. *Journal of Cleaner Production, 222*, 499–510.

Nas, P. J., & Jaffe, R. (2004). Informal waste management. *Environment, Development and Sustainability, 6*(3), 337–353.

Nilsson, M., Zamparutti, T., Petersen, J. E., Nykvist, B., Rudberg, P., & McGuinn, J. (2012). Understanding policy coherence: Analytical framework and examples of sector-environment policy interactions in the EU. *Environmental Policy and Governance, 22*, 395–423.

Nzeadibe, T. C. (2009). Solid waste reforms and informal recycling in Enugu urban area, Nigeria. *Habitat International, 33*, 93–99.

Oyebode, O. J. (2013). Solid waste management for sustainable development and public health: A case study of Lagos State in Nigeria. *Universal Journal of Public Health, 1*(3), 33–39.

Oyebode, O. J. (2018a). Evaluation of municipal solid waste management for improved public health and environment in Nigeria. *European Journal of Advances in Engineering and Technology, 5*(8), 525–534.

Oyebode, O. J. (2018b). Impact of environmental laws and regulations on Nigerian environment. *World Journal of Research and Review, 7*(3), 262587.

Oyebode, O. J. (2021). Hazard assessment and implementation strategies for mitigating of flood and erosion in Nigeria. In IOP Conference Series: Materials Science and Engineering (Vol. 1036, No. 1, p. 012066). IOP Publishing.

Scarlet, N., Motola, V., Dallemand, J. F., Monforti-Ferrario, F., & Mofor, L. (2015). Evaluation of energy potential of municipal solid waste from African urban areas. *Renewable and Sustainable Energy Reviews, 50*, 1269–1286.

Wilson, D. C., Velis, C. A., & Rodic, L. (2013). Integrated sustainable waste management in developing countries. *Waste Resource Management, 166*, 52–68. Available online http://www.icevirtuallibrary.com/doi/abs/10.1680/warm.12.00005. Accessed 1 Dec 2016.

Biomedical Waste Management Practices in Sub-Saharan Africa: Insights of Its Impacts and Strategies for Its Mitigation

O. J. Oyebode, A. O. Coker, M. K. C. Sridhar, and N. M. Ogarekpe

1 Introduction

The handling and overall management of biomedical wastes are vital for the preservation of public health, wealth and environment in most nations of the world. A lot of people in developing countries are at risk of infection and diseases due to biomedical waste. The assertion that people die unnecessarily each year from pitiable practices in the management of biomedical wastes has been reported (Chand et al., 2020).

Biomedical waste management is a critical issue and of great importance as it is a potential threat to the environment. Biomedical engineering with a sub-section in clinical engineering deals with hospital, medicinal, clinical, pharmaceutical, diagnosis services, scientific research treatment and other healthcare services.

O. J. Oyebode (✉)
Civil and Environmental Engineering Department, Afe Babalola University, Ado-Ekiti (ABUAD), Ekiti State, Nigeria

A. O. Coker
Civil Engineering Department, Faculty of Technology, University of Ibadan, Ibadan, Nigeria

M. K. C. Sridhar
Department of Environmental Health Sciences, Faculty of Public Health, College of Medicine, University of Ibadan, Ibadan, Nigeria

N. M. Ogarekpe
Department of Civil Engineering, Cross River University of Technology, Calabar, Nigeria
e-mail: nkpaogarekpe@crutech.edu.ng

A. O. Ayeni et al. (eds.), *Advanced Manufacturing in Biological, Petroleum, and Nanotechnology Processing*, Green Energy and Technology,
https://doi.org/10.1007/978-3-030-95820-6_21

2 Waste Generation in Urban Centres

There are a lot of solution that can be adopted for waste management such as applying resource efficiency, circular economy, recycling system, effective solid waste management and many other notable techniques (US Congress Office of Technology Assessment, 1988; Rao, 2008; Coker et al., 2009; Hossain et al., 2011).

Urbanization increases the incidence of non-communicable diseases in most nations of the world especially in sub-Saharan Africa (Bickler et al., 2018). Recent development, climate change and conflicts affect economic aspect and safety of transportation and health sector (Porter, 2014).

The African continent has the capacity to bear greater burden of pandemic for the identical reasons it suffers from other communicable diseases: ecology, social and economic circumstances, shortage of water and hygiene infrastructures and pathetic health systems (Evans et al., 2020). Sub-Saharan Africa is a main province within the world's second largest continent and is almost non-existent in majority information systems research, teaching, biomedical practice and development (Datta et al., 2005). There is an urgent need to build assurance in the health sector through improved healthcare delivery, sensitization, risk assessment and waste management (Musheke et al., 2013). Diseases are rampant among countries in sub-Saharan Africa and other parts of the world. They result in ability loss, morbidity and loss of quality life (Gyasi & Phillips, 2020).

There have been scrupulous concerns for sub-Saharan Africa, as the main factors that drive high level of infectious diseases, such as water scarcity, environmental issues and socio-economic conditions and inadequate health systems (Evans et al., 2020). There are leadership issues in sub-Saharan Africa due to lack of attention to public health, economic crisis, debt, unstable political system and other governance-related issues (Bräutigam & Knack, 2004).

Transport services can affect medical intervention and research work in rural sub-Saharan Africa, with reference to the crucial significance of transport services for reducing poverty and encouraging growth. Improved well-being is vital for the generation of direct employment, addressing health problems, effects on agricultural activities, health and education (Porter, 2014). Groundwater is the major and most significant water resource in Africa (MacDonald et al., 2012). It is often more dependable, in closer proximity to users, less vulnerable to pollution and more resilient to climate variability than surface water (MacDonald et al., 2011; Lapworth et al., 2013). Right to secure and hygienic water is vital for recuperating health and livelihood for low-income villages and cities in Africa (Hunter et al., 2010).

Biomedical waste is generated from biological and medical sources and activities, such as the diagnosis, prevention or treatment of diseases. Common producers of biomedical waste include nursing homes; hospitals; research laboratories; clinics; emergency medical services; offices of physicians, dentists and veterinarians; healthcare; and morgues. You can also call these waste clinical wastes. In larger aspects of sub-Saharan Africa, there is continuous expansion of urban and rural settlements. A recent study (UNPFA, 2007) estimates that between 2000 and 2030,

Africa's urban population, compared to rural population, will double and become the majority. Overall, 37% of Africa's population are presently urbanized, and the urban proportion is rising (World Bank, 2016; World Population, 2016). Africa is estimated to have 67,740 health facilities and produce approximately 282,447 tons of medical waste every year (Tulokhonova & Ulanova, 2013). Previous studies indicated that municipal solid waste management in sub-Saharan African (SSA) cities are inefficient and ineffective (Komakech, 2014; Laner et al., 2012).

3 Waste Management Approach in the Study Area

The methodology adopted includes reconnaissance survey, secondary data from literatures, interview and interactions with doctors in selected part of sub-Saharan Africa and selected hospitals within the study area. Several articles and blog posts were studied and analysed for quantitative and qualitative information. Figure 1 indicates sub-Saharan Africa map.

Figure 4 indicates the collection and disposal vehicles for biomedical wastes. Table 1 gave the estimated volume of biomedical wastes in public health facilities in part of sub-Saharan Africa.

Figure 2 presented the segregation of hospital biomedical wastes, while Fig. 3 gave the hospital biomedical waste disposal to environment (Fig. 4).

Incinerators also are inherently inefficient from an energy standpoint, particularly when handling wastes with high water content. Figure 5 presented the strategies for the mitigation of wastes by hierarchy. Figure presented the mechanical equipment for incineration.

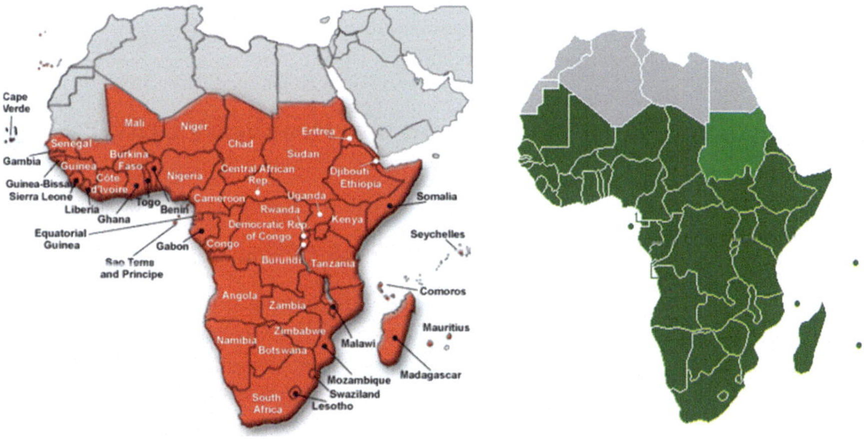

Fig. 1 Map of sub-Saharan Africa

Table 1 Estimated volume of biomedical wastes in public health facilities in typical African country

Type of health facilities	Number of health facilities	Quantity (liter/day)	Quantity (1,000 liters/day)
Hospitals	34	1200	40.8
Health centers	89	300	26.7
Health posts	1240	30	37.2
DPC	76	30	23
Health cases	1722	10	17.2
Total			124.2

Source: Shamim et al. (2015)

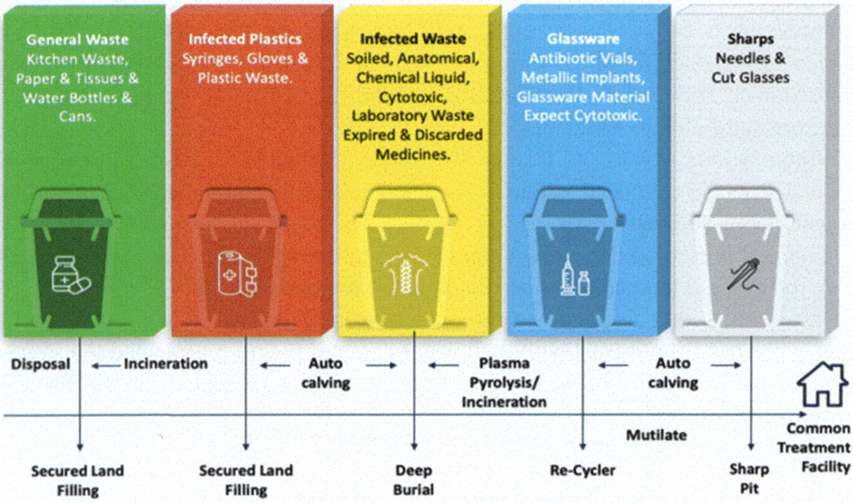

Fig. 2 Segregation of hospital biomedical wastes. (Source: Patan & Mathur, 2015)

4 Biomedical Waste Management and Healthcare Delivery

This study indicates the fact that sub-Saharan Africa has not been able to deal with the quantity of hazardous wastes generated annually and segregation of medical waste has not been given the high priority that it demands. Healthcare waste management is still a knotty issue in developing countries because it has not been given adequate attention. HIV infection and waste management-related problems led to health hazards in not just the public but health workers as well, although improvements have been made in biomedical waste management in 2010.

Fig. 3 Hospital biomedical waste disposal to environment

Fig. 4 Incineration of biomedical waste in the southwestern Nigeria

5 Conclusions

Biomedical waste management in sub-Saharan Africa is unarguably very poor in its sense of practice judging from statistics taking in comparison with the stipulated World Health Organization standards. There is an urgent need to address it for public health and effective sanitation. Holistic approach is needed by engineers, health workers, doctors, government, stakeholders and other experts in biomedical waste

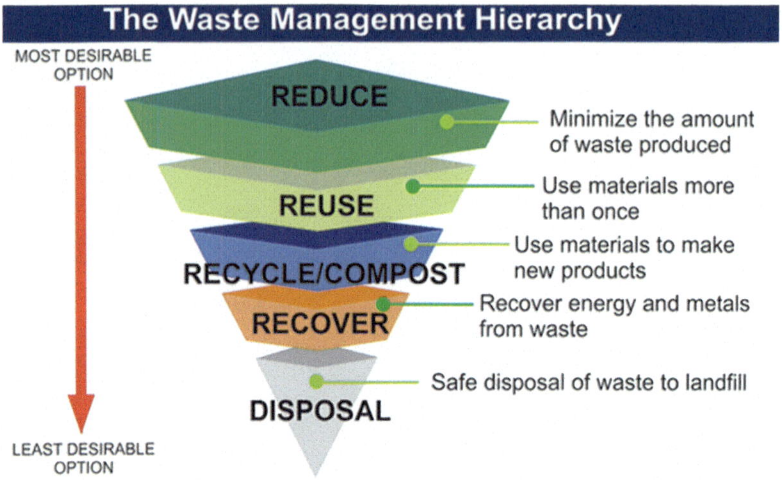

Fig. 5 Strategies for the mitigation of wastes by hierarchy. (Source: Fischer et al., 2011)

management. Biomedical waste management has great impacts on the environment and requires urgent mitigation strategies for public health and hygienic environment. Findings are common across SSA, and, therefore, the mitigation strategies extend beyond Nigerian borders and need implementing in other developing nations. The issue of biomedical waste management has assumed great significance in recent times, and it should be addressed without any delay for the health and benefit of mankind. Causes of mismanagement of biomedical wastes were traceable to lack of sensitization, inadequate regulation, technological limitations, lack of maintenance and inefficient disposal systems for biomedical waste.

Mitigation strategies, legal framework and government policies are needed for the effective management of biomedical wastes. Biomedical waste management practices should be seen as a team effort and process. No one man or woman can perform all this and keep the entire hospital environment clean.

There are many severe issues in the sub-Saharan African medical waste disposal system and the potential troubles it may bring with possible outbreaks and danger to medical waste personnel as well as scavengers, so it should also be considered that with relatively enough effort, it is possible to reform and ensure the creation of a much better and safer system.

6 Recommendations

Based on the findings of this study, the following recommendations are made to improve the MWM practices not only in Lagos but in the southwest state as a whole:

(i) There should be proper planning, construction of sanitary landfills and management practices for an effective and proper biomedical waste management in sub-Saharan Africa to alleviate environmental contamination, biological hazards and health risks.

(ii) Adequate legal framework, efficient structure, more healthcare facilities and considerable amount of money should be devoted biomedical waste management in all concerned nations.

(iii) With the aid of mandatory standards and schedules in healthcare sectors, growth and improvement in biomedical waste management are possible, as well as a reduction in the exposure of the environment and people to risk.

(iv) Also, financial aids for the healthcare system to aid the availability of necessary waste disposal materials (e.g. incinerators and chemicals) will help improve the quality of biomedical waste management.

(v) Further studies should be done on this to study specific impacts on various countries for health, wealth and environment.

References

Bickler, S. W., Wang, A., Amin, S., Halbach, J., Lizardo, R., Cauvi, D. M., & De Maio, A. (2018). Urbanization in sub-Saharan Africa: Declining rates of chronic and recurrent infection and their possible role in the origins of non-communicable diseases. *World Journal of Surgery, 42*(6), 1617–1628.

Bräutigam, D. A., & Knack, S. (2004). Foreign aid, institutions, and governance in sub-Saharan Africa. *Economic Development and Cultural Change, 52*(2), 255–285.

Chand, S., Shastry, C. S., Hiremath, S., Joel, J. J., Bhat, C. K., & Mateti, U. V. (2020). Water, sanitation, hygiene and biomedical waste disposal in the healthcare system: A review. *Biomedicine, 40*(1), 14–19.

Coker, A., Sangodoyin, A., Sridhar, M., Booth, M. C., Olomolaiye, P., & Hammond, F. (2009). Medical waste management in Ibadan, Nigeria: Obstacles and prospects. *Waste Management, 29*(2), 804–811.

Datta, P., Byrd, T. A., Okoli, C., & Mbarika, V. W. (2005). The neglected continent of IS research: A research agenda for Sub-Saharan Africa. *Journal of the Association for Information Systems, 6*(5), 6.

Evans, M. V., Garchitorena, A., Rakotonanahary, R. J., Drake, J. M., Andriamihaja, B., Rajaonarifara, E., & Rakotonirina, J. (2020). Reconciling model predictions with low reported cases of COVID-19 in Sub-Saharan Africa: Insights from Madagascar. *Global Health Action, 13*(1), 1816044.

Fischer, T. B., Potter, K., Donaldson, S., & Scott, T. (2011). Municipal waste management strategies, strategic environmental assessment and the consideration of climate change in England. *Journal of Environmental Assessment Policy and Management, 13*(04), 541–565.

Gyasi, R. M., & Phillips, D. R. (2020). Aging and the rising burden of non communicable diseases in sub-Saharan Africa and other low and middle-income countries: A call for holistic action. *The Gerontologist, 60*(5), 806–811.

Hossain, M. S., Santhanam, A., Norulani, N. A. N., & Omar, A. K. M. (2011). Clinical solid waste management practices and its impact on human health and environment – A review. *Waste Management, 31*(4), 754–766.

Hunter, P. R., MacDonald, A. M., & Carter, R. C. (2010). Water supply and health. *PLoS Medicine, 7*(11), e1000361.

Komakech, A. J. (2014). *Urban waste management and the environmental impact of organic waste treatment systems in Kampala*. Uganda in PhD dissertation, D.o.E.a.T. Faculty of

Natural Resources and Agricultural Sciences, Uppsala, Sweden and D.o.A.a.B.E. College of Agricultural and Environmental Sciences, Makerere University, Kampala, Uganda Editors. 2014: Sweden.

Laner, D., et al. (2012). A review of approaches for the long-term management of municipal solid waste landfills. *Waste Management, 32*(3), 498–512.

Lapworth, D. J., MacDonald, A. M., Tijani, M. N., Darling, W. G., Gooddy, D. C., Bonsor, H. C., & Araguás-Araguás, L. J. (2013). Residence times of shallow groundwater in West Africa: Implications for hydrogeology and resilience to future changes in climate. *Hydrogeology Journal, 21*(3), 673–686.

MacDonald, A. M., Bonsor, H. C., Calow, R. C., Taylor, R. G., Lapworth, D. J., Maurice, L., Tucker, J., & O'Dochartaigh, B. E. (2011). *Groundwater resilience to climate change in Africa*. British Geological Survey Open Report, OR/11/031, BGS, Keyworth, UK.

MacDonald, A. M., Bonsor, H. C., O'Dochartaigh, B. E. O., & Taylor, R. G. (2012). Quantitative maps of groundwater resources in Africa. *Environmental Research Letters, 7*. https://doi.org/10.1088/1748-9326/7/2/024009

Musheke, M., Ntalasha, H., Gari, S., Mckenzie, O., Bond, V., Martin-Hilber, A., & Merten, S. (2013). A systematic review of qualitative findings on factors enabling and deterring uptake of HIV testing in Sub-Saharan Africa. *BMC Public Health, 13*(1), 220.

Patan, S., & Mathur, P. (2015). Assessment of biomedical waste management in government hospital of Ajmer city – A study. *International Journal of Research in Pharmacy and Science, 5*(1), 6–11.

Porter, G. (2014). Transport services and their impact on poverty and growth in rural sub-Saharan Africa: A review of recent research and future research needs. *Transport Reviews, 34*(1), 25–45.

Rao, P. H. (2008). Report: Hospital waste management – Awareness and practices: A study of three states in India. *Waste Management and Research, 26*, 297–303.

Shamim, A., Mursheda, A. K., & Rafiq, I. (2015). E-waste trading impact on public health and ecosystem services in developing countries. *International Journal of Waste Resources, 5*(4), 1–18.

Tulokhonova, A., & Ulanova, O. (2013). Assessment of municipal solid waste management scenarios in Irkutsk (Russia) using a life cycle assessment-integrated waste management model. *Waste Management & Research, 31*(5), 475–484.

U. S. Congress Office of Technology Assessment. (1988). *Issues in medical waste management: Background Paper, OTA-BP-O-49*. United States Government Printing Office.

Udofia, E. A., & Nriagu, J. (2013). Health-care waste in Africa: A silent crisis? *Global Perspectives in Health, 1*(1), 3–10.

UNFPA. (2007). *State of world population 2007: Unleashing the potential of urban growth*. United Nations Population Fund (UNPFA).

World Bank. (2016). *Urban development*. http://data.worldbank.org/topic/urban-development. Accessed June 2016.

World Population. (2016). *Urban data sets*. http://www.worldpop.org.uk/data/. Accessed May 2016.

Bioplastic: A Sustainable Remedy to Manage Environmental Waste

S. A. Afolalu, O. O. Yusuf, M. E. Emetere, S. O. Ongbali, and F. Ademuyiwa

1 Introduction

Environmental nuisance attributed to the disposal of hydrocarbon plastics is a major concern globally. The bulk population of hydrocarbon plastics is derived from crude oil. They are synthetic and, when disposed of, cause land and water pollution, endangering our health and the ecosystem in general (Andrady & Neal, 2009). Plastics possess properties which make them seem almost indestructible; thus, they are utilized in a broad range of applications and are pretty much irreplaceable (DeArmitt, 2011). Since the world cannot do without plastic products, it became imperative that eco-friendly alternatives to fossil-based plastics be produced. Globally, the production of plastics grew from 1.7 million tonnes in 1950 to approximately 275 million tonnes by 2010 and accumulated exponentially to 396 million tonnes in 2018 (Afolalu et al., 2018a, b, c, d). The estimated value disposed into the water bodies annually ranges between 4.8 million and 12.7 million tonnes. This is majorly as a result of the one-off use of most plastics which are immediately discarded carelessly into the environment. "Bio-plastic" simply refers to a type of plastic that is either partially or fully bio-based and/or biodegradable. Bioplastics are of

S. A. Afolalu (✉) · S. O. Ongbali · F. Ademuyiwa
Department of Mechanical Engineering, Covenant University, Ota, Nigeria
e-mail: sunday.afolalu@covenantuniversity.edu.ng;
samson.ongbali@covenantuniversity.edu.ng; faith.ademuyiwa@covenantuniversity.edu.ng

O. O. Yusuf
Department of Microbiology, Obafemi Awolowo University, Ile-Ife, Nigeria
e-mail: omolola.yusuf@oau.edu.ng

M. E. Emetere
Department of Physics, Covenant University, Ota, Nigeria
e-mail: moses.emetere@covenantuniversity.edu.ng

A. O. Ayeni et al. (eds.), *Advanced Manufacturing in Biological, Petroleum, and Nanotechnology Processing*, Green Energy and Technology,
https://doi.org/10.1007/978-3-030-95820-6_22

two classes, namely, bio-based and biodegradable plastic (Tokiwa et al., 2009). Bio-based plastics are those that are either completely or partly renewably sourced, while biodegradable plastics undergo microbial action and are reduced to water (H_2O), carbon dioxide (CO_2), methane (CH_4), and biomass under defined time and environmental conditions. A noteworthy fact is that not all bioplastics or degradable plastics are biodegradable (Vert et al., 2012). European Bioplastics (2019) classified bioplastics into three main groups which are technical performance polymers, like polytrimethylene terephthalate and thermoplastic polyester elastomers, which are bio-based, as well as non-biodegradable plastics which are partially or completely bio-based (like bio-based polypropylene, polyethylene, and polyethylene terephthalate); plastics that are both biodegradable and bio-based, like poly (butane) succinate, polyhydroxyalkanoates, and poly(lactic) acid; and biodegradable fossil-sourced plastics, like polycaprolactone and polybutylene adipate terephthalate. However, biodegradable plastics have received a lot of attention as they go in line with sustainable environment creation by broadening the scope of plastic waste management. As opposed to landfilling, which has been banned in a number of countries, and incineration, which generates greenhouse gases, biodegradable plastics are considered to be eco-friendly (Emadian et al., 2017). Hence, attention should be given to the enormous production of biodegradable bioplastics whose possibility could be explored from bio-based polymers, also known as "biopolymers". Biopolymers are materials produced from renewable resources and are used for the commercial production of bioplastics. Biopolymers are of two main types (Díez-Pascual, 2019), those derived from living organisms and those from renewable sources but require polymerization. Biopolymers are also categorized with respect to their sources or processes through which they are synthesized, which are microorganisms, biotechnology, petroleum sources, and agro-resources (Karthik & Rathinamoorthy, 2018).

2 Biodegradation Mechanisms of Bioplastics

According to Lucas et al. (2008), biodegradation occurs in three major processes which are biodeterioration, biofragmentation, and assimilation. Biodeterioration results from microbial growth on the surface or within the structure of the polymers. This leads to alterations in the chemical, mechanical, and physical properties of the polymers. It is essential to note that biodeterioration is also contributed to by abiotic factors, which are thermal, mechanical, chemical, and light related. Forces such as tension, compression, and shearing cause mechanical property change (Peng et al., 2011; Saito & Isogai, 2004). Energy transfer – by either heat, thermal radiation, luminescence, photoionization, or fluorescence – contributes to thermal and photodegradation (Afolalu et al., 2018a). Atmospheric conditions contribute to oxidation and hydrolysis, which is due to either acid or alkali contact. The cytoplasmic membranes and cell walls that microorganisms possess are impenetrable by polymers, which are macromolecular. The microbes secrete free radicals and enzymes that engender polymeric biofragmentation, an enzymatic reaction that could be carried out through either enzymatic hydrolysis or enzymatic oxidation (Afolalu et al., 2020). Enzymatic hydrolysis could be due to anaerobic or aerobic microorganism activity, while enzymatic oxidation is strictly due to aerobic microorganism activity (Amalia et al., 2020). This reaction promotes the depolymerization of the polymer into monomers, dimers, and oligomers.

Assimilation is the process through which there is a cellular integration of atoms resulting from the fragmentation within the microorganisms. This is necessary, as elements, electrons, and energy required for cell structure development are supplied through these atoms to engender microorganism growth and reproduction. Molecules which cannot permeate membranes are further transformed into products that are either assimilable or not (Afolalu et al., 2020). The assimilated molecules undergo catabolism through either respiration (anaerobic or aerobic) or fermentation to produce cell constituents like adenosine triphosphate (ATP) and the like which are necessary for cell functions. Mineralization, the release of various oxidized salts along with molecules of water (H_2O), carbon dioxide (CO_2), methane (CH_4), and nitrogen (N_2) into the environment, occurs as the end product of intracellular metabolism (European Bioplastics, 2019) (Fig. 1).

Abioye et al. (2019) described biodegradation as a two-step process comprising physical disintegration and chemically induced biodegradation. These two must occur simultaneously to engender complete decomposition which is a prerequisite

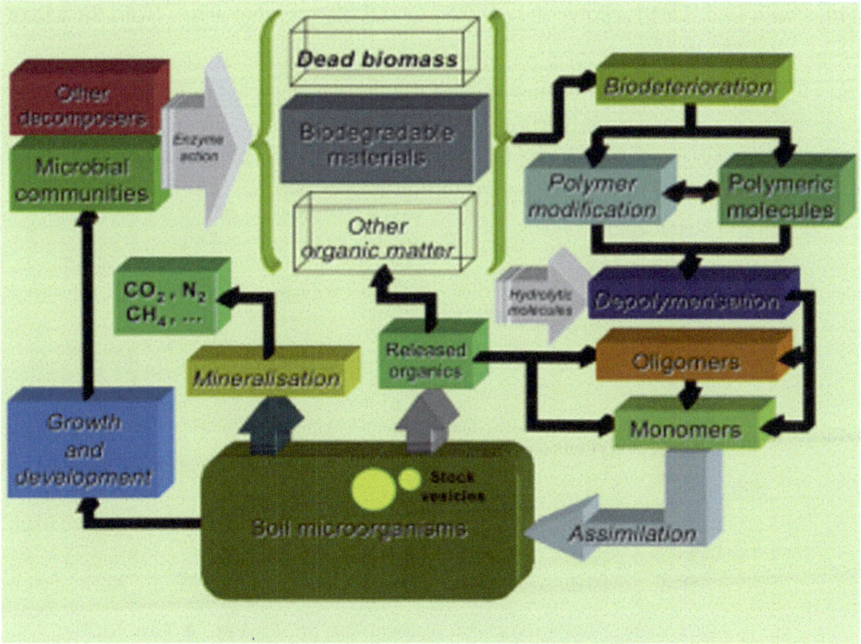

Fig. 1 A typical bioplastic biodegradation scheme. (European Bioplastics, 2019)

for mineralization to occur. According to Arutchelvi et al. (2008), physical and chemical degradation are separate from and have higher rates than biodegradation, with all major factors present. Mineralization occurs due to biodegradation, which is facilitated by the initial chemical and physical degradation of the biopolymers.

2.1 Production of Bioplastics

The demand for bioplastics is rising as new products and applications arise, giving hope for the bioplastic market to grow from its approximate 1% level in the global plastic market (European Bioplastics, 2019). With all things being equal, production is predicted to rise from 2019's 2.114 million tonnes to 2.426 million tonnes in 2024 (Fig. 2).

2.2 Production of Bioplastics from Agro-resources

The production of bioplastics from biopolymers sourced from agro-resources, especially starch and cellulose, is an increasing trend. The potential of starch as a major player in the production of eco-friendly and renewable plastics is huge. Great interest has been put into creating and examining the properties of biomaterials made with starch in order to resolve the environmental distress that arises from the use of

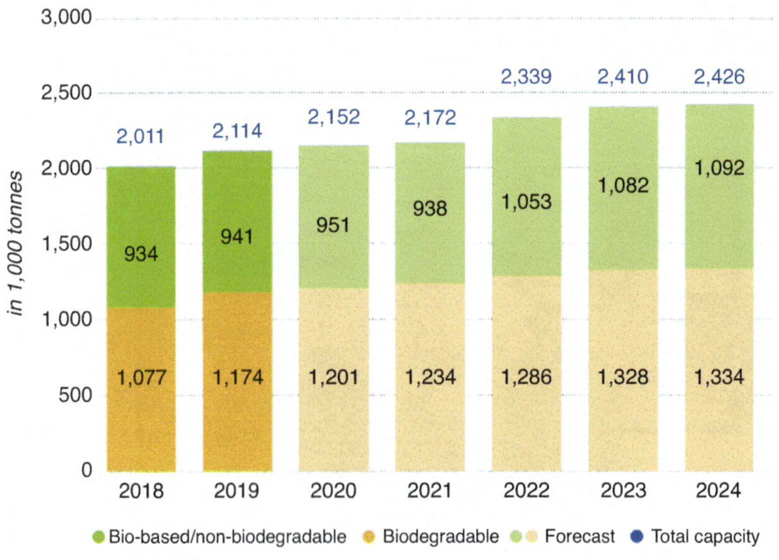

Fig. 2 Global production capacities of bioplastics. (European Bioplastics, 2019)

synthetic polymers (Mohan et al., 2016). A huge amount of research work has gone into upgrading the biodegradable polymers with the use of starch-based products. To this end, two states of starch – granular and thermoplastic – have been used in different studies. Granular starch can be either modified or of pure form, while thermoplastic starch is starch that has been plasticized (Afolalu et al., 2019a, b). Research has been made on polymer blends that are composed of starch and polymers (either biologically or synthetically derived), and their properties have been seen to be affected by a host of variable factors which are included but not limited to particle sizes of starch, the amylopectin and amylose contents, methods of processing, plasticizer, compatibilizer type, and content (Din et al., 2017). Properties of a blend can be varied by a mere adjustment of starch type and contents in a blend. The resultant polymer blends have been found to have pleasant characteristics, some of which are eco-friendliness and vast applicability. For example, in the medical field, the use of starch in items for the transmission of medicine has been seen to engender tremendous results in the hindrance of microbial activity (Karthik & Rathinamoorthy, 2018). Cellulose is a biodegradable and hydrophobic linear polysaccharide compound found in algae and plants. It has a melting point of 467 °C and contact angles ranging from 20 to 30 °C (Abioye et al., 2019; Krumm et al., 2016). While it is not dissolved by water and most organic solvents, it dissolves in solutions like cupriethylenediamine and Schweizer's reagent (Abioye et al., 2019). The matrix is made up of microfibrils which possess higher crystallinity than starch and high tensile strength, which in turn strengthens the matrices in which they are incorporated. The fibrils can be treated mechanically or by hydrolysis to achieve individualization into nanoparticles with lengths ranging from a few 100 nm to 1 μm (Afolalu et al., 2019a; Udo et al., 2018), which have many applications. This paper focuses on managing environmental issues that result from synthetic plastics through scaling up the production of bioplastics.

2.3 Biodegradable Plastics from Starches, Cellulose, and Their Blends

Several studies were reported on the synthesis of bioplastics from starch, cellulose, and other agricultural materials. Studies on blends of starch from agricultural crops such as *Zea mays*, *Oryza sativa*, and jackfruit have been made, and the increase in mechanical properties was observed (Afolalu et al., 2019b). Amalia et al. (2020) and Marichelvam et al. (2019) reported tensile strength of 11.7164 MPa and 12.5 MPa, respectively. Lower solubility and moisture content were also reported. Biodegradation experiments by exposure to biodegrading microorganisms were conducted as well, and both authors reported 78% and 48% biodegradation, respectively. Starch-cellulose composites synthesized from agricultural products have also been studied, and their blends have been shown to have higher mechanical properties than when used singly (Afolalu et al., 2018b, c, 2020). Agustin (2014) assessed the mechanical properties from cellulose nanocrystals obtained from rice straws and garlic straws in addition to corn starch. The study obtained better properties for rice straw nano-cellulose-filled composites (tensile strength = 26.0 MPa) than garlic stalk nano-cellulose-filled composites (tensile strength = 15.6 MPa) compared with the control sample (tensile strength = 10 MPa). It could also be predicted from the study that the source agricultural materials and the quantity of the synthesized material determine the properties of the bioplastics. Abdulahi et al. (2019) studied the effects of varying starch quantities on the properties on bioplastic sheets using native potato starch and plasticized with glycerol. The bioplastics with the highest quantity of starch to bioplastics (3.5:1) showed the highest tensile strength density and contact angle and the fastest enzymatic degradation by microorganisms (Amalia et al., 2020).

These studies have displayed that the utilization of bioplastics can be explored in different sectors and poses to be more economical and prospective in less environmental pollution over a long time. This is also dependent on the strategies deployed to manage its waste and its disposal (Marichelvam et al., 2019).

3 Applications and Market Sectors

The number of sectors in which bioplastics are applied is on the increase (Abdullah et al., 2019). Although they are currently used in automotive, consumer electronic, textile, and agriculture/horticulture industries, their largest application is the packaging sector. According to Babu et al. (2013), PHAs and their copolymers and blends are utilized majorly in items for biomedical implants like cardiovascular patches, repair patches, sutures and their fasteners, and bone plates. PLA fibres and blends are used in textile manufacture, bioresorbable scaffolds, bone support splints, and a lot more areas. PBS and its blends are used in packaging, pharmaceuticals, and medical delivery systems, while bio-based polyethylene is used in packaging, toys, and mulch films. Starches and their blends are used in implant devices, bone cement adhesives, and food applications, while celluloses and their compounds are applied in textiles, separation membranes, drug formulations, and wound dressings. These are just a few applications of biopolymers; Fig. 3 shows the industries for which biopolymers were produced in 2019 (Karthik & Rathinamoorthy, 2018; Santana et al., 2017).

3.1 Product Names of Bioplastics in the World Market

There are quite a number of bioplastics commercially available for various purposes and under different trade names. Bio-based polymer products include NatureFlex™ and Celuluzon® (cellulose polymers), NatureWorks® PLA and Lacty (polylactic acid), and Mirel™ and Nodax™ (polyhydroxyalkanoates) (Glaser, 2019; Siraccusa,

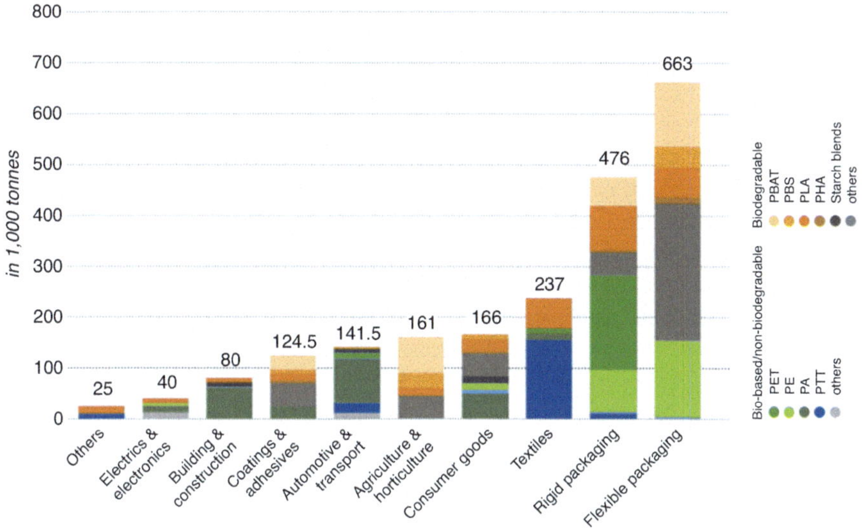

Fig. 3 Data showing the production and use of bioplastics in different sectors. (Karthik & Rathinamoorthy, 2018)

2019; Afolalu et al., 2018d; Abioye et al., 2018; Arutchelvi et al., 2008; Lucas et al., 2008; Kale et al., 2007; Briassoulis, 2005). Some synthetic biopolymers are present as Sorona® (polytetramethylene adipate-co-terephthalate), TONE™ and Placcel® (polycaprolactone), POVAL and Elvanol® (polyvinyl alcohol), and Ecoflex® and Eastar Bio™ (polybutylene adipate-co-terephthalate). Several polymers have come about from blending different bio-based polymers with other bio-based polymers or synthetically sourced polymers. Mater-Bi® is a product of starch and polycaprolactone blended at different levels, Ecovio® is a product of polybutylene adipate-co-terephthalate and polylactic acid, and Cereplast Compostables™ is a product of blending polylactic acid with other bio-based polymers (Afolalu et al., 2018d; Abioye et al., 2018; Arutchelvi et al., 2008).

According to European Bioplastics (Babu et al., 2013), polyhydroxyalkanoates and the 100% bio-based polypropylene are currently the best received and most demanded bioplastics in the market. This is grossly due to the properties they possess, which make them widely applied across sectors. It is expected that their rates of production increase exponentially in the coming years (Afolalu et al. 2015). All the non-biodegradable bio-based plastics constitute more than 44% of the global production of bioplastics (European Bioplastics, 2019). Although the capacities of the bio-based production of polyethylene terephthalate (PET) that were forecasted have not yet been achieved, a new polymer which is 100% bio-based, polyethylene furanoate (PEF), has taken the attention as its properties are commensurate with those of PET. The biodegradable plastics accounted for 55.5% of the total production with starch blends, polylactic acid, and polybutylene adipate terephthalate taking the lead (Chen, 2014).

4 Conclusion

Several studies, as reported in this study, showed that bioplastics have properties that have been researched at an adequate measure, most especially their biodegradability, and can be recommended globally as a sustainable product to be utilized. Furthermore, the rate of the production of bioplastics is on the increasing side annually, and their utilization in the market worldwide has revealed that their use is being embraced and accepted. This is expected to reduce environmental pollution in the aquatic and terrestrial environment. However, legislation should be enforced for adequate disposal, and more studies should be conducted on the recycling of bioplastics.

Acknowledgements We acknowledge the financial support offered by Covenant University in the actualization of this research work for publication.

References

Abdullah, A. H. D., Pudjirahar, S., Karina, M., Dwi Putri, O., & Fauziyyah, R. H. (2019). Fabrication and characterization of sweet potato starch-based bioplastics plasticized with glycerol. *Journal of Biological Sciences, 19*(1), 57–64.

Abioye, O. P., Abioye, A. A., Afolalu, S. A., Ongbali, S. O., & Akinlabi, S. A. (2018). A review of biodegradable plastics in Nigeria. *International Journal of Mechanical Engineering and Technology, 9*(10), 1172–1185.

Abioye, A. A., Obuekwe, C., Fasanmi, O., Oluwadare, O., Abioye, O. P., Afolalu, S. A., Akinlabi, S. A., & Bolu, C. A. (2019). Investigation of biodegradation speed and biodegradability of polyethylene and Manihot Esculenta starch blends. *Journal of Ecological Engineering, 20*(2), 65–72.

Afolalu, S. A., Adejuyigbe, S. B., Adetunji, O. R., & Olusola, O. I. (2015). Production of cutting tools from recycled steel with palm kernel shell as carbon additives. *International Journal of Innovation and Applied Studies, 12*(1), 110.

Afolalu, S. A., Orenuga, O., Samson Ongbali, O., Abioye, A. A., Okokpujie, P. I., & Olamilekan, R. O. (2018a). The study of the impact of nano carbon additives on ASTM A53 mild steel during machining. In *IOP conference series materials science and engineering* (Vol. 413, No. 2, p. 012028). IOP Publishing.

Afolalu, S. A., Abioye, A. A., Udo, M. O., Adetunji, O. R., Ikumapayi, O. M., & Adejuyigbe, S. B. (2018b). Data showing the effects of temperature and time variances on Nano-additives treatment of mild steel during machining. *Data in Brief, 19*, 2456–2461.

Afolalu, S. A., Samson Ongbali, O., Abioye, A. A., Udo, O. M., & Akintayo, C. T. (2018c). Experimental investigation of the effects of bi-nano additives on the mechanical properties of AISI 5130 mild steel during machining. *International Journal of Mechanical Engineering and Technology, 9*(12), 264–274.

Afolalu, S. A., Ongbali, S. O., Abioye, A. A., Udo, O. M., & Akintayo, C. T. (2018d). Experimental investigation of the effects of bi-nano additives on the mechanical properties of AISI 5130 mild steel during machining. *International Journal of Mechanical Engineering and Technology, 9*(12), 264–273.

Afolalu, S. A., Efekodha, G. E., Ongbali, S. O., Abioye, A. A., Salawu, E. Y., Ajayi, O. O., & Oluwabunmi, A. P. (2019a). Experimental analysis of the effect of tri-nano additives on wear rate of mild steel during machining. *Procedia Manufacturing, 35*, 395–400.

Afolalu, S. A., Oladipupo, S., Bose, M. E., Abioye, A. A., Adejuyigbe, S. B., Ajayi, O. O., & Ongbali, S. O. (2019b). Agro waste a sustainable source for steel reinforcement-review. In *Journal of Physics: Conference Series* (Vol. 1378, No. 3, p. 032032). IOP Publishing.

Afolalu, S. A., Samuel, O. D., & Ikumapayi, O. M. (2020). Development and characterization of nano-flux welding powder from calcined coconut shell ash admixture with FeO particles. *Journal of Materials Research and Technology, 9*(4), 9232–9241.

Agustin, M. B., Leon, E. R. P., De Buenaobra, J. L., Alonzo, S. M. M., Patriana, F. M., Hirose, F., & Ahmmad. (2014). Starch based bioplastics reinforced with cellulose nanocrystals from agricultural residues. In *International conference on advances in engineering and technology (ICAET'2014) March 29–30, 2014 Singapore* (pp. 593–597).

Amalia, D., Saleh, D., & Djonaedi, E. (2020). Synthesis of biodegradable plastics using corn starch and corn husk as the fillers as well as chitosan and sorbitol. *Journal of Physics: Conference Series, 1442*, 012007.

Andrady, A. L., & Neal, M. A. (2009). Applications and societal benefits of plastics. *Philosophical Transactions of the Royal Society, B: Biological Sciences, 364*(1526), 1977.

Arutchelvi, J., Sudhakar, M., Arkatkar, A., Doble, M., Bhaduri, S., & Uppara, P. V. (2008). Biodegradation of polyethylene and polypropylene. *Indian Journal of Biotechnology, 7*(February), 9–22.

Babu, R. P., O'Connor, K., & Seeram, R. (2013). Current progress on bio-based polymers and their future trends. *Progress in Biomaterials, 2*(8), 1–16.

Briassoulis, D. (2005). The effects of tensile stress and the agrochemical Vapam on the ageing of low density polyethylene (LDPE) agricultural films. Part I. Mechanical behaviour. *Polymer Degradation and Stability, 88*(3), 489–503.

Chen, Y. J. (2014). Bioplastics and their role in achieving global sustainability. *Journal of Chemical and Pharmaceutical Research, 6*(1), 226–231.

DeArmitt, C. (2011). Applied plastics engineering handbook. In *Applied plastics engineering handbook* (Issue January 2011).

Díez-Pascual, A. M. (2019). Synthesis and applications of biopolymer composites. *International Journal of Molecular Sciences, 20*(9), 2321.

Din, Z., Xiong, Z. H., & Fei, P. (2017). Physical and chemical modification of starches: A review. *Critical Reviews in Food Science and Nutrition, 57*(12), 2691–2705.

Emadian, S. M., Onay, T. T., & Demirel, B. (2017). Biodegradation of bioplastics in natural environments. *Waste Management, 59*, 526–536.

European Bioplastics. (2019). *Bioplastics market data 2019.* https://docs.european-bioplastics.org/publications/market_data/Report_Bioplastics_Market_Data_2019.pdf

Glaser, J. A. (2019). Biological degradation of polymers in the environment. In *Plastics in the environment: Vol. i* (p. 13). IntechOpen.

Kale, G., Kijchavengkul, T., Auras, R., Rubino, M., Selke, S. E., & Singh, S. P. (2007). Compostability of bioplastic packaging materials: An overview. *Macromolecular Bioscience, 7*(3), 255–277.

Karthik, T., & Rathinamoorthy, R. (2018). Sustainable biopolymers in textiles: An overview. In *Handbook of ecomaterials* (pp. 1–27, 53–1). Springer.

Krumm, C., Pfaendtner, J., & Dauenhauer, P. J. (2016). Millisecond pulsed films unify the mechanisms of cellulose fragmentation. *Chemistry of Materials, 28*(9), 3108–3114.

Lucas, N., Bienaime, C., Belloy, C., Queneudec, M., Silvestre, F., & Nava-Saucedo, J. E. (2008). Polymer biodegradation: Mechanisms and estimation techniques – A review. *Chemosphere, 73*(4), 429–442.

Marichelvam, M. K., Jawaid, M., & Asim, M. (2019). Corn and rice starch-based bio-plastics as alternative packaging materials. *Fibres, 7*(4), 32.

Mohan, S., Oluwafemi, O. S., Kalarikkal, N., Thomas, S., & Songca, S. P. (2016). Biopolymers – Application in nanoscience and nanotechnology. In *Recent advances in biopolymers* (Issue March 2018, pp. 47–72).

Peng, B. L., Dhar, N., Liu, H. L., & Tam, K. C. (2011). Chemistry and applications of nanocrystalline cellulose and its derivatives: A nanotechnology perspective. *The Canadian Journal of Chemical Engineering, 89*(5), 1191–1206.

Saito, T., & Isogai, A. (2004). TEMPO-mediated oxidation of native cellulose. The effect of oxidation conditions on chemical and crystal structures of the water-insoluble fractions. *Biomacromolecules, 5*(5), 1983–1989.

Santana, R. F., Bonomo, R. C. F., Gandolfi, O. R. R., Rodrigues, L. B., Santos, L. S., dos Santos Pires, A. C., de Oliveira, C. P., da Costa Ilhéu Fontan, R., & Veloso, C. M. (2017). Characterization of starch-based bioplastics from jackfruit seed plasticized with glycerol. *Journal of Food Science and Technology, 55*(1), 278–286.

Siracusa, V. (2019). Microbial degradation of synthetic biopolymers waste. *Polymers, 11*(6), 1066.

Tokiwa, Y., Calabia, B., Ugwu, C., & Aiba, S. (2009). Biodegradability of plastics. *International Journal of Molecular Sciences, 10*(9), 3722–3742.

Udo, M., Esezobor, D., Afolalu, S., Onovo, H., Ongbali, S., & Okokpuji, I. (2018). Investigation of balling characteristics of mixture of iron oxide bearing wastes and iron ore concentrates. In *IOP conference series materials science and engineering* (Vol. 413, No. 2, p. 012042). IOP Publishing.

Vert, M., Doi, Y., Hellwich, K. H., Hess, M., Hodge, P., Kubisa, P., Rinaudo, M., & Schué, F. (2012). Terminology for biorelated polymers and applications (IUPAC recommendation 2012). *Pure and Applied Chemistry, 84*(2), 377–410.

A Short Review on Land/Soil Pollution: The Pollutants and the Treatment Techniques

F. B. Elehinafe, O. G. Olomukoro, A. O. Ayeni, and O. B. Okedere

1 Introduction

Environmental pollution through land, water and air are a result of human-induced activities (Hou & Al-Tabbaa, 2014). Numerous pollutants are released into the environment by various human activities. Some of these pollutants are particulate matters, heavy metals, polycyclic aromatic hydrocarbons, persistent organic pollutants, volatile organic compounds, etc. Particulate matters are produced mainly through the burning of organic materials, manufacturing and road construction. Harmful elements contained in particulate matter may include calcium, sulphur, aluminium, silicon, lead, zinc, nickel, potassium, chromium, copper, etc. When in the human body, these substances cause health problems. Soil pollutants that contain chlorinated compounds are high in toxicity (Hou & Li, 2017).

Land is a valuable resource and vital factor of production. It offers support to human activities. Land pollution has become a worldwide concern, particularly in developing nations. It is even more common in countries undergoing rapid industrialization (Hou & Al-Tabbaa, 2014; Elehinafe et al., 2020). The highest form of land pollution is through high population density and urbanization. Land can be contaminated through the release of sludge, sewage, pesticides and synthetic organic

F. B. Elehinafe (✉) · O. G. Olomukoro · A. O. Ayeni
Department of Chemical Engineering, College of Engineering, Covenant University, Ota, Ogun State, Nigeria
e-mail: francis.elehinafe@covenantuniversity.edu.ng, augustine.ayeni@covenantuniversity.edu.ng

O. B. Okedere
Department of Chemical Engineering, College of Engineering, Osun State University, Osogbo, Nigeria
e-mail: oyetunji.okedere@uniosun.edu.ng

A. O. Ayeni et al. (eds.), *Advanced Manufacturing in Biological, Petroleum, and Nanotechnology Processing*, Green Energy and Technology,
https://doi.org/10.1007/978-3-030-95820-6_23

chemicals (Wuana & Okieimen, 2011). Millions of hectares of agricultural lands in Europe are polluted. Mining activities have led to a decline in the quality of land in China (Lestan et al., 2008).

Soil is a component of land where agricultural activities are carried out. Soil pollution occurs when the soil contains contaminants (harmful chemicals) in concentrations that are significant enough to pose a threat to man and environment. The emission of heavy metals to the atmosphere in the form of aerosols is later deposited on the ground and hence contaminates the soil. The rate of pollution in soils through heavy metals relies on properties such as organic matter, grain size and mineralogy of the soil (Soriano et al., 2012). Man is susceptible to dangerous contaminants through the consumption of food crop grown on contaminated farmlands or soils.

The causes of soil pollution include industrial wastes (toxic chemicals and gases), insecticides, pesticides, harmful irrigation activities, sewage leakages and acid rain production through mixture of industrial smoke with rain. The effects of polluted soil result in the following: a reduction in yield of the affected soil, soil erosion, disruption in the ecosystem of the soil and high soil salinity (Uddin et al., 2017).

The review identifies the different causes of land pollution, category of land pollutants and separation techniques applied in the control of land pollution, defines gaps in literature and gives recommendations for the development of novel separation techniques for land pollution analysis and control. This will give a convincing knowledge on whether the present separation techniques are adequate in the analysis and control of land pollution.

2 Methodology

This work was put together through scientific literature review procedures. Some electronic sources for scientific peer-reviewed journals were explored. They include ScienceDirect, Springer, Scopus and Web of Science. The terminologies used during the search were 'separation techniques and land pollution', 'land pollutants and their effects', 'separation methods in soils' and 'different land contaminants'. Over 20 journals/articles relating to soil pollutants/contaminants and removal techniques were obtained using the above search terms from different scientific data base. Information were also retrieved from the United States Environmental Protection Agency (US EPA). The search was limited to English-based peer-reviewed papers.

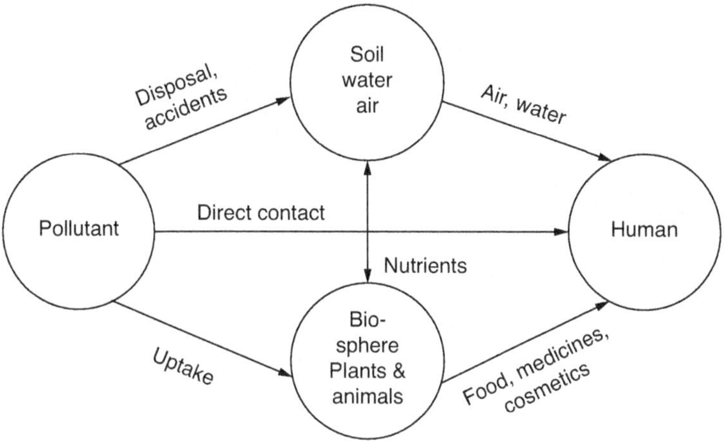

Fig. 1 Movement of pollutants through direct and indirect routes. (Source: Mohamed & Antia, 1998)

3 Land/Soil Pollutants

Pollutants have affected the different components of the ecosystems such as the biosphere, hydrosphere, atmosphere and lithosphere. The motions of these pollutants (gaseous, liquid and solid pollutants) through the ecosystems have a great influence on its bioavailability (Olukanni & Adeoye, 2012). The pollutant enters through any of the ecosystem and is distributed quickly especially when there is high phase mobility, such as the atmosphere. There are several chemical products manufactured from a wide range of industries such as medicine, energy and agriculture, among others, which are utilized by man on a constant basis. Most of these manufactured chemical products are toxic to the environment, humans and the ecosystems. The movement of pollutants through the ecosystem and eventually to man is shown in Fig. 1 (Mohamed & Antia, 1998).

Soils are components of the ecosystem which vary in characteristics, constituents and structure (Mohamed & Paleologos, 2018). They are distinguished by particle size scattering. Biomass is formed in soils through the decomposition of plants and organic matter. Soil comprises water, solid matter, air and other organisms, all interacting together. The composition of sulphides, silicates, organic matter and carbonates will determine the complexity of the chemical reactions occurring in the soils (Mohamed & Paleologos, 2018). Clay and silt are made up of very fine particles of soil, which in most cases possess the largest amount of pollutants. These fine particles are made up of natural organic constituent of soil and, hence, take in organic pollutants. Fine particles contain the largest surface area per unit volume. Coarse soil particles are distinguished mainly by their surface roughness which is favoured by the formation of oxides, disintegration or breakdown and precipitation of salt. This coarseness of soil particles enables the penetration of pollutants. Stones and pebbles are very coarse particles that contain a low surface area which makes them low in porosity to pollutants. Most polluted soils contain a mixture of fine, coarse and natural organic constituents (Gräfe et al., 2014).

4 Separation Techniques Applied in the Control of Land/Soil Pollution

The separation of contaminants from soil is through remediation methods applied in the treatment of soils. Remediation methods in treating polluted soils are classified according to the United States Environmental Protection Agency (USEPA, 1997) into source control and containment remedies. Source control consists of in situ and ex situ treatment techniques. For in situ technique, the polluted soil is treated without being excavated or removed, i.e. the soil is in its original location. Ex situ technique refers to a polluted soil that is excavated or removed from its original location for treatment. Containment remedies refer to erection of barriers or walls to divert the flow of contaminated particles (Wuana & Okieimen, 2011). In situ treatment methods include electrokinetics, soil flushing, stabilization and phytoremediation (USEPA, 1997).

4.1 Physical Separation

Mechanical separation is a size selection process that separates smaller contaminated particles from clean bigger particles. The size of the particle and contaminant rate are required to characterize this method of separation (Filgueiras et al., 2002). Small-scale tests are carried out to analyse different separation methods which include hydrocyclones, fluidized bed, flotation and gravimetric settling techniques. Hydrocyclones utilize centrifugal force in separating larger particles (>10–20 μm) from smaller ones. Fluidized bed separation removes lighter particles (<50 μm) from the top in a countercurrent manner by flotation and gravimetric settling techniques which are being utilized in the processing of mineral ore. A frothing agent is used in the froth flotation process whereby contaminants are made to float. Magnetic separation through the magnetic properties of metals separates ferrous materials (Mulligan et al., 2001a, b).

4.2 Pyrometallurgical Separation

In this method of separation, soil contaminants (metals) are made to evaporate (volatilization) at temperatures between 200 and 700 °C. The metals can be recovered after the volatilization process (Filgueiras et al., 2002). This separation technique applies to mercury as it can be simply transformed to its metal at elevated temperatures. Other metals such as cadmium, lead, chromium and arsenic can also be recovered from the soil. It is an ex situ separation method employed in soils with a high amount of contaminants. Soil washing method is carried out before the pyrometallurgical separation is carried out. Equipment utilized in carrying out this separation process include arc furnace or rotary kilns. This equipment successfully extracts heavy metals alongside waste matter known as slag (Mulligan et al., 2001a, b).

4.3 Soil Washing

This is a method of separating metal pollutants from soils and sediments through chemical, physical or biological processes (Dermont et al., 2010). This process separates pollutants from the soil matrix by washing them in a solution (e.g. alcohols, bases, acids and chelating agents) in which they are soluble. Soil washing is an ex situ process. Soils contaminated with metals and oil can be treated by the use of surfactants. These surfactants are added to washing water to release, scatter and solubilize the contaminants in the excavated soil (Mulligan et al., 1999).

4.4 Electrokinetic Separation

This method involves the passage of direct current through electrodes (anode and cathode) mounted in the soil. The pollutants migrate as ionic particles. In this method, the movement of pollutants is directed towards the centre of the polluted region of the soil (USEPA, 1997). According to Mulligan et al. (2001a, b), during the electrokinetic process, there is charge difference (electric gradient) which causes movement of the charged particles (pollutants) and electrolysis reaction. The metals are extracted through surface pumping, electrode precipitation or ion exchange resins (Smith et al., 1995). Electrokinetics is an in situ process but can also be applied in excavated soil (Mulligan et al., 2001a, b).

4.5 Froth Flotation

Froth flotation is a chemical method in the removal of metal contaminants from the solid particles of the soil. It is a soil washing process that is applied in mining industry for the processing of minerals, wastewater treatment and paper recycling. In the removal of metals from polluted soils, froth flotation works by compressing the size of the metal particles. It is applicable to polluted soils with particle size between 10 and 300 μm (Dermont et al., 2010). A combined study carried out by Vanthuyne and Maes (2002) revealed the separation of metals from soils and sediments. In these studies, froth flotation method was applicable for the treatment of fine particles of soil. Froth flotation operates by two major particle movement mechanisms, namely, the true flotation and the mechanical entrainment mechanism. The true flotation deals with characteristics such as chemical reactivity, mineralogy and hydrophobicity. This mechanism involves a bond between a bubble and particle and then a desorption process. The mechanism entrainment deals with the spread of elements in the particle, soil texture and hydrodynamic flotation (Dermont et al., 2010).

4.6 The Use of Chelating Agents

A chelant is a chemical compound (ligand) that forms bond with metal ions. It binds tightly to the metallic ions since they possess electron donor groups (Zhu et al., 2013). Chelating agents are used as soil remediation measures for the removal (extraction) of toxic metals from contaminated soils. These agents extract contaminants through soil flushing, phytoextraction and chelant-enhanced electrokinetic extraction (Lestan et al., 2008). Chelating agents and acids are usually applied in extracting heavy metals from polluted soils. Ethylenediaminetetraacetic acid (EDTA) is a well-known chelating agent that forms complexes with different metallic ions. Other chelating agents include citranox, gluconate, ammonium acetate and oxalate. According to Wuana and Okieimen (2011), metals in soils are set in motion through the generation of complexes, soil acidity alteration and alteration in the concentration of ions in solution. As the components of the polluted soil gradually change (age), the metals trapped in the soil will decline in mobility due to the formation of more complexes (Zhu et al., 2013). The agents are divided into biodegradable and non-biodegradable.

The earlier mentioned chelating agents are not easily biodegradable and can result in the leaching of heavy metal. There is a need for biodegradable chelating agents such as organic phosphonic acid and amino carboxylic acid chelators for an effective phytoremediation in soils. Chelating agents are categorized into natural and synthetic chelating agents (Shuang et al., 2021).

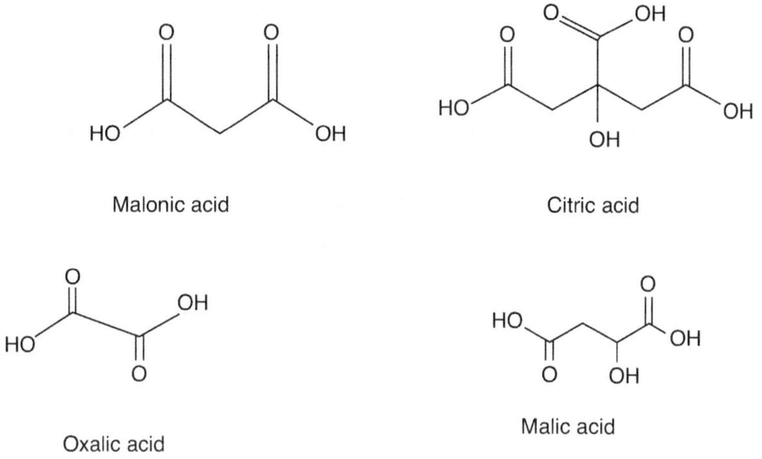

Fig. 2 Chemical structures of natural chelating agents. (Source: Shuang et al., 2021)

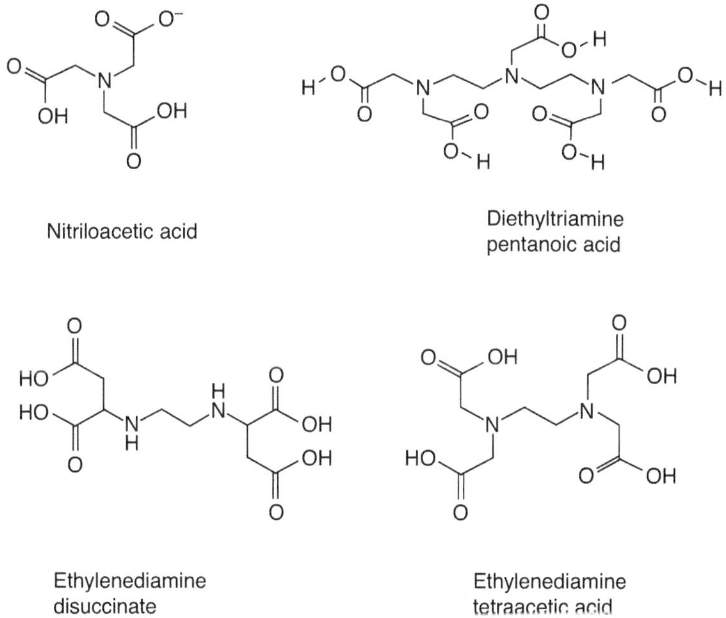

Fig. 3 Chemical structures of synthetic chelating agents. (Source: Shuang et al., 2021)

4.6.1 Natural Chelating Agents

They are low molecular weight chelating agents that are easily degradable. They are not toxic to the ecosystem. They readily dissolve heavy metals, hence enabling their mobility. Examples include organic acids such as malonic acid, citric acid, oxalic acid, malic acid, etc. The structures of these acids are shown in Fig. 2 (Shuang et al., 2021).

4.6.2 Artificial/Synthetic Chelating Agents

These chelating agents can remove heavy metals from the soil by reacting with the metals to produce complexes. Synthetic chelants have low biodegradable ability and are quite expensive (Shuang et al., 2021). They are liable to cause contamination in underground water and soils. Examples include nitrilotriacetic acid (NTA), diethylenetriaminepentaacetic acid (DTPA), ethylenediamine disuccinate (EDDS), ethylenediaminetetraacetic acid (EDTA), etc. Some of the structures of synthetic chelants are shown in Fig. 3.

5 Conclusion

Land/soil contamination is a global concern, and the ways of controlling the spread of these pollutants are reviewed in this article. The selection of the different separation technique may be peculiar to location. A proper evaluation of soil pH, porosity and presence of moisture are usually conducted to determine the most relevant technique to the target soil. It is recommended that more studies should be carried out to separate contaminants from land/soil as to obtain sustainable use of soil and more research is required in site hydrology so as to hinder the mobility of pollutants to unwanted locations.

Acknowledgement The authors would like to thank the Covenant University Centre for Research, Innovation and Discovery (CUCRID), Ota, Nigeria, for its support in making the publication of this research possible.

References

Dermont, G., Bergeron, M., Richer-Lafleche, M., & Mercier, G. (2010). Remediation of metal-contaminated urban soil using flotation technique. *The Science of the Total Environment, 408*, 1199–1211.

Elehinafe, F. B., Mamudu, A. O., Okedere, O. B., & Ibitioye, A. (2020). Risk assessment of chromium and cadmium emissions from the consumption of premium motor spirit (PMS) and automotive gas oil (AGO) in Nigeria. *Heliyon*. https://doi.org/10.1016/j.heliyon.2020.e05301

Filgueiras, A. V., Lavilla, I., & Bendicho, C. (2002). Chemical sequential extraction for metal partitioning environmental solid samples. *Journal of Environmental Monitoring, 4*, 823–857.

Gräfe, M., Donner, E., Collins, R. N., & Lombi, E. (2014). Speciation of metalloids in environmental samples by X-ray absorption spectroscopy: A critical review. *Analytica Chimica Acta, 822*, 1–22.

Hou, D., & Al-Tabbaa, A. (2014). Sustainability: A new imperative in contaminated land remediation. *Environmental Science & Policy, 39*, 25–34.
Hou, D., & Li, F. (2017). Complexities surrounding China's soil action plan. *Land Degradation and Development, 28*, 2315–2320.
Lestan, D., Luo, C., & Li, X. (2008). The use of chelating agents in the remediation of metal-contaminated soils: A review. *Environmental Pollution, 153*, 3–13.
Mohamed, A. M. O., & Antia, H. (1998). *Geoenvironmental engineering* (1st ed., p. 82).
Mohamed, A. M. O., & Paleologos, E. K. (2018). Fate and effects of pollutants on the land environment. *All Works, 903*.
Mulligan, C. N., Yong, R. N., & Gibbs, B. F. (1999). On the use of biosurfactants for removal of heavy metals from oil contaminated soil. *Process Safety Progress, 18*(1), 50–54.
Mulligan, C. N., Yong, R. N., & Gibbs, B. F. (2001a). Heavy metal removal from sediments by biosurfactants. *Journal of Hazardous Materials, 85*, 111–125.
Mulligan, C. N., Yong, R. N., & Gibbs, B. F. (2001b). Remediation technologies for metal-contaminated soils and groundwater: An evaluation. *Engineering Geology, 60*, 193–207.
Olukanni, D. O., & Adeoye, D. O. (2012). Heavy metal concentrations in road side soils from selected locations in the Lagos Metropolis, Nigeria. *International Journal of Engineering and Technology, 2*(10), 1743–1752.
Shuang, C., Qing, H., & Tianyi, Z. (2021). Overview of leaching remediation of heavy metal contamination in soil. *Web of Conferences, 245*, 1–4.
Smith, L. A., Means, J. L., Chen, A., Alleman, B., Chapma, C. C., Tixier, J. S., Brauning, S. E., Gavaskar, A. R., & Royer, M. D. (1995). *Remedial option for metals-contaminated sites*. Lewis.
Soriano, A., Pallares, S., Pardo, F., Vicente, A. B., Sanfeliu, T., & Bech, J. (2012). Deposition of heavy metals from particulate settleable matter in soils of an industrialised area. *Journal of Geochemical Exploration, 113*, 36–44.
U.S. Environmental Protection Agency. (1997). *Recent developments for in situ treatment of metal contaminated soils*. USEPA.
Uddin, M. J., Sagar, G. A., & Jagdeeshwar, J. (2017). Soil pollution and soil remediation techniques. *International Journal of Advance Research, Ideas and Innovations in Technology, 3*(1), 582–593.
Vanthuyne, M., & Maes, A. (2002). The removal of heavy metals from contaminated soil by a combination of sulphidisation and flotation. *Science of the Total Environment, 290*, 69–80.
Wuana, R. A., & Okieimen, F. E. (2011). Heavy metals in contaminated soils: A review of sources, chemistry, risks and best available strategies for remediation. *International Scholarly Research Network Ecology, 2011*, 1–21.
Zhu, J., Pigna, M., Cozzolino, V., Caporale, A. G., & Violante, A. (2013). Higher sorption of arsenate versus arsenite on amorphous Al-oxide, effect of ligands. *Environmental Chemistry Letters, 2013*(11), 289–294.

The Effect of Air-Fuel Ratio on Tailpipe Exhaust Emission Measurement of Motorcars

O. A. Odunlami, F. A. Akeredolu, K. O. Oderinde, A. A. Ayoola, A. A. Busari, and V. E. Efeovbokhan

1 Introduction

On-road vehicles such as bikes, cars and trucks are the major source of emissions. Traffic emissions contribute majorly to atmospheric pollution. Among other compounds, CO, HC, and CO_2 are mostly present in vehicular emissions. Studies have shown that exposure to CO and HC has adverse health effects on the respiratory organs which can increase mortality and morbidity (Chen et al., 2017; Dockery et al., 1993). Studies have shown that 6.7% of deaths worldwide is from ambient air pollution (Chowdhury et al., 2020; Mannucci et al., 2015; Song et al., 2019). Economic development has increased energy demand, and the burning of fossil fuels has brought about large amounts of CO, CO_2 and HC emissions. Statistics have shown that energy consumption due to transportation needs in the cities had doubled in last 30 years and by 2050 will increase much more because the population of people in cities would have grown by nearly 70% of the present

O. A. Odunlami (✉) · A. A. Ayoola · V. E. Efeovbokhan (✉)
Department of Chemical Engineering, Covenant University, Ota, Ogun State, Nigeria
e-mail: olayemi.odunlami@covenantuniversity.edu.ng;
ayodeji.ayoola@covenantuniversity.edu.ng;
vincent.efeovbokhan@covenantuniversity.edu.ng

F. A. Akeredolu
Department of Chemical Engineering, Obafemi Awolowo University, Ile Ife, Osun State, Nigeria

K. O. Oderinde
Department of Chemistry, Lead City University, Ibadan, Oyo State, Nigeria

A. A. Busari
Department of Civil Engineering, University of Oye Ekiti, Oye-Ekiti, Ekiti State, Nigeria
e-mail: ayobami.busari@covenantuniversity.edu.ng

A. O. Ayeni et al. (eds.), *Advanced Manufacturing in Biological, Petroleum, and Nanotechnology Processing*, Green Energy and Technology,
https://doi.org/10.1007/978-3-030-95820-6_24

value (Dulac, 2012). In developing countries, outdoor air pollution is a major problem that requires an urgent attention because of daily increase in the number of vehicles (Odunlami et al., 2018a, b). Vehicular emission standards are technical policy tools for the alleviation of pollutants from vehicles. Recent studies show that as much as 45% of the pollutants released in the United States are from vehicular emissions (Cheng et al., 2013; Bektaş & Laporte, 2011). Many of the developed nations have expended so much resources on science, technology and regulations to address automotive emissions (Williams et al., 2012; Haines et al., 2009).

2 Air-Fuel Ratio Lambda (λ)

The air-fuel ratio is an indicator of the combustibility of a mixture; it determines the amount of energy released and the pollutants produced from the vehicles. Combustion takes place in the cylinders of the internal combustion gasoline engines (Jia et al., 2016; Manente et al., 2010). Heat energy from combustion process is converted to mechanical work (Al-Himyari et al., 2014). An engine needs air, fuel and spark to make power. To improve the performance of the engine, there is a need for the appropriate mixture of air and fuel. Fuel mixes with air for combustion to take place. AFR shows the right proportions of air and fuel to be mixed. For example, a 14.7:1 AFR shows mixture of 14.7 parts air to one part fuel. This is the stoichiometric or ideal ratio for motor cars with gasoline engine. There is a complete combustion for an ideal AFR. AFR is the ratio of the mass of air to mass of fuel in gasoline combustion process; it is an indicator for controlling the performance of the engine. AFR is a very effective measure for the control of fuel in the combustion chamber; it therefore reduces vehicular emissions more than any of the engine control factors like spark timing, fuel injection timing, engine torque, air intake and engine speed (Williams et al., 2012; Haines et al., 2009; Jia et al., 2016). AFR numbers less than stoichiometric are called the rich mixtures, while AFR numbers greater than stoichiometric are called lean mixtures. A lambda value less than 1 gives rich mixture, while lambda value greater than 1 gives a lean mixture. When more air (oxygen) is present than necessary to combust the fuel, resulting in leftover oxygen, it is called a lean mixture. When more than the required fuel is used for combustion, it results into an incomplete combustion leaving unburned fuel in the chamber; this is known as a rich mixture. This increases the power of the engine, and reduces the fuel economy therfore, in the rich mixture, an amount of fuel is leftover after the combustion process leading to an increase in the concentration of emissions of CO and hydrocarbon in the ambient air. The rich mixture can also result into overheating of the internal combustion engine. In the lean mixture, oxygen is in excess resulting into an increase in fuel efficiency.

3 Methodology

The exhaust emission tests were carried out in Abeokuta, Ogun State, Nigeria. Hand-held, Kane automotive 4-gas analyser with detector tube (Model Auto 4-1) was used to measure the vehicular emissions. It measures four exhaust gases and calculates air-fuel ratio and lambda. The machine could measure CO (a resolution of 0.01% with an accuracy of ±0.5% volume), HC (a range of 1.0 ppm), CO_2 (a resolution of 0.1% with an accuracy of ±0.5% volume) and O_2 (a resolution of 0.1% with an accuracy of ±0.5% volume). The analyser could be used on petrol, LPG or CNG powered engines. All measured and calculated parameters can be printed on the optional infrared printer or saved to the analyser's memory. The analyser was switched on and set within the allowable range (especially oxygen, O_2, which should be between 20.9% and 21%). For petrol engine, the engine speed was raised to 2500 rpm. This was maintained for 30 seconds to warm and precondition the engine. Then the engine was returned to idle by taking the foot off the accelerator. The exhaust was observed for a while to ensure the steadiness of the smoke, and then the analyser's probe was fully inserted into the exhaust pipe and clamped. The concentration of CO was recorded at the maximum level. The concentrations of the tail pipe pollutants measured were reported in parts per million (ppm) or per cent (%) (where 1% = 10,000 ppm). Raw exhaust samples were taken from the tested vehicles.

4 Results and Discussion

The results of this study give the relationship between air-fuel ratio, lambda (λ) and the gasoline-fuelled vehicle exhaust emission concentrations, relating the actual field measurements with theoretical relation between air-fuel ratio, lambda (λ) and the gasoline-fuelled vehicle exhaust emission concentrations. It was observed from the results on Tables 1, 2, 3, 4, and 5 that about 39% of the sampled vehicles were within the standard value for lambda by the catalytic technology for exhaust gas emission, which is 1 (±5%). This showed that most of the sampled vehicles had the equivalence air-fuel ratio or lambda (λ) to be more than the expected value. This could result into the emission of very high concentrations of carbon monoxide (CO), hydrocarbon (HC) and carbon dioxide (CO_2) and can be attributed to bad maintenance culture of most Nigerian drivers and the absence of major emission control devices in many of fairly used vehicles regularly imported into the country. Regular maintenance of vehicles is needed to regulate the air-fuel ratio, thereby reducing the exhaust emission. Poorly maintained vehicles consume more fuel and emit higher levels of CO and VOC than those that are regularly serviced. Most commercial drivers use substandard engine oils to lubricate their engines in addition to postponing vehicle maintenance to maximise profits.

Tables 1, 2, 3, 4, and 5 show the pollutants from vehicular emissions, AFR and the lambda values from petrol engine (Table 6).

Table 1 Vehicular emission data, AFR and lambda values

Fuel type	Vehicle use	CO%	CO_2%	O_2%	HC (ppm)	AFR	Lambda (λ)
Petrol	Private	0.01	14.5	0.19	59	14.847	1.01
Petrol	Private	0.02	13.7	1.54	1	15.847	1.07
Petrol	Private	0.02	14.4	0.81	11	15.273	1.03
Petrol	Commercial	0.03	14.9	0.27	99	14.818	1.01
Petrol	Private	0.03	7.7	10.36	82	28.503	1.93
Petrol	Private	0.03	13.5	1.28	6	15.685	1.06
Petrol	Commercial	0.04	13.72	2.32	67	16.464	1.12
Petrol	Private	0.05	11.1	3.58	165	17.934	1.22
Petrol	Private	0.05	10.6	6.96	38	21.477	1.46
Petrol	Commercial	0.05	15.2	0.56	157	14.965	1.02
Petrol	Commercial	0.06	14.8	0.33	98	14.847	1.01
Petrol	Commercial	0.07	14.3	1.06	11	15.45	1.05
Petrol	Private	0.07	12.6	4.24	103	18.037	1.23
Petrol	Private	0.07	14.2	1.17	88	16.109	1.10
Petrol	Commercial	0.08	14.9	0.3	59	14.876	1.01
Petrol	Commercial	0.08	14.1	0.66	98	15.112	1.03
Petrol	Private	0.08	9.1	8.56	288	23.825	1.96
Petrol	Official	0.09	14.6	0.17	1	14.347	0.98
Petrol	Private	0.09	7.9	10.93	233	28.415	1.93
Petrol	Private	0.09	12	4.53	19	18.507	1.26

Table 2 Vehicular emission data, AFR and lambda values

Fuel type	Vehicle use	CO%	CO_2%	O_2%	HC (ppm)	AFR	Lambda (λ)
Petrol	Commercial	0.10	13.1	2.61	104	16.67	1.13
Petrol	Official	0.10	14.4	0.49	5	15.023	1.02
Petrol	Private	0.11	5.9	4.62	98	22.388	1.52
Petrol	Private	0.11	15	0.18	150	14.70	1.00
Petrol	Commercial	0.12	14.2	1.19	47	15.415	1.05
Petrol	Private	0.12	13.5	2.04	263	16.082	1.09
Petrol	Private	0.12	15	0.77	1	15.185	1.03
Petrol	Commercial	0.13	13	2.81	308	16.64	1.13
Petrol	Official	0.13	14.5	0.2	98	14.788	1.01
Petrol	Private	0.14	10.7	0.05	90	14.627	1.00
Petrol	Private	0.14	9.7	6.26	446	20.845	1.42
Petrol	Private	0.15	13.9	1.4	279	15.479	1.05
Petrol	Commercial	0.15	13.5	1.6	505	15.523	1.06
Petrol	Private	0.16	13.2	3.76	300	17.346	1.18
Petrol	Private	0.16	10.4	6.6	354	20.668	1.41
Petrol	Private	0.17	11.8	3.98	180	17.963	1.22
Petrol	Private	0.17	13.3	3.03	94	16.905	1.15
Petrol	Private	0.18	8.8	8.91	553	24.02	1.63
Petrol	Private	0.18	12.3	0.76	201	13.789	0.94
Petrol	Private	0.19	14.5	0.18	116	14.729	1.00

Table 3 Vehicular emission data, AFR and lambda values

Fuel type	Vehicle use	CO%	CO_2%	O_2%	HC (ppm)	AFR	Lambda (λ)
Petrol	Private	0.20	14.5	0.2	29	14.759	1.00
Petrol	Official	0.21	12.9	2.98	378	16.714	1.14
Petrol	Official	0.22	13.8	1.8	155	15.876	1.08
Petrol	Private	0.23	14.3	0.6	10	15.068	1.03
Petrol	Commercial	0.24	10.8	5.48	200	19.566	1.33
Petrol	Official	0.24	11.2	2.23	195	16.435	1.12
Petrol	Private	0.25	11.3	5.44	212	19.213	1.31
Petrol	Private	0.25	13.7	4.51	92	17.875	1.22
Petrol	Private	0.26	14.4	0.14	1	14.715	1.00
Petrol	Official	0.27	13.1	2.93	487	16.493	1.12
Petrol	Commercial	0.28	12.5	2.7	322	16.538	1.13
Petrol	Private	0.29	13.3	1.49	282	15.553	1.06
Petrol	Commercial	0.3	11.3	5.23	492	18.743	1.28
Petrol	Official	0.3	10.7	5.82	86	19.933	1.36
Petrol	Official	0.31	13.7	0.81	108	15.097	1.03
Petrol	Private	0.32	13.5	2.15	147	16.097	1.10
Petrol	Private	0.32	14	1.52	44	15.626	1.06
Petrol	Private	0.33	13.3	1.06	253	15.17	1.03
Petrol	Private	0.35	12.7	3.49	132	17.067	1.16
Petrol	Official	0.35	14.8	1.24	174	15.347	1.04

Table 4 Vehicular emission data, AFR and lambda values

Fuel type	Vehicle use	CO%	CO_2%	O_2%	HC (ppm)	AFR	Lambda (λ)
Petrol	Private	0.39	12.7	2.6	93	15.552	1.06
Petrol	Private	0.39	13.2	1.51	248	15.523	1.06
Petrol	Private	0.4	13.7	1.77	211	15.7	0.94
Petrol	Private	0.42	9.2	8.55	26	23.564	1.60
Petrol	Private	0.42	12.8	1.84	278	15.758	1.07
Petrol	Private	0.43	14.1	0.93	93	15.126	1.03
Petrol	Private	0.44	13.5	1.22	59	15.406	1.05
Petrol	Private	0.45	13.6	2.37	232	16.126	1.10
Petrol	Private	0.46	11.6	3.82	82	17.669	1.20
Petrol	Commercial	0.47	13.8	1.98	287	15.744	1.07
Petrol	Commercial	0.48	13.5	1.24	106	15.376	1.05
Petrol	Official	0.48	13.7	1.58	109	15.597	1.06
Petrol	Private	0.5	10.2	6.16	16	20.374	1.39
Petrol	Private	0.51	13.7	1.69	130	15.611	1.06
Petrol	Private	0.52	14	0.91	193	15.009	1.02
Petrol	Private	0.53	8	10.25	126	26.563	1.81
Petrol	Private	0.54	11	5.51	122	19.228	1.31
Petrol	Private	0.54	12.1	3.9	96	17.581	1.20
Petrol	Private	0.55	11	6.24	161	19.874	1.35
Petrol	Commercial	0.56	12.5	1.7	177	15.656	1.15

Table 5 Vehicular emission data, AFR and lambda values

Fuel type	Vehicle use	CO%	CO_2%	O_2%	HC (ppm)	AFR	Lambda (λ)
Petrol	Official	0.58	13.7	1.28	64	15.362	1.04
Petrol	Commercial	0.59	13.5	1.65	108	15.582	1.06
Petrol	Official	0.60	13.6	1.75	108	15.685	1.07
Petrol	Official	0.71	12.7	2.44	98	16.214	1.10
Petrol	Private	0.71	12.3	3.11	104	16.817	1.14
Petrol	Official	0.73	14.1	0.96	116	15.038	1.02
Petrol	Private	0.75	13.6	1.99	108	15.714	1.07
Petrol	Private	0.75	10.6	4.69	437	18.125	1.23
Petrol	Private	0.78	14.3	1.03	102	15.362	1.05
Petrol	Private	0.79	10.9	5.63	98	19.139	1.30
Petrol	Official	0.8	13.9	1.61	243	15.332	1.04
Petrol	Official	0.82	13.4	1.59	159	15.42	1.05
Petrol	Commercial	0.83	14.2	1.08	156	14.994	1.02
Petrol	Private	0.84	8.1	9.54	216	24.99	1.70
Petrol	Commercial	0.85	14.4	0.48	96	14.656	1.00
Petrol	Private	0.86	10.6	4.94	368	18.316	1.26
Petrol	Private	0.87	14	3.8	413	16.758	1.14
Petrol	Official	0.88	13.2	0.63	109	14.7	1.00
Petrol	Commercial	0.9	6	2.31	19	14.95	1.02
Petrol	Private	0.90	13.9	1.54	285	15.273	1.04

Table 6 Nomenclature

Symbol	Meaning
AFR	Air-fuel ratio
AFR_{actual}	Actual air-fuel ratio
AFR_{ideal}	Ideal/stoichiometric air-fuel ratio
λ	Lambda
CO	Carbon monoxide
CO_2	Carbon dioxide
VOC	Volatile organic compound
HC	Hydrocarbon
O_2	Oxygen
IC	Internal combustion
NOx	Nitrogen oxides
LPG	Liquefied petroleum gas
CNG	Compressed natural gas
Ppm	Parts per million

5 Conclusion

The actual air-fuel ratio was higher than the ideal air to fuel ratio in most of the data obtained from the vehicular emission analysis; therefore, lambda (λ) values were greater than 1. This shows that most of these vehicles were operating with rich mixtures because the fuel used in the combustion process was more than required and it altered the expected air-fuel ratio. This study hereby concludes that an alteration in the air-fuel ratio in the combustion process of motorcars will lead to high concentrations of the emitted environmental pollutants like CO, CO_2, HC and VOC which have adverse effects on human health, plants, animals and the environment.

Acknowledgement The authors acknowledge Covenant University for the financial support offered in the publication of this research.

References

Al-Himyari, B. A., Yasin, A., & Gitano, H. (2014). Review of air-fuel ratio prediction and control methods. *Asian Journal of Applied Sciences, 2*, 4.

Bektaş, T., & Laporte, G. (2011). The pollution-routing problem. *Transportation Research Part B: Methodological, 45*(8), 1232–1250.

Chen, S. Y., Chan, C. C., & Su, T. C. (2017). Particulate and gaseous pollutants on inflammation, thrombosis, and autonomic imbalance in subjects at risk for cardiovascular disease. *Environmental Pollution, 223*, 403–408.

Cheng, S., Lang, J., Zhou, Y., Han, L., Wang, G., & Chen, D. (2013). A new monitoring-simulation-source apportionment approach for investigating the vehicular emission contribution to the PM2. 5 pollution in Beijing, China. *Atmospheric Environment, 79*, 308–316.

Chowdhury, S., Pozzer, A., Dey, S., Klingmueller, K., & Lelieveld, J. (2020). Changing risk factors that contribute to premature mortality from ambient air pollution between 2000 and 2015. *Environmental Research Letters, 15*(7), 074010.

Dockery, D. W., Pope, C. A., Xu, X., Spengler, J. D., Ware, J. H., Fay, M. E., & Speizer, F. E. (1993). An association between air pollution and mortality in six US cities. *New England Journal of Medicine, 329*(24), 1753–1759.

Dulac, J. (2012). *Global transport outlook to 2050 energy analyst*. Energy Technology Policy Division International Energy Agency Targets and Scenarios for a Low-Carbon Transport Sector. OECD/International Energy Agency.

Haines, A., McMichael, A. J., Smith, K. R., Roberts, I., Woodcock, J., Markandya, A., Armstrong, B. G., Campbell-Lendrum, D., Dangour, A. D., Davies, M., & Bruce, N. (2009). Public health benefits of strategies to reduce greenhouse-gas emissions: Overview and implications for policy makers. *The Lancet, 374*(9707), 2104–2114.

Jia, M., Gingrich, E., Wang, H., Li, Y., Ghandhi, J. B., & Reitz, R. D. (2016). Effect of combustion regime on in-cylinder heat transfer in internal combustion engines. *International Journal of Engine Research, 17*(3), 331–346.

Manente V, Zander CG, Johansson B, Tunestal P, Cannella W. An advanced internal combustion engine concept for low emissions and high efficiency from idle to max load using gasoline partially premixed combustion. SAE Technical Paper; 2010.

Mannucci, P. M., Harari, S., Martinelli, I., & Franchini, M. (2015). Effects on health of air pollution: A narrative review. *Internal and Emergency Medicine, 10*(6), 657–662.

Odunlami, O. A., Elehinafe, F. B., Oladimeji, T. E., Fajobi, M. A., Okedere, O. B., & Fakinle, B. S. (2018a). Implications of lack of maintenance of motorcycles on ambient air quality. *IOP Conference Series: Materials Science and Engineering, IOP Publishing, 413*, 012055.

Odunlami, O. A., Elehinafe, F. B., Okorie, C. G., Abioye, O. P., Abioye, A. A., & Fakinle, B. S. (2018b). Assessment of in-tricycle exposure to carbon monoxide emission on roads in Nigerian urban centres. *International Journal of Mechanical Engineering and Technology (IJMET), 9*(9), 671–684.

Song, J., Lu, M., Lu, J., Chao, L., An, Z., Liu, Y., & Wu, W. (2019). Acute effect of ambient air pollution on hospitalization in patients with hypertension: A time-series study in Shijiazhuang, China. *Ecotoxicology and Environmental Safety, 170*, 286–292.

Jiang, Z., McDonald, B. C., Worden, H., Worden, J. R., Miyazaki, K., Qu, Z., Henze, D. K., Jones, D. B., Arellano, A. F., Fischer, E. V., & Zhu, L. (2018). Unexpected slowdown of US pollutant emission reduction in the past decade. *Proceedings of the National Academy of Sciences, 115*(20), 5099–5104.

Williams, J. H., DeBenedictis, A., Ghanadan, R., Mahone, A., Moore, J., Morrow, W. R., & Torn, M. S. (2012). The technology path to deep greenhouse gas emissions cuts by 2050: The pivotal role of electricity. *Science, 335*(6064), 53–59.

Part III
Process Modelling and Simulation

Design of Water Retaining Structures and Application of Environmental Engineering for Sustainable Environment

O. J. Oyebode

1 Introduction

Civil engineers are saddled with copious responsibilities such as design of structures, foundation investigations, effective planning, supervision of projects, proper delivering of projects on time, sustainable engineering practices, efficient management, and maintenance of most private, state, and federal government infrastructures. Civil and environmental engineers have contributed immensely to the sustainability, design, and enhancement of water retaining structures and other infrastructures in built environment. One of the reasons for water retaining structure design is to conserve water for various purposes such navigation, recreation, pollution abatement, flood control, and provision of adequate water (Oyebode, 2020).

There are deficits of sustainable and environmentally friendly facilities for water and wastewater systems in Nigeria (Oyebode, 2017). There is urgent need for developing nations to overcome challenges of high-quality water for residential for the achievement of public health and hygienic environment (Oyebode et al., 2019).

There are sufficient reasons for flood control and management of water resources and reuse of excess water for irrigation and civil engineering construction works (Oyebode, 2018).

Water retaining structure is a structure built to retain the water. It is a wall built across the rivers and small streams where water is stored. The various types of water retaining structures are overhead water tanks in buildings, underground water tanks, underground water reservoirs, elevated surface reservoirs, etc. Each is unique in

O. J. Oyebode (✉)
Civil and Environmental Engineering Department, Afe Babalola University,
Ado-Ekiti, Ekiti State, Nigeria

A. O. Ayeni et al. (eds.), *Advanced Manufacturing in Biological, Petroleum, and Nanotechnology Processing*, Green Energy and Technology,
https://doi.org/10.1007/978-3-030-95820-6_25

design aspect and can be made of different materials ranging from clay or plain concrete to steel and glass.

Sustainability of water retaining structures entails making these structures serve their purpose structurally but not at the expense of either its social, economic, or environmental purpose. The importance and need of sustainable water structures can never be overemphasized. Water plays a crucial role in the environment which ranges from agriculture to domestic use (Basunia et al., 2015).

Water will be useless if not stored and channeled properly with the effective use of water retaining structures. This hydraulic structure is relevant for water storage, irrigation purposes, power generation, etc. All these are necessary for an enhanced environment. A lot of factors come into play when designing water retaining structures. These factors are the design methods to be used, the design code to be adopted, the concrete mix, external loads, materials to be used, and most importantly the sustainability of the structure (Mosley et al., 2012).

Thus for a water retaining structure to be sustainable, the design should be done using the required amount of resources available and most importantly adhering to a standard building code. The code should be followed because the effectiveness of these water retaining structures strongly promotes the enhancement of the society. These water retaining structures are those which are required for a particular purpose, and these structures are water towers, storage tanks, and reservoirs. These problems can be solved by using good quality building materials, by improving life cycle performance, and lastly through the strategic use of alternative building materials.

1.1 BS 8007:1987 for the Water Retaining Structure Design

BS 8007 code can be used to design structures based on the probabilities of the structure becoming unfit in any use case. Civil engineers must perform life cycle assessment and provide dynamic and innovative solutions to construction challenges (ASCE, 2001; BS 8007, 1987).

The limit state recommendations of the code stress that the size and amount of reinforcements used should be checked. It accesses factor of safety based on its resistance when subjected to groundwater pressure; it also accesses the cracking based on the serviceability crack width limit, thus determining whether it should be pre-stressed or tensioned, and lastly deflection is also accessed and checked by the introduction of calculated allowances.

Geo-environmental impact assessment and remediation can be applied to contaminated soil and groundwater (Ojuri et al., 2014). Constant rise in price of cement and other binders has caused a constant surge in cost of construction, rehabilitation, and maintenance of roads (Ola et al., 2020). Infrastructures in most cases can be constructed on soils that are unsaturated especially hot areas of Nigeria. Adequate soil mechanics tests and foundation investigations must be done despite difficulties that might be encountered (Ola et al., 2019).

Dams are the most common example of water retaining structures. Importance of these structures includes electrical generation, renewable, clean energy, flood control, water storage, irrigation, navigation, recreation, and so on. Classifications of concrete dams are based on the designs used to resist the stress due to reservoir water pressure. Example of dams are gravity dams, butress and arch. Figure 1 indicates different types of retaining structures, Fig. 2 presents typical gravity retaining wall, Fig. 3 describes a typical embankment dam, Fig. 4 indicates gravity dam, Fig. 5 indicates buttress dam, Fig. 6 presents arch dam, and Fig. 7 gives typical construction site for retaining wall with reinforcement arrangement and concrete works.

Concrete arch dams are thin dams. The forces acting on it are transferred directly to the abutments, and it looks like the segment of a circle. Figure 6 indicates an arch dam.

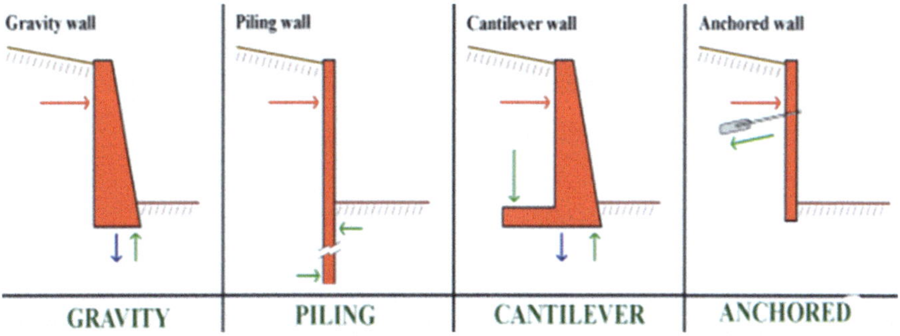

Fig. 1 Types of retaining walls. (Source: Khan & Sikder, 2004)

Fig. 2 Typical gravity retaining wall. (Source: Field Study, 2021)

Fig. 3 Typical embankment dam. (Source: Field Study, 2021)

Fig. 4 Gravity dam. (Source: Field Study, 2021)

Fig. 5 Buttress dam. (Source: Field Study, 2021)

Fig. 6 Arch dam. (Source: Field Study, 2021)

Fig. 7 Typical construction site for retaining wall with reinforcement arrangement and concrete works. (Source: Field Study, 2021)

Design and safety aspects must be taken into consideration, and all reinforcements should be arranged according to civil engineering drawing. Example of construction site with reinforcement arrangement and concrete works is indicated in Fig. 7.

2 Water Retaining Structures and Sustainable Development

Sustainability of water retaining structures simply means meeting the demands of the society economically, socially, and environmentally; thus the design of the structures needs to be structurally sound.

For this work, the method to be used includes accessing the structural engineer as a determining factor for the sustainability of water retaining structures, thus addressing the global environmental impact, the solutions for today, and the challenges for the future. Figure 8 presented a typical flexural cracking.

Calculation of crack widths should be done with care. The required surface width of cracks shouldn't be more than the specified in order to support the limit state of cracking. Figure 9 gave typical detailing of spacer reinforcement. Appropriate

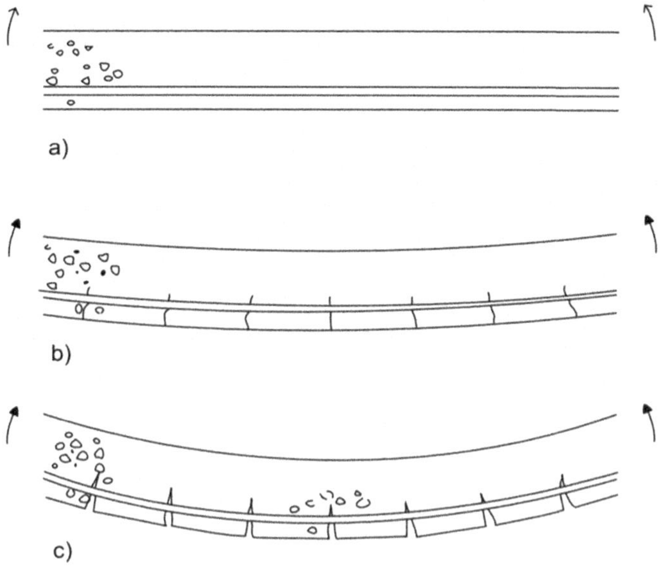

Fig. 8 Flexural cracking. (**a**) Uncracked concrete with low steel stress, (**b**) fine cracks and increased steel stress, and (**c**) wide cracks and high steel stress. (Source: Soltani et al., 2013)

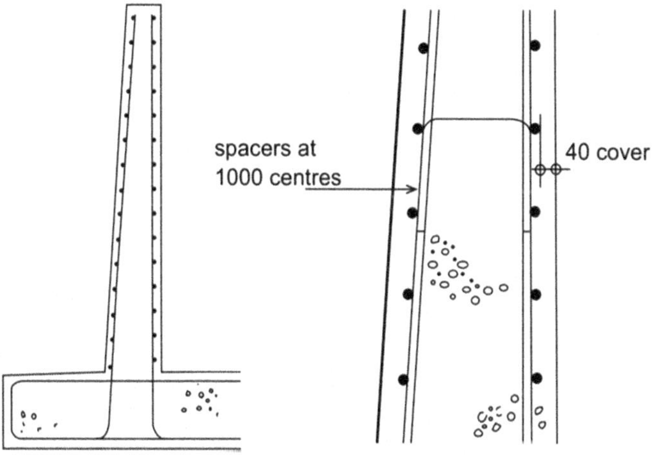

Fig. 9 Detailing of spacer reinforcement

Table 1 Maximum bar spacing

Steel stress (MPa)	BS EN 1992–2001	BS EN 1992–1993
160	32	–
200	25	35
240	16	22
280	12	16
320	10	12
360	8	9
400	6	6
450	5	5

calculations and codes of practice must be followed during design and detailing of structural members to prevent failure of structures (Table 1).

The problem of sustainability can be narrowed to global environmental impact, meaning all the non-environmental practices being done structurally contribute wholly to the large-scale depletion of resources. Here the emissions that come out of steel and concrete are a major shortcoming of sustainability in structures.

3 Retaining Structure Design and Construction

The graph below shows the world production of cement and steel and thus shows how far the world is in sustaining the structures (Fig. 10).

Just like every other problem, there are various solutions to the environmental issues facing retaining structures. Since sustainability is a long-term investment, most of the solutions are based on green practices and awareness. Below are some ways in which the issue of poor design and sustainability can be solved:

3.1 Improve Life Cycle Performance

Life cycle performance is an important factor when it comes to sustainability because it entails the capability of a structure to function over a long period of time. Life cycle costs of water retaining structures are usually overlooked because of how huge the initial cost is, thus making the structural engineer to save initial cost, thereby increasing the life cycle cost of the structure. Water tanks can be constructed in a variety of shapes such as circular (cylindrical), rectangular, oval, octagonal, etc. Furthermore, different materials such as reinforced concrete, steel, high-density plastics, reinforced polymers, etc. have been successfully used in the construction

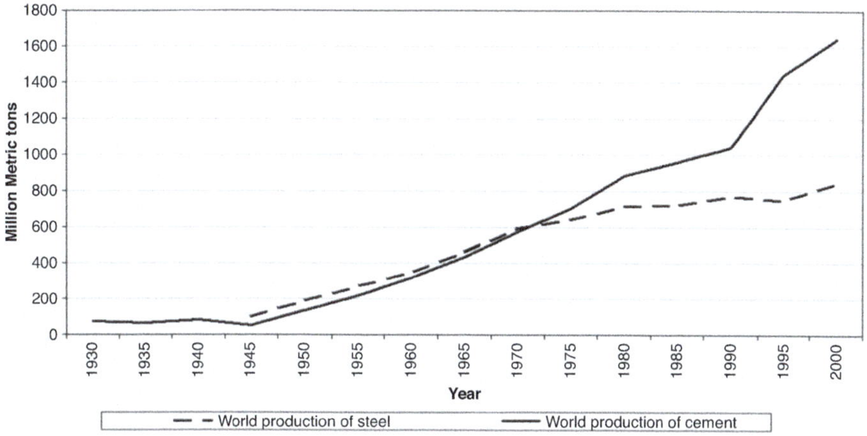

Fig. 10 Graph showing the world production of cement and steel. (Source: Gan et al., 2017)

Fig. 11 Concrete materials for water retaining structures. (Source: Field Study, 2021)

of water tanks. Various materials can be used for the construction of water retaining structures. Hydrostatic forces and earth pressures should be taken into consideration in the design of retaining structures. Figure 11 presented concrete and steel sections for water tanks, while Fig. 12 indicated coated steel sections for water tanks to handle corrosion effects. Figure 13 also presented reinforced concrete for

Fig. 12 Coated steel sections for water tanks. (Source: Field Study, 2021)

Fig. 13 Reinforced concrete for swimming pool construction. (Source: Field Study, 2021)

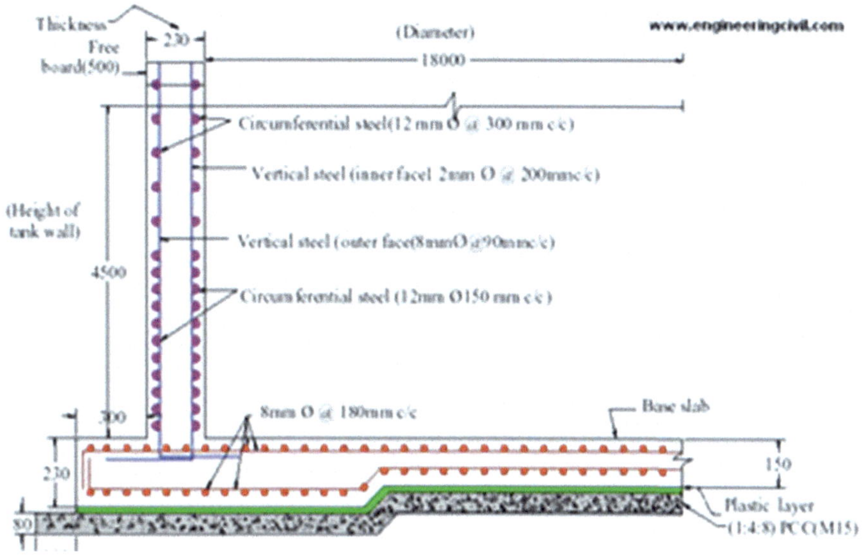

Fig. 14 Typical detailing of reinforced water tank walls and rigid base. (Source: Khan & Sikder, 2004)

swimming pool construction, while Fig. 14 presented a typical detailing of reinforced water tank walls and rigid base.

Since the usual building materials are causing the reduction of the quality of life thus downgrading the social, economic, and environmental issues of the society. And all these are as a result of the predominant use of non-environmentally friendly materials.

4 Conclusion

Design of water retaining structures should be done for safety, accuracy, quality materials, experienced workmanship, integrity of environmental engineering works, and sustainable environment. Over the years, the role of structural engineers has been linked to sustainability; thus all the main points from this research revolve around adequate and standard structural engineering. It is the role of the structural engineer to ensure that whatever material is being used in the water retaining structure, it will not make it structurally unfit. Recent trends and development have to be incorporated into the design of water retaining structures to avert flooding, natural disaster, and other issues that can affect our environment. Proper checks have to be done on analysis, design, and drawings coupled with appropriate geotechnical tests on dams, water retaining, and other hydraulic structures. Computer software and application can be used to validate issues linked with settlements, stability, and deflection. It has been concluded that legislations, engineering law, managerial economy, policies, standards, and guidelines should be formulated, structured, implemented, and monitored for better design and sustainability of water retaining structures for enhanced environment. Design of retaining wall must be sustainable.

5 Recommendations

Recommendations made include the following:

(i) Members of engineering family should create healthy environments by our engineering construction, consultancy, and other activities. We must not contribute to the failure of retaining walls and other structures in our country.
(ii) There is a need for appropriate supervision and selection of quality materials.
(iii) Design of retaining structures should be done with the use of latest codes of practice with appropriate checks against sliding and overturning moments.
(iv) Adoption of standard design code of practice approved by building regulations, Council for the regulation of engineering practice in Nigeria (COREN) and other regulatory bodies.
(v) Awareness campaign of the importance of sustainability in the society.
(vi) Engineer must ensure innovation, latest technology, and thoroughness in design and supervision of retaining wall. Engineers, project managers, and clients should give safety of retaining structures adequate priority above the cost.
(vii) The topography of the soil, overturning moment, specific peculiarities of locations, earth pressure, and factor of safety should be carefully checked for safety of life and for enhanced environment.
(viii) Appropriate geotechnical tests and checks for factor of safety should be carried out.
(ix) Quality control and appropriate tests should be carried out on all materials of construction.

References

American Society of Civil Engineers. (2001). The role of the civil engineer in sustainable development. *ASCE Policy Statement, 418*.

Basunia, Z. A., Hossain, M. A., Sayed, A., Ali, M., & Rahman, G. A. (2015). Evaluation of ground water resources in Mymensingh Sadar Upazilla, Bangladesh. *Asian Journal of Agricultural Extension, Economics and Sociology*, 1–10.

Civil Engineering and Building Structures Standards Committee. (1987). *BS 8007:1987 code of practice for design of concrete structures for retaining aqueous liquids*. BSI.

Field Study (2021). Studies carried out on the field done by researcher during site investigations in 2021.

Gan, V. J., Cheng, J. C., Lo, I. M., & Chan, C. M. (2017). Developing a CO2-e accounting method for quantification and analysis of embodied carbon in high-rise buildings. *Journal of Cleaner Production, 141*, 825–836.

Khan, A. J., & Sikder, M. (2004). Design basis and economic aspects of different types of retaining walls. *Journal of Civil Engineering (IEB), 32*(1), 17–34.

Mosley, W. H., Hulse, R., & Bungey, J. H. (2012). *Reinforced concrete design: To Eurocode 2*. Macmillan International Higher Education.

Ojuri, O. O., Ola, S. A., Fadugba, O. G., & Uduebor, M. A. (2014). Site remediation in Nigeria: Proven and innovative technologies (recovery of free hydrocarbon from soil/groundwater). *GEOMATE-Brisbane*, 1–7.

Ola, S. A., Braimoh, A. S., & Fadugba, O. G. (2019). Measurement and estimation of soil water characteristic curve for four unsaturated tropical soils. *International Journal of Engineering*, 1–11.

Ola, S. A., Harris, F. I., & Fadugba, O. G. (2020). Verification of the effect of lateralite on soil stabilization for highway construction. *Annals of the Faculty of Engineering Hunedoara, 18*(1), 171–179.

Oyebode, O. J. (2017). Design of engineered sanitary landfill for efficient solid waste management in Ado–Ekiti, South-Western Nigeria. *Journal of Multidisciplinary Engineering Science Studies, 3*(9), 2144–2160.

Oyebode, O. J. (2018). Strategy for tackling flood and water scarcity for adequate water supply and public health in Nigeria. *European Journal of Advances in Engineering and Technology, 5*(8), 656–664.

Oyebode, O. J. (2020). Design and sustainability of water retaining structures for enhanced environment *1*(2), 3–4. Conference on Engineering Research, Technology, Innovation and Practice "2020" e-Conference Organized by Mechanical Engineering in the University of Nigeria, Nsukka.

Oyebode, O. J., Adebanjo, A. S., & Nd-ezuma, S. R. (2019). Conditions of available sources of water for domestic use in selected communities in Ado-Ekiti, Ekiti State, Nigeria. *African Journal of Environmental Science and Technology, 13*(2), 84–90.

Soltani, A., Harries, K. A., & Shahrooz, B. M. (2013). Crack opening behavior of concrete reinforced with high strength reinforcing steel. *International Journal of Concrete Structures and Materials, 7*(4), 253–264.

Development of Robust Fractional-Order Controllers to Provide Effective Economic Load Dispatch Management in Interconnected Area Networks

J. Okoronkwo, A. Mati, Y. Jibril, G. Olarinoye, and A. S. Abubakar

1 Introduction

The integration of renewable energy sources into the grid affects the dynamic operation, as well as the response of the power system-to-system disturbances in terms of change in power flow patterns and sensitivity of the system to faults. Thus reducing frequency and tie-line power deviation within accepted operating standard in multiple areas of interconnected power systems is paramount. Load frequency controller (LFC) and automatic voltage regulator (AVR) play an important role in maintaining steady frequency and voltage to ensure reliability of electric power systems (Kroposki et al., 2017) (Grigsby, 2017). Load frequency control is one of the functions of automatic generation control (AGC) (Soundarrajan et al., 2012). The present challenge is in situations when there are integrated power systems of two or more areas with cascading multiple disturbances.

2 Theoretical Background

When considering renewable energy systems and other energy systems to be integrated in a multi-area interconnected microgrid power system, limiting frequency fluctuation by compensating for the deviation between generation and demand is optimal, (Khooban et al., 2018; Lal et al., 2018). This functionality and capability are often referred to as load frequency control (LFC). The load frequency control

J. Okoronkwo (✉) · A. Mati · Y. Jibril · G. Olarinoye · A. S. Abubakar
Department of Electrical Engineering, Ahmadu Bello University, Zaria, Kaduna State, Nigeria

A. O. Ayeni et al. (eds.), *Advanced Manufacturing in Biological, Petroleum, and Nanotechnology Processing*, Green Energy and Technology,
https://doi.org/10.1007/978-3-030-95820-6_26

(LFC) is to sustain and maintain system frequency of each area and tie-line power flow between areas during normal operating situations and variations in load demands (Chidambaram & Paramasivam, 2013; Lal et al., 2018). A functional LFC can not only guarantee the frequency stability of the MG but also increase its fuel saving efficiency (Khooban et al., 2018). To improve the response of LFC, many controllers including conventional PID control (Khooban et al., 2018), intelligent control (Bevrani et al., 2012), adaptive control (Xu et al., 2018), robust control (Liao & Xu, 2018), and MPC control (Ersdal et al., 2016) have been applied to the DGs of MAIMGs (Khooban et al., 2018).

The MG's low inertia capability compared to the grid remains a challenge maintaining the active and reactive power balances between the supply and utilization (Khooban et al., 2018), especially in the presence of intermittent RES and frequent load variations (Roslan et al., 2019). Few works have been carried out to address frequency control of isolated MG power systems; however, interconnected MG systems have not received much attention (Bevrani et al., 2012; Cau et al., 2014; Khooban et al., 2018).

Since the operating conditions of the LFC can change instantaneously (Khooban et al., 2018), the controller tuned for nominal conditions cannot work properly when exposed to other conditions. The performance of the controller primarily depends on its parameters; effective optimization of these parameters can play a controlling role and significant impact in promoting the output performance of the LFC control. So, to solve this problem, the control parameters were tuned according to the setoff point (Abdulkhader et al., 2018; Khooban et al., 2018).

As a measure of conformity to the control performance standards, a first compliance factor knows as ACE (area control error) is applied to the power system operations (Bevrani & Hiyama, 2017). To match the dynamic economic dispatch to load, system operators pick out trajectories of the generation units repeatedly or resort to manual outputs (Ross & Kim, 1980). The economic dispatch function provides set economic exponentials for the generation units. The tie schedules combine to obtain the net optimal AGC response tracks desired by unit generation as requirements for all units on AGC over the economic trajectories (Ross & Kim, 1980).

3 Material and Methods

3.1 Materials

The materials used for the research work are discussed in this subsection. They include all hardware and software used for the implementation and realization of the set objectives.

3.1.1 Hardware Platform

The research is simulated on a Dell laptop with specifications (operating system, Microsoft Windows 10; processor, 1.7 GHz Intel Core i5; memory, 4 GB 1600 MHz DDR3).

3.1.2 Software Platform

MATLAB ("MATrix LABoratory") software is used throughout this work for performing modeling, mathematical computations, algorithm development, and simulations.

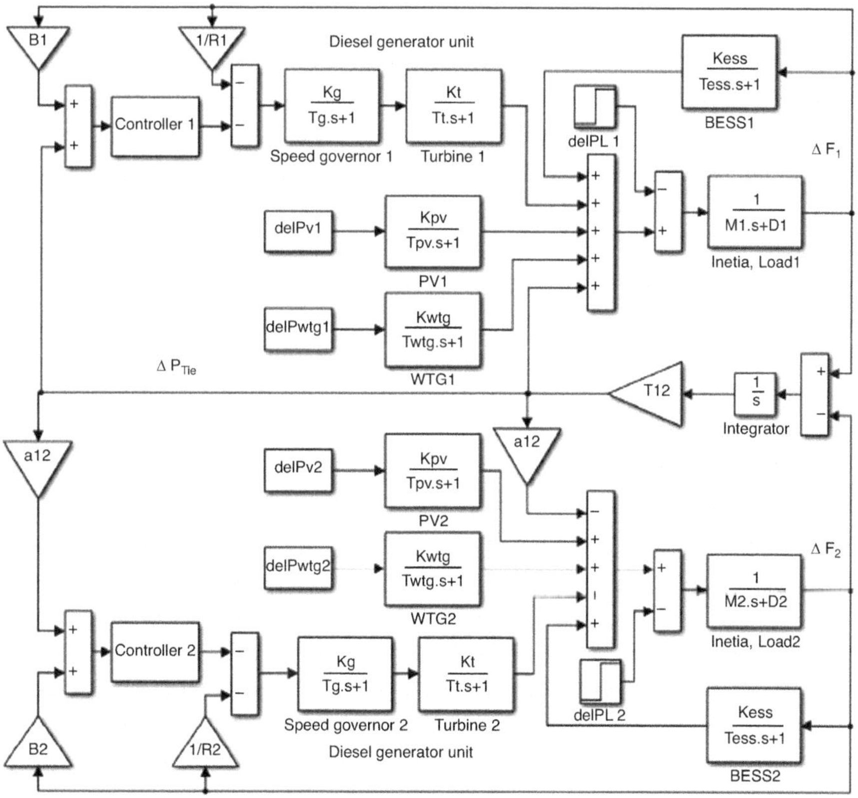

Fig. 1 Block diagram of two-area interconnected microgrid system. (Lal et al., 2018)

3.2 Methods

The steps of the methodology used in achieving the set objectives are as follows:

3.2.1 Two-Area Interconnected Microgrid Power System

The schematic diagram of a two-area interconnected microgrid network using energy generation units illustrated in Fig. 1 was implemented in MATLAB Simulink.

3.3 Performance Criteria for Optimization-Based Tuning of Controller Parameters

The controller parameters were optimized using the IAE, ITAE, ISE, and ITSE performance criteria, and the performance criteria are defined in Eqs. 1, 2, 3, 4, 5, 6, 7, 8 and 9:

$$J = \text{IAE} = \int_{t=0}^{t_{\text{sim}}} \left(|\Delta F_1| + |\Delta F_2| + |\Delta P_{\text{tie}}| \right) dt \tag{1}$$

$$J = \text{ITAE} = \int_{t=0}^{t_{\text{sim}}} t \left(|\Delta F_1| + |\Delta F_2| + |\Delta P_{tie}| \right) dt \tag{2}$$

$$J = \text{ISE} = \int_{t=0}^{t_{\text{sim}}} \left(\Delta F_1^2 + \Delta F_2^2 + \Delta P_{\text{tie}}^2 \right) dt \tag{3}$$

$$J = \text{ITSE} = \int_{t=0}^{t_{\text{sim}}} t \left(\Delta F_1^2 + \Delta F_2^2 + \Delta P_{\text{tie}}^2 \right) dt \tag{4}$$

where ΔF_1 and ΔF_2 are the system frequency deviations of areas 1 and 2, respectively (Lal & Barisal, 2019); ΔP_{tie} is the incremental change in tie-line power; and t is the simulation time. Hence, the optimization problem is stated as (Chen et al., 2019):

$$\textit{Minimize}\, J \tag{5}$$

subject to

$$\left.\begin{array}{l} K_P^{\min} \le K_P \le K_P^{\max} \\ K_I^{\min} \le K_I \le K_I^{\max} \\ K_D^{\min} \le K_D \le K_D^{\max} \end{array}\right\} \text{for PID controller} \tag{6}$$

$$\left.\begin{array}{l} K_1^{\min} \le K_1 \le K_1^{\max} \\ K_2^{\min} \le K_2 \le K_2^{\max} \\ K_3^{\min} \le K_3 \le K_3^{\max} \\ K_4^{\min} \le K_4 \le K_4^{\max} \end{array}\right\} \text{for fuzzy } PID \text{ controller} \tag{7}$$

$$\left.\begin{array}{l} K_P^{\min} \le K_P \le K_P^{\max} \\ K_I^{\min} \le K_I \le K_I^{\max} \\ K_D^{\min} \le K_D \le K_D^{\max} \\ \lambda_{\min} \le \lambda \le \lambda_{\max} \\ \mu_{\min} \le \mu \le \mu_{\max} \end{array}\right\} \text{for FOPID controller} \tag{8}$$

$$\left.\begin{array}{l} K_1^{\min} \le K_1 \le K_1^{\max} \\ K_2^{\min} \le K_2 \le K_2^{\max} \\ K_3^{\min} \le K_3 \le K_3^{\max} \\ K_4^{\min} \le K_4 \le K_4^{\max} \\ \lambda_{\min} \le \lambda \le \lambda_{\max} \\ \mu_{\min} \le \mu \le \mu_{\max} \end{array}\right\} \text{for fuzzy FOPID controller} \tag{9}$$

where K_1, K_2, K_3, K_4, K_P, K_I, K_D are the controller gains and λ, μ are fractional order of the FOPID controller. These are independent variables which need to be optimally selected in order to control the load frequency of the system.

4 Results and Discussion

The fuzzy logic good performance attained in specific member functions was enhanced by the introduction of the FOPID controller to avoid the need for additional retuning or online auto-tuning, even in unstable cases (Pan & Das, 2016). The FOPID controller also shows high robustness properties concerning parameter variation in nonlinear rate constraint on feedback elements and on disconnection of some components (Pan & Das, 2012). The FOPID controllers offer very good robustness, and the performance does not degrade appreciably even when there are changes in system parameters (Pan & Das, 2016). The ruggedness of the FOPID controller in the feedback loop is evident because it consistently keeps the controller gains especially ISE at lower values compared to PID and fuzzy PID structures. The

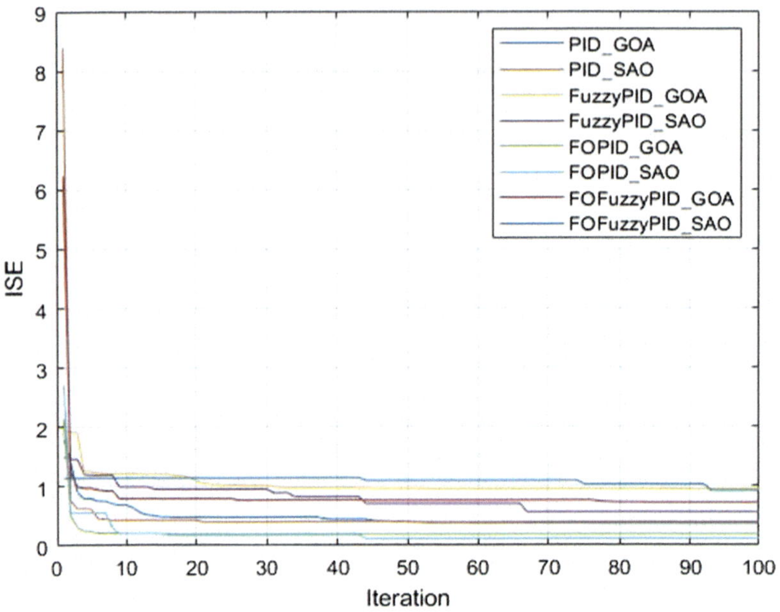

Fig. 2 Convergence plot of ISE criterion

robust features are due to the use of gain margin, phase margin, and iso-damping property. The parameters of the controllers were optimized using the grasshopper optimization algorithm and smell agent optimization as described above. The ISE, IAE, ITSE, and ITAE performance criteria were utilized in the optimization process. Each of the performance criteria was applied to different controllers, and a comparison of their performances was carried out to determine the best performing criterion. The ISE performance criterion was found to be the best performing (Kocaarslan & Çam, 2005) (Figs. 2 and 3, Table 1).

Four performance criteria were used to develop the controllers for this work. The results obtained for the IAE performance criterion are shown in Table 2; the FOPID controller outperformed other controllers with a minimum fitness value of 18.2555 when optimized with the SAO, followed by the fuzzy PID controller optimized with the GOA with a minimum fitness value of 27.0761 as shown in Figs. 4 and 5. In Fig. 4, the controllers converge to values less than 100 after the first ten iterations, but after the 60th iteration, they all converge to stable values less than 50. This shows that the worst IAE criterion fitness value was obtained from the FO fuzzy PID controller optimized with SAO. The results also show that the performance of PID and FO fuzzy controllers was generally poor when optimized for the IAE criterion.

The results obtained for the ITSE performance criterion are shown in Table 3; the FOPID controller outperformed other controllers with minimum fitness values of 0.0325 and 0.089336 when optimized with the GOA and SAO, respectively. This is

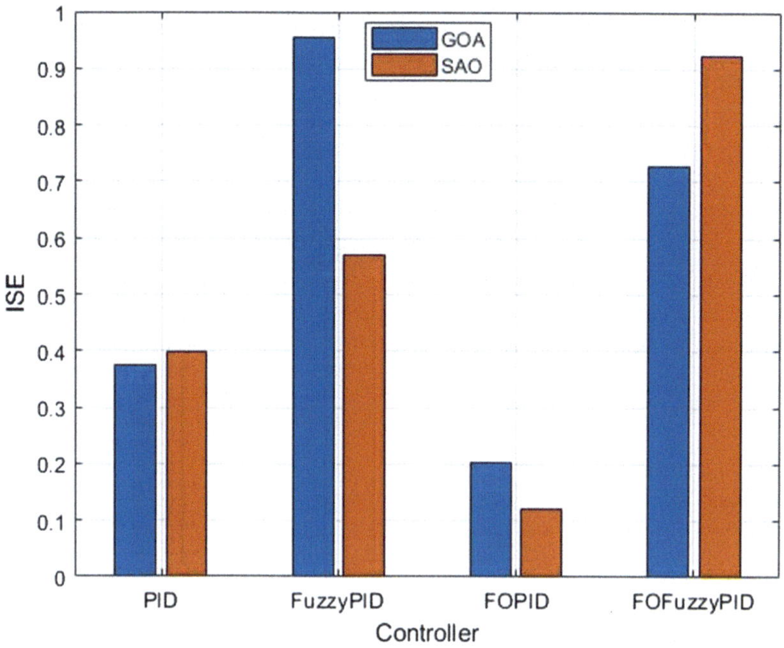

Fig. 3 Optimal fitness values of ISE criterion

Table 1 Optimal fitness values of ISE performance criterion

Controller structure	Optimization algorithm	ISE
PID	GOA	0.37474
	SAO	0.3981
Fuzzy PID	GOA	0.95426
	SAO	0.57044
FOPID	GOA	0.2034
	SAO	0.1215
FO fuzzy PID	GOA	0.7276
	SAO	0.92324

shown in Fig. 6 similar to previous convergence plots of Fig. 4; this converges to values between 0 and 0.5 after five iterations. The worst fitness value was obtained from the fuzzy PID controller optimized with GOA. The results also show that the performance of fuzzy controllers was generally poor when optimized for the ITSE criterion especially using GOA (Fig. 7).

The results obtained for the ITAE performance criterion are shown in Table 4; the fuzzy FOPID controller outperformed other controllers with minimum fitness values of 12.213 and 13.3172 when optimized with the SAO and GOA, respectively, as shown in Figs. 8 and 9. The signals in Fig. 8 converge to values between 0 and 50

Table 2 Optimal fitness values of IAE performance criterion

Controller structure	Optimization algorithm	IAE
PID	GOA	35.0241
	SAO	36.0429
Fuzzy PID	GOA	27.0761
	SAO	24.5477
FOPID	GOA	29.9658
	SAO	18.2555
FO fuzzy PID	GOA	37.852
	SAO	33.5883

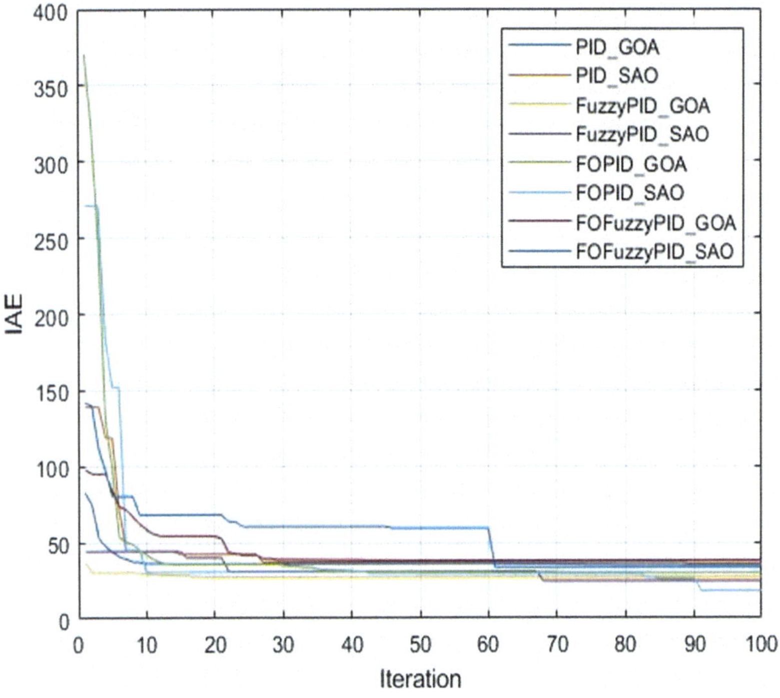

Fig. 4 Convergence plot of IAE criterion

after 30 iterations. The worst fitness value was obtained from the FO fuzzy PID controller optimized with GOA as the convergence pattern takes the form of an exponentially decaying function. The results also show that the performance of PID controllers was generally poor when optimized for the ITAE criterion especially using GOA.

In summary, the performance of developed FOPID and fuzzy FOPID controllers was compared with basic PID and fuzzy PID controllers using the convergence

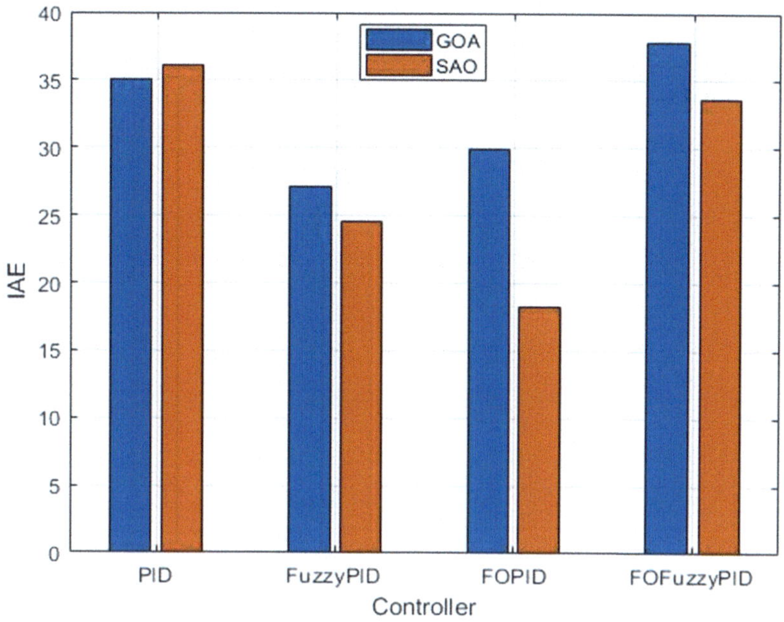

Fig. 5 Optimal fitness values of IAE criterion

Table 3 Optimal fitness values of ITSE performance criterion

Controller structure	Optimization algorithm	ITSE
PID	GOA	0.13694
	SAO	0.1586
Fuzzy PID	GOA	0.34457
	SAO	0.092525
FOPID	GOA	0.0325
	SAO	0.089336
FO fuzzy PID	GOA	0.27617
	SAO	0.22589

characteristics and optimal fitness values of performance criteria. The results obtained showed that the FOPID controller outperformed other controllers as shown along with the optimal fitness values of the ITSE performance criterion. This is evident that the developed controller is able to match the dynamic dispatch of the economic generation to load relating to load management between control areas.

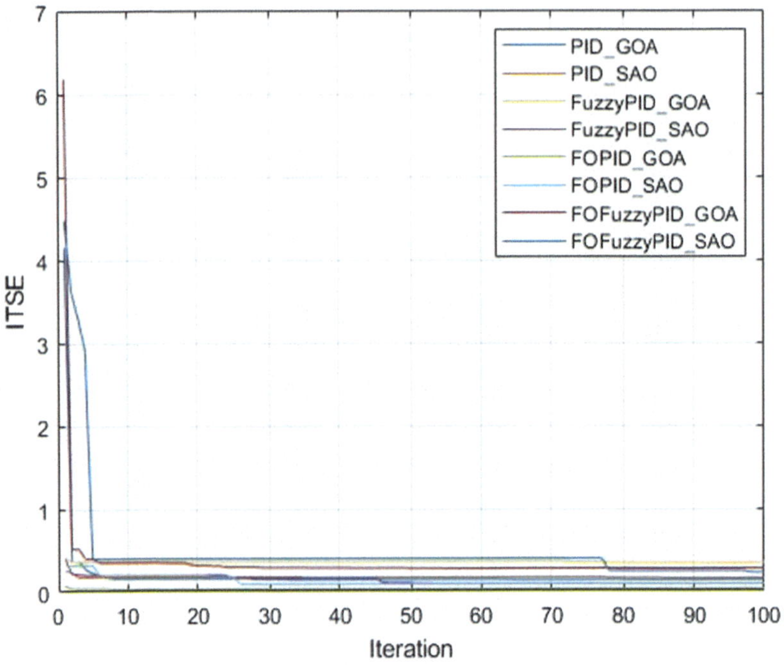

Fig. 6 Convergence plot of ITSE criterion

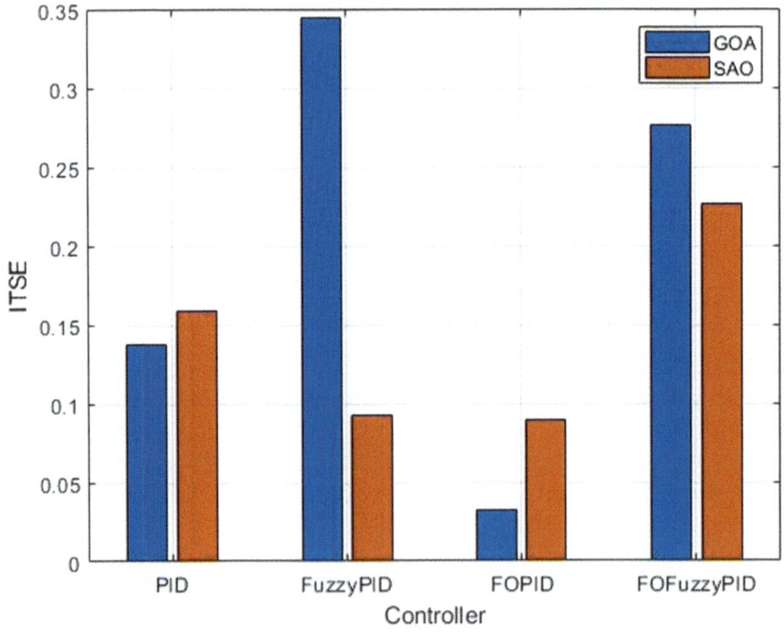

Fig. 7 Optimal fitness values of ITSE criterion

Table 4 Optimal fitness values of ITAE performance criterion

Controller structure	Optimization algorithm	ITAE
PID	GOA	40.5866
	SAO	42.8153
Fuzzy PID	GOA	13.3172
	SAO	12.213
FOPID	GOA	23.3432
	SAO	23.105
FO fuzzy PID	GOA	56.1118
	SAO	23.0759

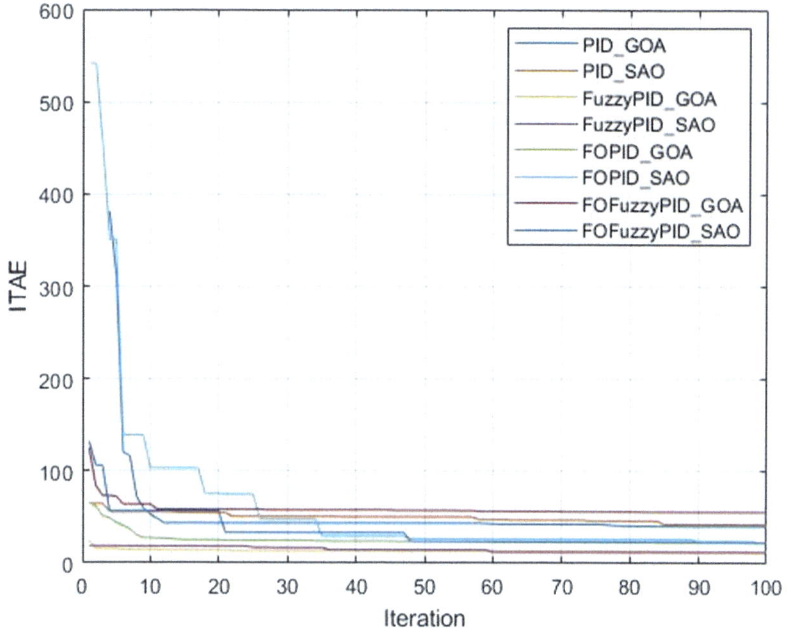

Fig. 8 Convergence plot of ITAE criterion

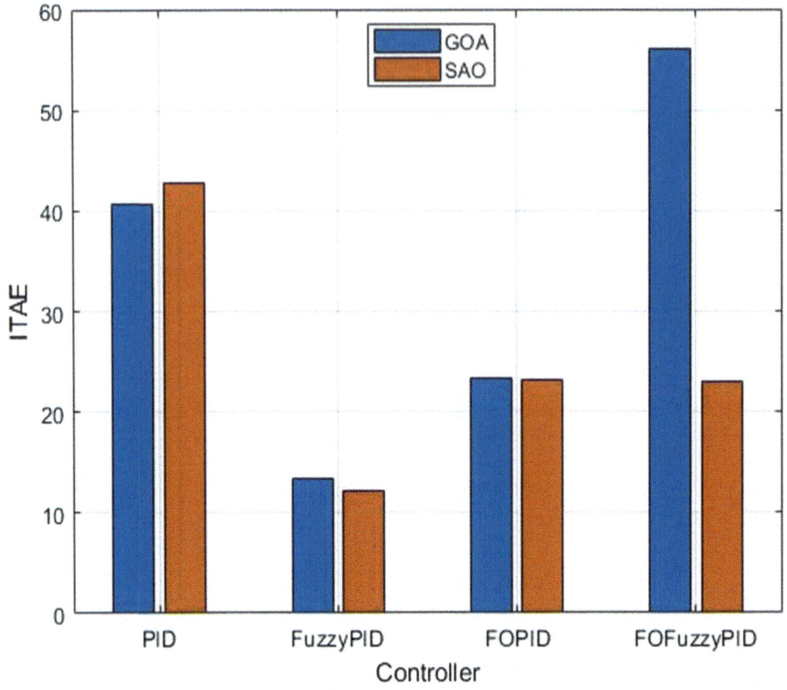

Fig. 9 Optimal fitness values of ITAE criterion

References

Abdulkhader, H. K., Jacob, J., & Mathew, A. T. (2018). Fractional-order lead-lag compensator-based multi-band power system stabiliser design using a hybrid dynamic GA-PSO algorithm. *IET Generation, Transmission & Distribution, 12*(13), 3248–3260. https://doi.org/10.1049/iet-gtd.2017.1087

Bevrani, H., & Hiyama, T. (2017). *Intelligent automatic generation control*. CRC Press.

Bevrani, H., Habibi, F., Babahajyani, P., Watanabe, M., & Mitani, Y. (2012). Intelligent Frequency Control in an AC Microgrid: Online PSO-Based Fuzzy Tuning Approach. *IEEE Transactions on Smart Grid, 3*(4), 1935–1944. https://doi.org/10.1109/TSG.2012.2196806

Cau, G., Cocco, D., Petrollese, M., Knudsen Kær, S., & Milan, C. (2014). Energy management strategy based on short-term generation scheduling for a renewable microgrid using a hydrogen storage system. *Energy Conversion and Management, 87*, 820–831. https://doi.org/10.1016/j.enconman.2014.07.078

Chen, G., Li, Z., Zhang, Z., & Li, S. (2019). An improved ACO algorithm optimized fuzzy PID controller for load frequency control in multi area interconnected power systems. *IEEE Access, 8*, 6429–6447.

Chidambaram, I. A., & Paramasivam, B. (2013). Optimized load-frequency simulation in restructured power system with Redox Flow Batteries and Interline Power Flow Controller. *International Journal of Electrical Power & Energy Systems, 50*, 9–24. https://doi.org/10.1016/j.ijepes.2013.02.004

Ersdal, A. M., Imsland, L., & Uhlen, K. (2016). Model predictive load-frequency control. *IEEE Transactions on Power Systems, 31*(1), 777–785. https://doi.org/10.1109/TPWRS.2015.2412614

Grigsby, L. L. (2017). *Power system stability and control*. CRC Press.

Khooban, M., Niknam, T., Shasadeghi, M., Dragicevic, T., & Blaabjerg, F. (2018). Load frequency control in microgrids based on a stochastic noninteger controller. *IEEE Transactions on Sustainable Energy, 9*(2), 853–861. https://doi.org/10.1109/TSTE.2017.2763607

Kocaarslan, I., & Çam, E. (2005). Fuzzy logic controller in interconnected electrical power systems for load-frequency control. *International Journal of Electrical Power & Energy Systems, 27*(8), 542–549.

Kroposki, B., Johnson, B., Zhang, Y., Gevorgian, V., Denholm, P., Hodge, B., & Hannegan, B. (2017). Achieving a 100% renewable grid: Operating electric power systems with extremely high levels of variable renewable energy. *IEEE Power and Energy Magazine, 15*(2), 61–73. https://doi.org/10.1109/MPE.2016.2637122

Lal, D. K., & Barisal, A. K. (2019). Combined load frequency and terminal voltage control of power systems using moth flame optimization algorithm. *Journal of Electrical Systems and Information Technology, 6*(1), 1–24.

Lal, D. K., Barisal, A. K., & Tripathy, M. (2018). Load frequency control of multi area interconnected microgrid power system using grasshopper optimization algorithm optimized fuzzy PID controller. *2018 Recent Advances on Engineering, Technology and Computational Sciences (RAETCS)*, 1–6. https://doi.org/10.1109/RAETCS.2018.8443847

Liao, K., & Xu, Y. (2018). A robust load frequency control scheme for power systems based on second-order sliding mode and extended disturbance observer. *IEEE Transactions on Industrial Informatics, 14*(7), 3076–3086. https://doi.org/10.1109/TII.2017.2771487

Pan, I., & Das, S. (2012). *Intelligent fractional order systems and control: An introduction* (Vol. 438). Springer.

Pan, I., & Das, S. (2016). Fractional order fuzzy control of hybrid power system with renewable generation using chaotic PSO. *ISA Transactions, 62*, 19–29. https://doi.org/10.1016/j.isatra.2015.03.003

Roslan, M. F., Hannan, M. A., Ker, P. J., & Uddin, M. N. (2019). Microgrid control methods toward achieving sustainable energy management. *Applied Energy, 240*, 583–607. https://doi.org/10.1016/j.apenergy.2019.02.070

Ross, D. W., & Kim, S. (1980). Dynamic economic dispatch of generation. *IEEE Transactions on Power Apparatus and Systems, PAS-99*(6), 2060–2068. https://doi.org/10.1109/TPAS.1980.319847

Soundarrajan, A., Sumathi, S., & Sivamurugan, G. (2012). Voltage and frequency control in power generating system using hybrid evolutionary algorithms. *Journal of Vibration and Control, 18*(2), 214–227.

Xu, D., Liu, J., Yan, X., & Yan, W. (2018). A novel adaptive neural network constrained control for a multi-area interconnected power system with hybrid energy storage. *IEEE Transactions on Industrial Electronics, 65*(8), 6625–6634. https://doi.org/10.1109/TIE.2017.2767544

Digital Technology and Sustainable Manufacturing: The Nexus

A. Noiki, F. Ademuyiwa, S. A. Afolalu, M. B. Edun, O. O. Yusuf, and M. E. Emetere

Abbreviation

SM	Sustainable manufacturing
DT	Digital technologies
IoT	Internet of things
SACPS	Stand-alone cyber-physical systems
BDA	Big data analytics
AI	Artificial intelligence
ML	Machine learning
DL	Deep learning
NLP	Natural learning processing
ICT	Information and communication technology
VR	Virtual reality
AR	Augmented reality
AM	Additive manufacturing
DT	Digital technology
VMware	Virtual machine ware

A. Noiki · F. Ademuyiwa (✉) · S. A. Afolalu · M. B. Edun
Department of Mechanical Engineering, Covenant University, Ota, Nigeria
e-mail: ayodeji.noiki@ogitech.edu.ng; faith.ademuyiwa@covenantuniversity.edu.ng; adeniran.afolalu@abuad.edu.ng; edun.mosun@ogitech,edu.ng

O. O. Yusuf
Department of Microbiology, Obafemi Awolowo University, Ile-Ife, Nigeria

M. E. Emetere
Department of Physics, Covenant University, Ota, Nigeria
e-mail: moses.emetere@covenantuniversity.edu.ng

A. O. Ayeni et al. (eds.), *Advanced Manufacturing in Biological, Petroleum, and Nanotechnology Processing*, Green Energy and Technology,
https://doi.org/10.1007/978-3-030-95820-6_27

1 Introduction

In this age of the Fourth Industrial Revolution, manufacturing organizations are under severe pressure to create sustainable manufacturing processes (Ghobakhloo, 2020; Sharma et al., 2021). With the recent emerging circular economy, ventures with sustainable and ethical practices are in high demand in our present world (Ngan et al., 2019). Research has shown that in achieving the objective of sustainable practices, digital technologies will play a paramount role (Kumar et al., 2020). The present manufacturing sector stands on the edge of the coming industrial revolution, the significant growth of the novel digital technology revolution termed Industry 4 (Li et al., 2017).

Recently, sustainable manufacturing (SM) is leading the transformation in the way businesses operate. SM is an indication of complete integration, collaborative manufacturing eco-friendly system that reacts to the continuous demands and the dynamic environment within the value chain (Bag et al., 2021a). The essentials of value creation and delivery in the manufacturing industry have been severely altered by SM through the technological revolution. The synergy between the physical and the digital world is the nucleus of sustainable manufacturing. This implies that the incorporation of information and digital technologies (DT) to every phase of manufacturing processes is paramount to modern-day manufacturers (Ngan et al., 2019). Manufacturers are progressively developing toward the global supply chains, made possible by the recent operational ties of digital technologies.

Moreover, digitalization of manufacturing processes and their wide adoption have characterized a new label of industrial transformation and its implementation in the manufacturing sector. The present-day acceptance of digital technology and its exploitation have significantly transformed business operations and impact business outcomes, most especially the manufacturing firms where implementation of digital technologies centered around interconnectivity among the manufacturing activities and stakeholders at various levels. This industrial transformation comprises numerous technologies that link and incorporate physical infrastructure, intelligent devices, and human factors driving this novel manufacturing processes (Ghadge et al., 2020). This emerging trend is a technology-based manufacturing shift that brought new issues at various stages of manufacturing processes and new opportunities as well both for new operational strategies and new business approach leading to an industrial revolution (Demeter et al., 2020). This shift necessitates the development of mechanisms, capacities, and strategies for its deployment (Ante, 2021). SM is termed as the making of manufactured products via a cost-effective process aimed at minimizing the negative environmental effects while energy consumption and resources have been conserved (Kuo de souse Jabbour et al., 2018). The significant growth of SM is due to twenty-first-century manufacturing under the prevalent environmental condition of energy and resource deficiency. Numerous efforts are made by several researchers to develop novel technologies for the new manufacturing paradigm shift (Lee et al., 2019). Present-day manufacturers are striving continuously to make their manufacturing practices progressively sustainable from an economic and environmental viewpoint. However, numerous manufacturing system models have been deployed for holistic assessment of various process configurations to ascertain their sustainability and the most preferred configuration (Fisher et al., 2021).

2 Components of Digital Technology

2.1 Internet of Things

Automation of manufacturing processes is implemented via the internet of things (IoT)-based technique and operations (Roy et al., 2020; Zheng et al., 2021). This covers the entire manufacturing process from the preliminary to the final stage of product delivery. Numerous data obtained from IoT-supported manufacturing are employed in addressing key issues associated with computation, a mechanism for control, and connectivity (Bosi et al., 2020). IoT-enabled gadgets can be utilized in the management of water systems, agro-allied and chemical manufacturing industries, power generating plants, and food and beverage manufacturing plants (Khatua et al., 2020). Sensors are utilized for the regulation and control of production flow lines. These are also used for the collection of real-time information regarding the process performance on the production floor (Zheng & Sivabalan, 2020). Furthermore, this helps in controlling production activities as well as improvement in product quality. Consequently, this enhances rapid production processes and aids the economic production cycle (Lakshmi et al., 2021). Since IoT is a development that facilitates autonomous communication between a series of activities, thus creating a volume of data that can be used for predictive analysis or decision-making (Sestino et al., 2020). IoT plays a paramount role in the digitalization of the supply chain management process. This implies an improvement in technological development connecting products and devices across the Internet. Several kinds of research have termed the utilization of IoT in product development as "Smart" (Benitez et al., 2021; Jasko et al., 2020). The emerging trends perceive real-time monitoring of manufacturing processes via the lens of sensor capability, whereby communication network permits equipment to be configured and for effective handling of production processes through selective operations (Jiao et al., 2020). The deployment of this novel technology concept has become important in facilitating the following: production line automation, process material management, product tracing and tracking, and energy and resource conservation (Margherita & Braccini, 2020; Leng et al., 2020). Consequently, resulting in an effective manufacturing sustainable supply chain process (Luthra & Mangla, 2018). It affords an exceptional level of prospect, responsiveness, and flexibility to deal with supply chain issues. IoT aids virtualization of the supply chain and thereby permits tracking and tracing of goods in transit along with planning and quality control checks (Tazhiyeva, 2018).

2.2 Stand-Alone Cyber-Physical System

Stand-alone cyber-physical systems (SACPS) are cutting-edge systems that are valuable in the reduction of cost and connected to the improvement of operational performance, the safety of humans in the workplace, and rapid production processes (Oks et al., 2017; Aheleroff et al., 2021). This system creates a connection between the physical and virtual worlds. This connotes the incorporation of computing with physical manufacturing processes which is a keystone for the implementation of Industry 4.0 (Lee et al., 2020; Pivoto et al., 2021). This system ensures real-time data and information transmission among components of virtual and physical settings, thereby permitting a high level of synchronization control, openness, and efficiency among supply chain stakeholders which leads to improvement in productivity and operational performance (Minerva et al., 2020). The supply chain management process involves the incorporation of technology in manufacturing operations, for example, assembly process regarding operations that ensue in logistics. This system initiates the foundation for serial manufacturing of items with a higher degree of quality and higher efficiency production processes (Rossit et al., 2019). Subsequently, traditional manufacturing processes are harmonized and characterized in the digital world (Peruzzini et al., 2020).

2.3 Big Data Analytics

Manufacturing supply chain processes are becoming sophisticated; it is anticipated that the volume of data generated for planning and assessment will become bigger. The management

of high volumes, high speeds, and multiplex data requires massive use of technology (Nascimento et al., 2019; Manavalan & Jayakrishna, 2019). The core of big data analytics (BDA) is identifying "what will happen" rather than "what has happened." The data represents a prediction of potential possibilities or unidentified occurrences (Sivri & Oztaysi, 2018). Due to the emerging trend, the use of big data has shifts from the collection of data to analytics of result processed data outcomes. The prevalent techniques for analyzing the supply chain resulted in three main branches which are the following: simulation, optimization, and statistical analysis (Akbari & Do, 2021). Research has shown that big data analytics at the buyer ends through the performance of reduction of operations can achieve numerous objectives such as mutual trust between manufacturers and suppliers, promotion of data sharing, reduced service utilization cost, and delegation of data sharing control (Dutta et al., 2020).

2.4 *Artificial Intelligence*

Artificial intelligence (AI) is an idea that assists machine learning (ML) which portrays the ability to learn without being trained. Mostly, ML is developed on neural networks such as deep learning (DL) and natural learning processing (NLP) (Ghoreishi & Happonen, 2020; Dubey et al., 2020). Recently, an average cyber-physical system does not incorporate AI tools that lead to training and solving a problem. This relatively serves as an available source of data, obtained from numerous types of machinery and other sources in a production plant and recirculated to users and stakeholders (Zhang et al., 2020). The integration of AI tools and systems is a significant step toward the utilization and generation of applications that result in efficient use of manufacturing resources despite a sudden change in production plans and as a response to production disruptions (Abubakr et al., 2020; Lu et al., 2020). Artificial intelligence aids the accomplishment of big data analytics via its computing capability and learning ability. Thus, it helps in processing data and identification of data pattern, as well as improving the manufacturing value chain (Osornio & Prieto, 2020).

2.5 *Cloud Computing*

Cloud computing (CC) is a noteworthy and evolving terminology in the information and communication technology (ICT) world that attains relevance in numerous sectors of the manufacturing industry (Arromba et al., 2020). This is a fast-rising emerging core area of the digital economy that allows the use of computer systems connected to the Internet (Taghipour et al. 2020). It has progressively changed its application and transformed how organizations manage their service and interactions between suppliers and customers (Primi & Toselli, 2020). Cloud technology typifies easy online storage of data and information and retrieval platforms via a web-based application without any installation. This technology permits convenient and cost-effective access to data (Sony & Aithal, 2020). Thus, it affords convenient planning and operations for the supply chain stakeholders, bringing about improvement in a firm's performance and efficiency (Li et al., 2020; Kayemuddin, 2019). Numerous cloud platforms have been in existence, for example, Amazon Elastic Compute Cloud, Microsoft Live Mesh, Azure, VMware, etc. Computational capabilities have been aligned with sustainability and operational efficiency for both small and large manufacturing firms by several researchers. Due to the challenge of cloud security, systems have been proposed for securing stored data in cloud computing via private-preserving public auditing system (Krishnamoorthy, 2021).

2.6 Virtual Reality and Augmented Reality

In the present-day manufacturing sector, virtual technologies are becoming prevalent; this includes maintenance, product development, product design, process development, etc. (Roblet et al., 2020). The virtualization process allows the virtual display of information providing an array of improved services to the firm through the utilization of virtual reality (VR) and augmented reality (AR) (Rejeb et al., 2020). The two concepts are not dissimilar; rather both complement human intellects to deliver all-inclusive information that cannot be comprehended by an end-user, for example, head-mounted display, smartphone camera, scanners lens, projection gadgets with 3D models, and spoken instructions. In addition, operational instructions in consecutive order can be displayed via the use of AR. However, in a maintenance system, augmented reality can facilitate useful instructions that are required to perform or complete a particular task. A few of the sustainable manufacturing enterprises have applied the use of augmented reality for facilitation of training and other services, product delivery, and identification inclusive (Baroroh et al., 2020).

2.7 Additive Manufacturing

Additive manufacturing (AM) unlike conventional machining processes where material removal is fundamental such as cutting, milling, and drilling operations. AM applications involved the use of computer-aided design popularly referred to as 3D printing. This technique aids decision-makers in creating prototypes. This is a sustainable manufacturing process that can bring about improved operational flexibility, offer reduced lead times and inventory reduction, and permit customization of the product. These advantages will enhance process performance and competitive edge over the traditional manufacturing processes (Li & Jiang, 2021). Thus, aids in creating a product with complex and intricate shapes that are difficult to produce using traditional manufacturing processes. This technique has a wide application in machining manufacturing organizations. Several manufacturers in the developed nation like Germany, Italy, the USA, and Japan have adopted the use of 3D printing for part manufacturing or prototyping automotive parts, marine component parts, defense parts, and motorsport components (Fig. 1).

3 Sustainable Manufacturing Approach

Sustainable manufacturing (SM) is a vital shift in which manufacturers produce items using a sustainable method while conserving global competitiveness and handling emerging challenges. SM is a system that takes into account the adverse effect

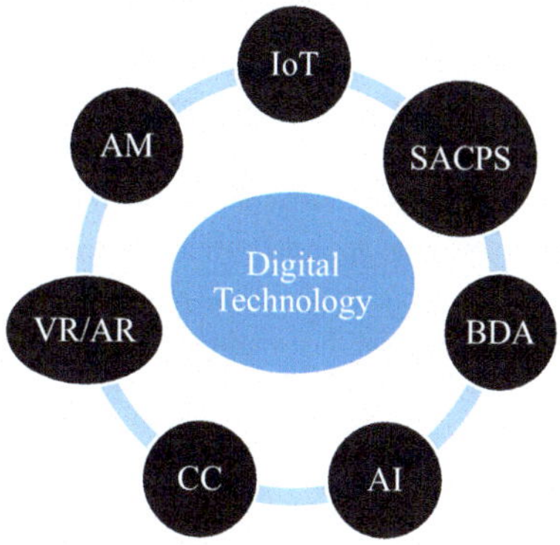

Fig 1 Components of digital technology

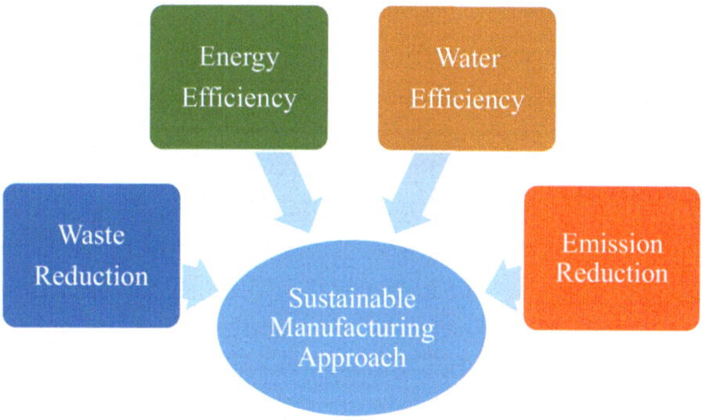

Fig 2 Sustainable manufacturing approach

of the manufacturing process on the natural resources, conservation of energy, the environment, consumers, and societal safety in an economically feasible procedure. Noteworthy financial and environmental benefits are gains of businesses conducted sustainably. However, sustainable manufacturing practice must be applied to the entire manufacturing processes and distribution process inclusive (Onu & Mbohova, 2021). Sustainable manufacturing is approached from the following perspectives: energy efficiency, water efficiency, emission reduction, and waste reduction (Fig. 2).

4 The Implications of Digitalization on Sustainable Manufacturing

Industry 4.0 has been adopted by several manufacturing firms to enhance and advance manufacturing capacities and by extension to meet their sustainable development goals. Industry 4.0 has brought development in technologies that are completely transforming the traditional manufacturing configuration (Bai et al., 2020; Leng et al., 2020). This encompasses advanced technologies for sensing, information processing, visualization, and network communication. Though there are kinds of digital systems and tools that result in improvement in planning and manufacturing operations. Aside from aiming at product customization and reduced throughput times, also resource and energy efficiency are improved. Digitalization and sustainability are both key drivers for change in manufacturing organizations. There is a strong synergy between sustainability and digitalization (Ozkan-Ozen et al., 2020). Digital systems and devices have brought significant improvement to the environmental performance of manufacturing-related processes. Furthermore, additional efforts required for the establishment of ICT must be taken into consideration (Thiede, 2021; Glatt et al., 2021).

4.1 Process Digitalization

Process and system efficiency in sustainable manufacturing can as well enhance manufacturing operations and other energy-related pointers. Simulation tools are often used to improve such kind of efficiency via physical asset diagnosis and evaluation (Bevilacqua et al., 2020). Specifically, the digital twin technology is presently attracting significant attention due to the technical functionality it provides at the category and phases of a work center. Additionally, it thus aids efficient control and decision-making via horizontal coordination and vertical integration (Lim et al., 2020). Also, the digital twin technology can influence the prevention of energy-associated inefficiency through the provision of functionalities necessary for the service structure of the functionalities that are outlined for improvement of process efficiency (Park et al., 2020). Through digital technology, there are several kinds of research addressing the life cycle analysis of products; this helps in improving the efficiency of simulation and modeling. However, there is limited research work that has examined the operations of DT (Dutta et al., 2020). There is no specific information as regards the implementation of DT using either a structured or an unstructured approach, since there no comprehensive definitions for the significant steps associated with the digital technology operations, for example, conception, synchronization, and deployment (Alexopoulos et al., 2020; Pacchini et al., 2019).

4.2 Quality Control Tools

In other to improve the flexibility, reliability of manufacturing processes, and the quality of the product, digital technologies are being applied (Shi et al., 2020). More so, they can be applied for examining and assessing the environmental effect of processes, improvement in the production quality control, and other supplementary processes that contribute to the environmental effect, thus providing vital performance indicators for efficiency of resources (Gillani et al., 2020). Through digital technology, an online monitoring system can be implemented based on the IoT. In the manufacturing sector, traditional database technology often encounters difficulties in the management of large volumes of data that are both structured and unstructured (Chiarini et al., 2020; Paiola & Gebauer, 2020). CC and BDA are significant tools that can be employed on historical data for the identification of product quality issues and thereby reduction in product rejects. Most manufacturers have adopted big data strategies for both monitoring and improvement of product quality (Gao et al., 2020). This technology utilizes a novel processing approach to obtain vital information from varying data sources and to have an in-depth understanding, gain insight, and make findings that will aid decision-making. Also, prediction analytics based on online data collected from the production line can aid in making the necessary adjustment to the production line to minimize the occurrence of product defects (Szalavetz, 2019; Muller & Voigt, 2018).

4.3 Production Efficiency

An energy-efficient manufacturing process can be achieved through the entire production stage with the application of effective data management. Energy data can be obtained through both on-site and off-site interfaces. This comprises water, power, and diesel data acquisition techniques, which can be employed as an on-site interface for the collection of real-time energy data (Ma et al., 2020). Consumption of energy can be monitored which includes the incorporation of special equipment. Multidimensional statistical tools are utilized for year-on-year analysis and product trend analysis for management assessment of energy efficiency. Consequently, optimization analyses of energy efficiency are executed in terms of process optimization, management energy evaluation, and audit (Ma et al., 2020). DT facilitates a major improvement in resource productivity through standard analytics where a large volume of data from a unit factory is utilized for the identification of "golden moments" where a product is produced using half the energy required or comparative analysis of carbon dioxide emission of the same product at the different time from different factories (Kong et al., 2020). Also, through the application of AI, AR, BDA, and IoT, improvement in resource productivity can be realized. Augmented reality is a key for driving resource productivity; this aids production engineers to visualize real-time flows of energy, waste, and water in the factory environment (Harikannan et al., 2020). One of the key benefits of utilizing digital technologies is to make visible hidden information that will contribute to a major reduction in total manufacturing costs (Sellitto et al., 2019).

4.4 Process Innovation and Sustainability

The manufacturing industry can design a reward or punishment system for each member of the production team based on the energy efficiency assessment of the team. This entails the evaluation of the production shop floor, production line, or machines which can deduce the variation in energy efficiency (Wheeldon et al., 2020). Thereafter, optimization analysis can be supplied for enterprise applications. The overall goal of a manufacturing system is to work within the capacity of the planet to deliver its resources such as energy, water, and other natural resources. Since the alternatives are not sustainable (Pontevedra et al., 2018). Numerous large and small manufacturing firms are engaging this ambition of sustainable manufacturing practices with great determination. However, it is evident that sustainability is not yet universal, but many leading organizations, for example, Unilever and Toyota, are taking giant strides toward achieving sustainable manufacturing practices (Sommer, 2017). Through the identification of input materials that locally available, they also make a critical examination of their products to assess products that can be produced and in a similar manner how to adjust their bill of material and quantities for product to be made through reuse, recycling, and remanufacturing (Leal et al., 2020). This can only be achieved due to flexibility arising from DT and process control technology and automation. Numerous manufacturing firms are utilizing the power of increasing data availability for product and manufacturing discovery and remanufacture such products (Oluyisola et al. 2020; Gohari et al., 2019).

4.5 Optimization Process

Manufacturing processes can be easily optimized with the incorporation of DT and a standard operating procedure, thus providing numerous alternatives to optimize business processes and remarkably reduce resources and lead times (Csalódi et al., 2021; Chiarini & Kumar, 2020). However, in an impulsive business environment, firms practicing recycling and remanufacturing of products do encounter challenges, for example, material supply-related bottlenecks, production downtime and losses, and excess inventory of spares which ultimately affect the efficiency of the overall operation. Optimization of energy and water efficiency can be achieved through a cloud platform. Consequently, this can be employed for the management of water and energy resources (Wamba, 2021). Recently, there are applications for running analysis of equipment, water, and energy warning, demand response, and optimization of water and energy parameters. These applications are a data-driven management system for advancing the sustainability of manufacturing practices (Meng et al., 2018). Analysis of equipment can automatically aid identification of runtime, downtime, and shutdown time of equipment (Ma et al., 2020).

4.6 Organization Flexibility and Improved Responsive Performance

The capability of an organization to perform its operations in an unstable business environment is termed "organization flexibility." The characteristics of organization flexibility encompass the ability to prompt adjustment of organization structure and quick response to change in the business environment (Wamba et al., 2020).This capability includes organizational structural change without any negative effects on the product or service quality. This also aids an organization in quick adaptation to ever-changing business situations and consequently staying ahead of competitors. SM and BDA are keenly dependent on organizational structure (Ali et al., 2020; Raut et al., 2019).

4.7 Environmental Performance

SM decision-making involves manufacturing system design and operations which are associated with the green design of both processes and products and management of the environmental impact of supply chain activities, respectively (Ma et al., 2020). Researchers have been able to identify and classify practices that tackle environmental and sustainable manufacturing decision-making (Machado et al., 2020; Malek & Desai, 2020; Hendiani et al., 2020; Raut et al., 2019). This entails designing for the environment which is termed product creation that demands fewer natural resources and harmful pollutants, waste minimization, and part reuse, recycle, and remanufacture (Ali et al., 2020). Also, cleaner production, which is established on material and production planning and process control geared toward less consumption of materials and waste generation. In addition, initiatives toward green supply chain management entail green logistics and green purchasing decisions to minimize the environmental effect and responsible waste disposal and recycling systems (De Giovanni & Cariola, 2020; Raut et al., 2019). To extend the product life cycle, reuse, recycle, and reduction have been recommended and utilization of clean technologies to reduce environmental pollution (Zhang et al., 2020). DT is an essential component of SM decisions; it is crucial to lowering production resources and consequently creating less harmful effects on the environment (Bag et al., 2021b). SM also necessitates critical analysis of production infrastructures and facilities that aids in manufacturing operations (Majeed et al., 2021).

5 Challenges

Despite the significant roles of digital technology in sustainable manufacturing, there are numerous challenges associated with this system due to its relative newness. Manufacturers are faced with difficulties such as a lack of technical know-how, financial limitations, and operational intricacies (Chirumalla, 2021). There is a need to develop a proper delivery system to overcome these challenges. The decision-making quality of managers is largely influenced by uncertainties. This as well can lead to a remarkable increase in expenditure, thereby reducing the profit margin of the manufacturing organization (Xu et al., 2020). Manufacturers are often keen about not losing customer's orders, thereby maintaining a slack inventory such as carrying all high lead time raw materials. Stocking these materials has a great effect on the working capital (Dbouk et al., 2020). Also, due to the increase in technological advancement, there is a probability of stock becoming outdated at some point which can lead to a huge financial loss (Ghisetti et al., 2017). More so, due to the state-of-the-art equipment required, the skill and training required by operating personnel are one of the key factors militating against the implementation of digital technology. Furthermore, there are no standard metrics for a comprehensive assessment of sustainable practices; the high cost of sustainable practices and technology cannot be overlooked as manufacturers are concerned about the financial implication (Srivastava et al., 2021). Thus manufacturers are reluctant in adopting this emerging technology.

6 Opportunities

Successful adoption of digital technology in achieving sustainable manufacturing can improve both the vertical and horizontal integration of manufacturing activities of organizations (Gupta et al., 2021). Adoption of DT and SM can bring about a 10% significant increase in the operational efficiency of the firm. Manufacturers can expect over a 10% reduction in operational cost via the adoption of sustainable manufacturing/digital technology integrated with the planning and scheduling of manufacturing activities (Felsberger et al., 2020; Paschou et al., 2020). Digital technology affords a significant reduction in machine downtime and production delays via predictive maintenance (Amrani et al., 2020). Manufacturers with a high level of digital technology adoption can deploy front-end and base technologies that will accomplish operational flexibility, improvement in operational effectiveness, and efficiency (Gillani et al., 2020; Frank et al., 2019). Manufacturing organizations play an important role in achieving sustainable development (Bhatt et al., 2020). However, the demand for high technology-oriented infrastructure and the financial implication remain a key issue. Integrating sustainability right from the product design stage to the product distribution stage can reduce the effect of global warming while permitting manufacturers to become competitive in the international market (Noiki et al., 2021).

7 Future Research

Optimization of manufacturing processes and operations via continuous improvement of operational efficiencies is attracting the attention of researchers. Achieving sustainable production and consumption pattern should involve the application of technological innovations such as automation, digital, and additive manufacturing technologies that will further reduce the manufacturing and labor costs and, consequently, result in greater operational efficiency. Besides, there is a need to increase energy efficiency to realize clean and renewable energy consumption and curtail energy usage in the manufacturing process. In addition, generating less emission and pollution and waste can reduce the cost of disposing waste, thereby creating a cleaner and safer environment (Noiki et al., 2021). This necessitates an industrial synergy approach to represent a sustainable cycle of an industrial system, whereby disposed material from various resources is reuse. Future sustainable manufacturing companies are expected to incorporate technological innovation into their businesses to enhance operations; this will as well necessitate an academic-industrial collaboration whereby research findings that will provide industrial solutions can be implemented. The future of sustainable manufacturing is all-inclusive, which requires a systematic approach, integrated infrastructure, application of designated tools, and interoperability measures. Since technological evolution is unending, sustainability should be pushed beyond the limit; technical readiness is a key factor.

8 Conclusion

In this era of Industry 4.0, digital technology plays a paramount role in advancing sustainability in the manufacturing processes. The synergy of sustainable manufacturing and digital technology is at crucial stages. Deploying DT is an effective and efficient way to advance sustainability. Therefore, this paper aimed to present a critical review of the digital technology-driven sustainable manufacturing process for realizing sustainable manufacturing and consequently contributing to the advancement of a sustainable society. In this study, components of digital technology and their deployment in the manufacturing sector were briefly discussed, such as IoT, big data analytics, cyber-physical systems, artificial intelligence, cloud computing, virtual and augmented reality, and additive manufacturing. Secondly, the various approaches to sustainable manufacturing were examined, for example, energy efficiency, water efficiency, waste reduction, and emission reduction. Thirdly, the implications of the integration of digitalization on the sustainable manufacturing processes bring about improvement in operational efficiency, process digitalization, sustainability, and optimization. Fourthly, challenges, opportunities, and future research associated with the implementation of digital technology in sustainable manufacturing practices were examined. However, the high cost of investment required for its implementation in developing countries remains a crucial factor.

Acknowledgments We acknowledge the financial support offered by Covenant University in the actualization of this research work for publication.

References

Abubakr, M., Abbas, A. T., Tomaz, I., Soliman, M. S., Luqman, M., & Hegab, H. (2020). Sustainable and smart manufacturing: An integrated approach. *Sustainability, 12*(6), 2280.

Aheleroff, S., Xu, X., Zhong, R. Y., & Lu, Y. (2021). Digital twin as a service (DTaaS) in industry 4.0: An architecture reference model. *Advanced Engineering Informatics, 47*, 101225.

Akbari, M., & Do, T. N. A. (2021). *A systematic review of machine learning in logistics and supply chain management: Current trends and future directions*. An International Journal.

Alexopoulos, K., Nikolakis, N., & Chryssolouris, G. (2020). Digital twin-driven supervised machine learning for the development of artificial intelligence applications in manufacturing. *International Journal of Computer Integrated Manufacturing, 33*(5), 429–439.

Ali, S. S., Kaur, R., Persis, D. J., Saha, R., Pattusamy, M., & Sreedharan, V. R. (2020). Developing a hybrid evaluation approach for the low carbon performance on sustainable manufacturing environment. *Annals of Operations Research*. https://doi.org/10.1007/s10479-020-03877-1

Amrani, M. A., Shameeri, M., Amrani, M. A., Al-Qubati, M., Alhomdi, M., Aswaidy, M. B., Ghaleb, A. M., Al-Qubati, M., & Amrani, M. A. (2020). Implementing an integrated maintenance management system for monitoring production lines: A case study for biscuit industry. *Journal of Quality in Maintenance Engineering*. https://doi.org/10.1108/JQME-06-2020-0049

Ante, L. (2021). Digital twin technology for smart manufacturing and industry 4.0: A bibliometric analysis of the intellectual structure of the research discourse. *Manufacturing Letters, 27*, 96–102.

Arromba, I. F., Martin, P. S., Ordoñez, R. C., Anholon, R., Rampasso, I. S., Santa-Eulalia, L. A., & Quelhas, O. L. G. (2020). Industry 4.0 in the product development process: Benefits, difficulties and its impact in marketing strategies and operations. *Journal of Business & Industrial Marketing*.

Bag, S., Pretorius, J. H. C., Gupta, S., & Dwivedi, Y. K. (2021a). Role of institutional pressures and resources in the adoption of big data analytics powered artificial intelligence, sustainable manufacturing practices and circular economy capabilities. *Technological Forecasting and Social Change, 163*, 120420. https://doi.org/10.1016/j.techfore.2020.120420

Bag, S., Gupta, S., & Kumar, S. (2021b). Industry 4.0 adoption and 10R advance manufacturing capabilities for sustainable development. *International Journal of Production Economics, 231*, 107844.

Bai, C., Dallasega, P., Orzes, G., & Sarkis, J. (2020). Industry 4.0 technologies assessment: A sustainability perspective. *International Journal of Production Economics, 229*, 107776.

Baroroh, D. K., Chu, C. H., & Wang, L. (2020). Systematic literature review on augmented reality in smart manufacturing: Collaboration between human and computational intelligence. *Journal of Manufacturing Systems*.

Benitez, G. B., Ferreira-Lima, M., Ayala, N. F., & Frank, A. G. (2021). Industry 4.0 technology provision: The moderating role of supply chain partners to support technology providers. *Supply Chain Management: An International Journal.*

Bevilacqua, M., Bottani, E., Ciarapica, F. E., Costantino, F., Di Donato, L., Ferraro, A., & Vignali, G. (2020). Digital twin reference model development to prevent operators' risk in process plants. *Sustainability, 12*(3), 1088.

Bhatt, Y., Ghuman, K., & Dhir, A. (2020). Sustainable manufacturing. Bibliometrics and content analysis. *Journal of Cleaner Production, 260*, 120988.

Bosi, F., Corradi, A., Modica, D., Foschini, L., Montanari, R., Patera, L., & Solimando, M. (2020, September). Enabling smart manufacturing by empowering data integration with industrial IoT support. In *2020 International Conference on Technology and Entrepreneurship (ICTE)* (pp. 1–8). IEEE.

Chiarini, A., & Kumar, M. (2020). Lean six sigma and industry 4.0 integration for operational excellence: Evidence from Italian manufacturing companies. *Production Planning & Control*, 1–18.

Chiarini, A., Belvedere, V., & Grando, A. (2020). Industry 4.0 strategies and technological developments. An exploratory research from Italian manufacturing companies. *Production Planning & Control, 31*(16), 1385–1398.

Chirumalla, K. (2021). Building digitally-enabled process innovation in the process industries: A dynamic capabilities approach. *Technovation*, 102256.

Csalódi, R., Süle, Z, Jaskó, S., Holczinger, T, & Abonyi, J. (2021). Industry 4.0-driven development of optimization algorithms: A systematic overview. Complexity.

Dbouk, W., Moussawi-Haidar, L., & Jaber, M. Y. (2020). The effect of economic uncertainty on inventory and working capital for manufacturing firms. *International Journal of Production Economics, 230*, 107888.

De Giovanni, P., & Cariola, A. (2020). Process innovation through industry 4.0 technologies, lean practices and green supply chains. *Research in Transportation Economics*, 23:22 100869.

de Sousa Jabbour, A. B. L., Jabbour, C. J. C., Foropon, C., & Godinho Filho, M. (2018). When titans meet–Can industry 4.0 revolutionise the environmentally-sustainable manufacturing wave? The role of critical success factors. *Technological Forecasting and Social Change, 132*, 18–25.

Demeter, K., Losonci, D., & Nagy, J. (2020). Road to digital manufacturing – A longitudinal case-based analysis. *Journal of Manufacturing Technology Management.*

Dubey, R., Gunasekaran, A., Childe, S. J., Bryde, D. J., Giannakis, M., Foropon, C., & Hazen, B. T. (2020). Big data analytics and artificial intelligence pathway to operational performance under the effects of entrepreneurial orientation and environmental dynamism: A study of manufacturing organisations. *International Journal of Production Economics, 226*, 107599.

Dutta, T., Barman, A., & Majumdar, G. (2020a). *Green and sustainable manufacturing of metallic, ceramic and composite materials*. Elsevier.

Dutta, G., Kumar, R., Sindhwani, R., & Singh, R. K. (2020b). Digital transformation priorities of India's discrete manufacturing SMEs – A conceptual study in perspective of Industry 4.0. *Competitiveness Review: An International Business Journal.*

Felsberger, A., Qaiser, F. H., Choudhary, A., & Reiner, G. (2020). The impact of Industry 4.0 on the reconciliation of dynamic capabilities: Evidence from the European manufacturing industries. *Production Planning & Control*, 1–24.

Fisher, O. J., Watson, N. J., Porcu, L., Bacon, D., Rigley, M., & Gomes, R. L. (2021). Multiple target data-driven models to enable sustainable process manufacturing: An industrial bioprocess case study. *Journal of Cleaner Production, 296*, 126242.

Frank, A. G., Mendes, G. H., Ayala, N. F., & Ghezzi, A. (2019). Servitization and Industry 4.0 convergence in the digital transformation of product firms: A business model innovation perspective. *Technological Forecasting and Social Change, 141*, 341–351.

Gao, R. X., Wang, L., Helu, M., & Teti, R. (2020). Big data analytics for smart factories of the future. *CIRP Annals, 69*(2), 668–692.

Ghadge, A., Kara, M. E., Moradlou, H., & Goswami, M. (2020). The impact of Industry 4.0 implementation on supply chains. *Journal of Manufacturing Technology Management*. 12:2, 2200.

Ghisetti, C., Mancinelli, S., Mazzanti, M., & Zoli, M. (2017). Financial barriers and environmental innovations: Evidence from EU manufacturing firms. *Climate Policy, 17*(sup1), S131–S147.

Ghobakhloo, M. (2020). Determinants of information and digital technology implementation for smart manufacturing. *International Journal of Production Research, 58*(8), 2384–2405.

Ghoreishi, M., & Happonen, A. (2020, May). Key enablers for deploying artificial intelligence for circular economy embracing sustainable product design: Three case studies. *AIP Conference Proceedings, 2233*(1), 050008.

Gillani, F., Chatha, K. A., Jajja, M. S. S., & Farooq, S. (2020). Implementation of digital manufacturing technologies: Antecedents and consequences. *International Journal of Production Economics, 229*, 107748.

Glatt, M., Kölsch, P., Siedler, C., Langlotz, P., Ehmsen, S., & Aurich, J. C. (2021). Edge-based Digital Twin to trace and ensure sustainability in cross-company production networks. *Procedia CIRP, 98*, 276–281.

Gohari, H., Berry, C., & Barari, A. (2019). A digital twin for integrated inspection system in digital manufacturing. *IFAC-PapersOnLine, 52*(10), 182–187.

Gupta, H., Kumar, A., & Wasan, P. (2021). Industry 4.0, cleaner production and circular economy: An integrative framework for evaluating ethical and sustainable business performance of manufacturing organizations. *Journal of Cleaner Production, 295*, 126253.

Harikannan, N., Vinodh, S., & Gurumurthy, A. (2020). Sustainable industry 4.0 – An exploratory study for uncovering the drivers for integration. *Journal of Modelling in Management.*

Hendiani, S., Liao, H., Bagherpour, M., Tvaronavičienė, M., Banaitis, A., & Antucheviciene, J. (2020). Analyzing the status of sustainable development in the manufacturing sector using multi-expert multi-criteria fuzzy decision-making and integrated triple bottom lines. *International Journal of Environmental Research and Public Health, 17*(11), 3800.

Jaskó, S., Skrop, A., Holczinger, T., Chován, T., & Abonyi, J. (2020). Development of manufacturing execution systems in accordance with Industry 4.0 requirements: A review of standard-and ontology-based methodologies and tools. *Computers in Industry, 123*, 103300.

Jiao, J., Zhou, F., Gebraeel, N. Z., & Duffy, V. (2020). Towards augmenting cyber-physical-human collaborative cognition for human-automation interaction in complex manufacturing and operational environments. *International Journal of Production Research, 58*(16), 5089–5111.

Kayemuddin. (2019). *Fareast International University Journal* (Vol. 3). Fareast International University.

Khatua, P. K., Ramachandaramurthy, V. K., Kasinathan, P., Yong, J. Y., Pasupuleti, J., & Rajagopalan, A. (2020). Application and assessment of internet of things toward the sustainability of energy systems: Challenges and issues. *Sustainable Cities and Society, 53*, 101957.

Kong, L., Zhao, J., Li, J., Lou, R., & Zhang, Y. (2020). Evaluating energy efficiency improvement of pulp and paper production: Case study from factory level. *Journal of Cleaner Production, 277*, 124018.

Krishnamoorthy, N. (2021). A review study on security issues, benefits, risks and various challenges in cloud computing platform and proposed model for enhancing security for sensitive data. *Turkish Journal of Computer and Mathematics Education (TURCOMAT), 12*(9), 1997–2012.

Kumar, R., Singh, R. K., & Dwivedi, Y. K. (2020). Application of Industry 4.0 technologies in Indian SMEs for sustainable growth: Analysis of challenges. *Journal of Cleaner Production, 257*, 124063.

Lakshmi, S. V., Janet, J., Rani, P. K., Sujatha, K., Satyamoorthy, K., & Marichamy, S. (2021). Role and applications of IoT in materials and manufacturing industries – Review. *Materials Today: Proceedings, 45*, 2925–2928.

Leal, J. M., Pompidou, S., Charbuillet, C., & Perry, N. (2020). Design for and from recycling: A circular ecodesign approach to improve the circular economy. *Sustainability, 12*(23), 9861.

Lee, H. T., Song, J. H., Min, S. H., Lee, H. S., Song, K. Y., Chu, C. N., & Ahn, S. H. (2019). Research trends in sustainable manufacturing: A review and future perspective based on

research databases. *International Journal of Precision Engineering and Manufacturing-Green Technology, 6*(4), 809–819.
Lee, J., Azamfar, M., Singh, J., & Siahpour, S. (2020). Integration of digital twin and deep learning in cyber-physical systems: Towards smart manufacturing. *IET Collaborative Intelligent Manufacturing, 2*(1), 34–36.
Leng, J., Ruan, G., Jiang, P., Xu, K., Liu, Q., Zhou, X., & Liu, C. (2020). Blockchain-empowered sustainable manufacturing and product lifecycle management in industry 4.0: A survey. *Renewable and Sustainable Energy Reviews, 132*, 110112.
Li, P., & Jiang, P. (2021). Enhanced agents in shared factory: Enabling high-efficiency selfsorganization and sustainability of the shared manufacturing resources. *Journal of Cleaner Production, 292*, 126020.
Li, G., Hou, Y., & Wu, A. (2017). Fourth Industrial Revolution: Technological drivers, impacts and coping methods. *Chinese Geographical Science, 27*(4), 626–637.
Li, J., Maiti, A., Springer, M., & Gray, T. (2020). Blockchain for supply chain quality management: Challenges and opportunities in context of open manufacturing and industrial internet of things. *International Journal of Computer Integrated Manufacturing, 33*(12), 1321–1355.
Lim, K., Zheng, P., & Chen, C. (2020). A state-of-the-art survey of Digital Twin: Techniques, engineering product lifecycle management and business innovation perspectives. *Journal of Intelligent Manufacturing, 31*(6), 1313–1337.
Lu, Y., Liu, C., Wang, K. I.-K., Huang, H., & Xu, X. (2020). Digital Twin-driven smart manufacturing: Connotation, reference model, applications and research issues. *Robotics and Computer-Integrated Manufacturing, 61*(July), 101837. https://doi.org/10.1016/j.rcim.2019.101837
Luthra, S., & Mangla, S. K. (2018). Evaluating challenges to industry 4.0 initiatives for supply chain sustainability in emerging economies. *Process Safety and Environmental Protection, 117*, 168–179.
Ma, S., Zhang, Y., Ren, S., Yang, H., & Zhu, Z. (2020a). A case-practice-theory-based method of implementing energy management in a manufacturing factory. *International Journal of Computer Integrated Manufacturing*, 1–15.
Ma, S., Zhang, Y., Liu, Y., Yang, H., Lv, J., & Ren, S. (2020b). Data-driven sustainable intelligent manufacturing based on demand response for energy-intensive industries. *Journal of Cleaner Production, 274*, 123155.
Machado, M. C., Vivaldini, M., & de Oliveira, O. J. (2020). Production and supply-chain as the basis for SMEs' environmental management development: A systematic literature review. *Journal of Cleaner Production*, 123141.
Majeed, A., Zhang, Y., Ren, S., Lv, J., Peng, T., Waqar, S., & Yin, E. (2021). A big data-driven framework for sustainable and smart additive manufacturing. *Robotics and Computer-Integrated Manufacturing, 67*, 102026.
Malek, J., & Desai, T. N. (2020). A systematic literature review to map literature focus of sustainable manufacturing. *Journal of Cleaner Production, 256*, 120345.
Manavalan, E., & Jayakrishna, K. (2019). A review of Internet of Things (IoT) embedded sustainable supply chain for industry 4.0 requirements. *Computers & Industrial Engineering, 127*, 925–953.
Margherita, E. G., & Braccini, A. M. (2020). Industry 4.0 technologies in flexible manufacturing for sustainable organizational value: Reflections from a multiple case study of italian manufacturers. *Information Systems Frontiers*, 1–22.
Meng, Y., Yang, Y., Chung, H., Lee, P. H., & Shao, C. (2018). Enhancing sustainability and energy efficiency in smart factories: A review. *Sustainability, 10*(12), 4779.
Minerva, R., Lee, G. M., & Crespi, N. (2020). Digital twin in the IoT context: A survey on technical features, scenarios, and architectural models. *Proceedings of the IEEE, 108*(10), 1785–1824.
Müller, J. M., & Voigt, K. I. (2018). Sustainable industrial value creation in SMEs: A comparison between Industry 4.0 and made in China 2025. *International Journal of Precision Engineering and Manufacturing-Green Technology, 5*(5), 659–670.

Nascimento, D. L. M., Alencastro, V., Quelhas, O. L. G., Caiado, R. G. G., Garza-Reyes, J. A., Rocha-Lona, L., & Tortorella, G. (2019). Exploring Industry 4.0 technologies to enable circular economy practices in a manufacturing context: A business model proposal. *Journal of Manufacturing Technology Management*.

Ngan, S. L., How, B. S., Teng, S. Y., Promentilla, M. A. B., Yatim, P., Er, A. C., & Lam, H. L. (2019). Prioritization of sustainability indicators for promoting the circular economy: The case of developing countries. *Renewable and Sustainable Energy Reviews, 111*, 314–331.

Noiki, A. A., Afolalu, S. A., Yusuf, O. O., Emetere, M. E., Ongbali, S. O., Oloyede, O. R., Joseph, O. O., & Banjo, S. O. (2021, April). Impact assessment of the current waste management practices in Nigeria. *IOP Conference Series: Materials Science and Engineering, 1107*(1), 012172.

Oks, S. J., Fritzsche, A., & Möslein, K. M. (2017). An application map for industrial cyber-physical systems. In *Industrial internet of things* (pp. 21–46). Springer.

Oluyisola, O. E., Sgarbossa, F., & Strandhagen, J. O. (2020). Smart production planning and control: Concept, use-cases and sustainability implications. *Sustainability, 12*(9), 3791.

Onu, P., & Mbohwa, C. (2021). Industry 4.0 opportunities in manufacturing SMEs: Sustainability outlook. *Materials Today: Proceedings, 44*, 1925–1930.

Osornio, R. A., & Prieto, M. D. (2020). New trends in the use of artificial intelligence for the industry 4.0. Heliyon. 22:24. 400.

Ozkan-Ozen, Y. D., Kazancoglu, Y., & Mangla, S. K. (2020). Synchronized barriers for circular supply chains in industry 3.5/industry 4.0 transition for sustainable resource management. *Resources, Conservation and Recycling, 161*, 104986.

Pacchini, A. P. T., Lucato, W. C., Facchini, F., & Mummolo, G. (2019). The degree of readiness for the implementation of Industry 4.0. *Computers in Industry, 113*, 103125.

Paiola, M., & Gebauer, H. (2020). Internet of things technologies, digital servitization and business model innovation in BtoB manufacturing firms. *Industrial Marketing Management, 89*, 245–264.

Park, K. T., Lee, D., & Do Noh, S. (2020). Operation procedures of a work-center-level digital twin for sustainable and smart manufacturing. *International Journal of Precision Engineering and Manufacturing-Green Technology, 7*(3), 791–814.

Paschou, T., Rapaccini, M., Adrodegari, F., & Saccani, N. (2020). Digital servitization in manufacturing: A systematic literature review and research agenda. *Industrial Marketing Management, 89*, 278–292.

Peruzzini, M., Wognum, N., Bil, C., & Stjepandic, J. (2020). Special issue on 'transdisciplinary approaches to digital manufacturing for industry 4.0'. *International Journal of Computer Integrated Manufacturing*.

Pivoto, D. G., de Almeida, L. F., da Rosa Righi, R., Rodrigues, J. J., Lugli, A. B., & Alberti, A. M. (2021). Cyber-physical systems architectures for industrial internet of things applications in Industry 4.0: A literature review. *Journal of Manufacturing Systems, 58*, 176–192.

Pontevedra, V., Chaim, O, & Carlos, S. (2018). Science Direct in Sustainable Manufacturing in the fourth industrial revolution: A positive Costing models for capacity optimization in Industry 4.0: Trade-off between used capacity and operational efficiency.

Primi, A., & Toselli, M. (2020). A global perspective on industry 4.0 and development: New gaps or opportunities to leapfrog? *Journal of Economic Policy Reform, 23*(4), 371–389.

Raj, G., Dwivedi, A., Sharma, A. B., de Sousa Jabbour, L., & Rajak, S. (2020). Barriers to the adoption of industry 4.0 technologies in the manufacturing sector: An inter-country comparative perspective. *International Journal of Production Economics, 224*. https://doi.org/10.1016/j.ijpe.2019.107546

Raut, R. D., Mangla, S. K., Narwane, V. S., Gardas, B. B., Priyadarshinee, P., & Narkhede, B. E. (2019). Linking big data analytics and operational sustainability practices for sustainable business management. *Journal of Cleaner Production, 224*, 10–24.

Rejeb, A., Keogh, J. G., Wamba, S. F., & Treiblmaier, H. (2020). The potentials of augmented reality in supply chain management: A state-of-the-art review. *Management Review Quarterly*, 1–38.

Roblek, V., Thorpe, O., Bach, M. P., Jerman, A., & Meško, M. (2020). The fourth industrial revolution and the sustainability practices: A comparative automated content analysis approach of theory and practice. *Sustainability, 12*(20), 8497.

Rossit, D. A., Tohme, F., & Frutos, M. (2019). Production planning and scheduling in Cyber-Physical Production Systems: A review. *International Journal of Computer Integrated Manufacturing, 32*(4–5), 385–395.

Roy, R. B., Mishra, D., Pal, S. K., Chakravarty, T., Panda, S., Chandra, M. G., & Misra, S. (2020). Digital twin: Current scenario and a case study on a manufacturing process. *The International Journal of Advanced Manufacturing Technology, 107*(9), 3691–3714.

Sellitto, M. A., Hermann, F. F., Blezs, A. E., Jr., & Barbosa-Póvoa, A. P. (2019). Describing and organizing green practices in the context of Green Supply Chain Management: Case studies. *Resources, Conservation and Recycling, 145*, 1–10.

Sestino, A., Prete, M. I., Piper, L., & Guido, G. (2020). Internet of Things and Big Data as enablers for business digitalization strategies. *Technovation*, 102173.

Sharma, M., Kamble, S., Mani, V., Sehrawat, R., Belhadi, A., & Sharma, V. (2021). Industry 4.0 adoption for sustainability in multi-tier manufacturing supply chain in emerging economies. *Journal of Cleaner Production, 281*, 125013.

Shi, Z., Xie, Y., Xue, W., Chen, Y., Fu, L., & Xu, X. (2020). Smart factory in Industry 4.0. *Systems Research and Behavioral Science, 37*(4), 607–617.

Sivri, M. S., & Oztaysi, B. (2018). Data analytics in manufacturing. In *Industry 4.0: Managing the digital transformation* (pp. 155–172). Springer.

Sommer, C. (2017). *Drivers and constraints for adopting sustainability standards in small and medium-sized enterprises (SMEs)* (Vol. No. 21/2017). Discussion Paper.

Sony, M., & Aithal, S. (2020). Transforming Indian engineering industries through Industry 4.0: An integrative conceptual analysis.

Srivastava, M., Narayanamurthy, G., Moser, R., Pereira, V., & Paille, P. (2021). Supplier's response to institutional pressure in uncertain environment: Implications for cleaner production. *Journal of Cleaner Production, 286*, 124954.

Szalavetz, A. (2019). Industry 4.0 and capability development in manufacturing subsidiaries. *Technological Forecasting and Social Change, 145*, 384–395.

Taghipour, M., Soofi, M. E., Mahboobi, M., & Abdi, J. (2020). Application of cloud computing in system management in order to control the process. *Management, 3*(3), 34–55.

Tazhiyeva, A. (2018). Challenges and opportunities of introducing Internet of Things and Artificial Intelligence applications into Supply Chain Management. IEOM 2::33.

Thiede, S. (2021). Digital technologies, methods and tools towards sustainable manufacturing: Does Industry 4.0 support to reach environmental targets?.

Varbanov, P. S., Jia, X., & Lim, J. S. (2021). Process assessment, integration and optimisation: The path towards cleaner production. *Journal of cleaner production, 281*, 124602.

Wamba, S. F., Dubey, R., Gunasekaran, A., & Akter, S. (2020). The performance effects of big data analytics and supply chain ambidexterity: The moderating effect of environmental dynamism. *International Journal of Production Economics, 222*, 107498.

Wheeldon, A., Shafik, R., Rahman, T., Lei, J., Yakovlev, A., & Granmo, O. C. (2020). Learning automata based energy efficient AI hardware design for IoT applications. *Philosophical Transactions of the Royal Society A, 378*(2182), 20190593.

Xu, X. L., Shen, T., Zhang, X., & Chen, H. H. (2020). The role of innovation investment and executive incentive on financial sustainability in tech-capital-labor intensive energy company: Moderate effect. *Energy Reports, 6*, 2667–2675.

Zhang, H., Yan, Q., & Zhenghua, W. (2020a). Information modeling for cyber-physical production system based on digital twin and Automation ML. *The International Journal of Advanced Manufacturing Technology, 107*(3–4), 1927–1945.

Zhang, X., Zhang, M., Zhang, H., Jiang, Z., Liu, C., & Cai, W. (2020b). A review on energy, environment and economic assessment in remanufacturing based on life cycle assessment method. *Journal of Cleaner Production, 255*, 120160.

Zheng, P., & Sivabalan, A. S. (2020). A generic tri-model-based approach for product-level digital twin development in a smart manufacturing environment. *Robotics and Computer-Integrated Manufacturing, 64*, 101958.

Zheng, T., Ardolino, M., Bacchetti, A., & Perona, M. (2021). The applications of Industry 4.0 technologies in manufacturing context: A systematic literature review. *International Journal of Production Research, 59*(6), 1922–1954. https://doi.org/10.1080/00207543.2020.1824085

Gasific: A Python Tool for Biomass Gasification Process

O. Oladokun, B. Nyakuma, W. S. Luing, E. Oladimeji, O. Abatan, A. O. Ayeni, O. Agboola, A. A. Ayoola, M. O. Ojewunmi, V. E. Efeovbokhan, L. Olagoke-Oladokun, and O. A. Odunlami

Code Metadata

Current code version	v1.0
Permanent link to code/repository used of this code version	https://github.com/goke-ai/gasific
Legal code license	MIT
Software code languages, tools, and services used	Python
Compilation requirements, operating environments, and dependencies	JupyterLab, NumPy, Pandas, SciPy, Matplotlib
If available, link to developer documentation/manual	https://github.com/goke-ai/gasific#readme
Support email for questions	olagoke.oladokun@covenantuniversity.edu.ng

O. Oladokun · E. Oladimeji · O. Abatan · A. O. Ayeni · O. Agboola · A. A. Ayoola · M. O. Ojewunmi · V. E. Efeovbokhan · O. A. Odunlami (✉)
Department of Chemical Engineering, Covenant University, Ota, Nigeria
e-mail: olagoke.oladokun@covenantuniversity.edu.ng; temitayo.oladimeji@covenantuniversity.edu.ng; olubunmi.abatan@covenantuniversity.edu.ng; augustine.ayeni@covenantuniversity.edu.ng; oluranti.agboola@covenantuniversity.edu.ng; ayodeji.ayoola@covenantuniversity.edu.ng; modupe.ojewunmi@covenantuniversity.edu.ng; vincent.efeovbokhan@covenantuniversity.edu.ng; olagoke.oladokun@covenantuniversity.edu.ng; olayemi.odunlami@covenantuniversity.edu.ng

B. Nyakuma
Department of Chemistry, Benue State University, Makurdi, Benue State, Nigeria

W. S. Luing
Universidad Rey Juan Carlos, Madrid, Spain
e-mail: kengyinnwong@utm.my

L. Olagoke-Oladokun
Department of Education and Development, Universiti Teknologi Malaysia, Johor Bahru, Malaysia

A. O. Ayeni et al. (eds.), *Advanced Manufacturing in Biological, Petroleum, and Nanotechnology Processing*, Green Energy and Technology,
https://doi.org/10.1007/978-3-030-95820-6_28

1 Motivation and Significance

Biomass gasification process is a good and clean source of hydrogen, a green and environmentally friendly fuel. And gasification is one of the thermochemical processes and takes place at a high temperature well above 800 °C in the presence of steam and little or no air. The gasification process is complex and expensive for empirical analysis, hence the need for the prediction of gasification gas products using modeling and simulation. Due to the biomass gasification model complexity, the existing simulation tools are proprietary, costly, and not user-friendly (Aspentech, 2021). The fact of these proprietary simulation tools is that they are very broad and bundles many other simulation modules together apart from the gasification model. It will be better to isolate the biomass gasification module, thereby reducing the time of execution and computing resources (Afanasyev et al., 2021). A self-contain biomass gasification model that is fully automated and open source will assist the researcher and renewable energy enthusiast in running simulations for optimal operating conditions that drive optimal gas yield without the usual financial constraint and need to run many experiments. The tools can also be used as a teaching aid for biomass gasification courses (Gartner et al., 2019).

In order to isolate, develop the model, and simulate the biomass gasification process, and expected to be accurate, robust, simple, and cost-effective. An extensive review of basic biomass thermochemical process models was reviewed, and a hybrid model was developed specifically for biomass gasification. Interestingly, there are myriad models available such as kinetic model (Sher et al., 2020), thermodynamic equilibrium model, stoichiometric equilibrium model (Oladokun et al., 2020), computational fluid dynamics (CFD) model (Zhong et al., 2020), artificial neural network (ANN) (Safarian et al., 2020), nonstoichiometric equilibrium model, and reaction kinetics and hydrodynamics model (Kardani et al., 2021; Basu & Kaushal, 2009). It is what states that all the mentioned models gave good predictions. However, some models are more complex and difficult to implement due to the considerations given to some phenomena such as kinetics and hydrodynamics of the system. Other models considered the effect of shapes, sizes, and dimensions of the reactors. Three of the mentioned models, namely, the thermodynamic equilibrium model, stoichiometric equilibrium model, and nonstoichiometric equilibrium model, are independent of the gasifier design and simply focus on the gas composition product from the gasification process. The three major gasification models namely: thermodynamic equilibrium model, stoichiometric equilibrium model, and nonstoichiometric equilibrium model can allow a quick determination of the gasification products with changes in the process parameter such as operating condition and biomass feedstock. The nonstoichiometric equilibrium model does not consider any reaction mechanism but on minimizing the total Gibbs free energy of the process components. However, the stoichiometric equilibrium model incorporates the chemical reactions and species involved and the equilibrium constants (Basu, 2013).

The Gasific application is based on the biomass stoichiometric equilibrium model (BSEM). The stoichiometric equilibrium model generates seven systems of nonlinear equations from one biomass reaction and four equilibrium chemical reactions in Boudouard reaction, water-gas or steam reaction, methanation reaction, and shift reaction. The mathematical system of nonlinear equations is derived from the atomic balance of carbon (C), hydrogen (H), oxygen (O), and nitrogen (N) over the biomass generalized equation Eq. 1, the atomic balance Eq. 2 and the balance of the nonlinear equations are derived from the reaction equilibrium balances Eq. 3. For a detailed explanation of the mathematical model and simulation, see one of the authors' papers (Oladokun et al., 2016).

$$\mathrm{CH}_a\mathrm{O}_b\mathrm{N}_c + d\mathrm{H_2O} + e\left(\mathrm{O_2} + 3.76\mathrm{N_2}\right) \rightarrow n_1\mathrm{C} + n_2\mathrm{H_2} + n_3\mathrm{CO} + n_4\mathrm{H_2O} + n_5\mathrm{CO_2} + n_6\mathrm{CH_4} + n_7\mathrm{N_2} \tag{1}$$

$$\begin{aligned} &\mathrm{C}: n_1 + n_3 + n_5 + n_6 = 1 \\ &\mathrm{H}: 2n_2 + 2n_4 + 4n_6 = a + 2d \\ &\mathrm{O}: n_3 + n_4 + 2n_5 = b + d + 2e \\ &\mathrm{N}: n_7 = c + 7.52e \end{aligned} \tag{2}$$

$$\begin{aligned} K_{e_1} &= \frac{y_{CO}^2}{y_{CO_2}} P, && R1 \\ K_{e_2} &= \frac{y_{CO}\, y_{\mathrm{H_2}}}{y_{\mathrm{H_2O}}} P, && R2 \\ K_{e_3} &= \frac{y_{CH_4}}{y_{\mathrm{H_2}}^2} \frac{1}{P}, && R3 \end{aligned} \tag{3}$$

This work designed, developed, and automated the biomass stoichiometric equilibrium model (BSEM) in Eqs. 1, 2 and 3 into the Gasific app, a Python application in a Jupyter Notebook file. The app will be an all-in-one and self-contain solution without the need for any third-party library apart from the Python standard scientific and mathematical packages such as NumPy, Pandas, SciPy, and Matplotlib for visualization.

2 Software Description

2.1 Software Architecture

We developed Gasific in Jupyter Notebook using Python 3.9 scripting language (Python 2021). The choice for Jupyter Notebook is to allow for clean and clear implementation allowing for easy code modification by future developers or extension for add-in (JupyterLab, 2021). The software entry file is in the gasific. ipynb; the file cells containing the code are run sequentially from the top to bottom. Gasific schematic flow is shown in Fig. 1; the flowchart consists of (1) collection of biomass ultimate analysis, (2) operating condition setup, (3) importing physical property data, (4) model property calculation, (5) equilibrium model simulation, and (6) composition visualization.

2.2 *Software Functionalities*

2.2.1 External Packages

Apart from Python 3.9 software and the JupyterLab (notebook) package installed for coding the Gasific app. Four other packages are required to successfully run the gasification simulation app without bugging out. The four packages include NumPy (2021), Pandas (2021), SciPy (2021), and Matplotlib (2021). The first cell installs all four packages using the Python installation package manager (pip). The pip command will tempt to install each package by downloading from the Internet and installing into the Python package folder, if not installed already. It is worth

Fig. 1 Gasific schematic flowchart

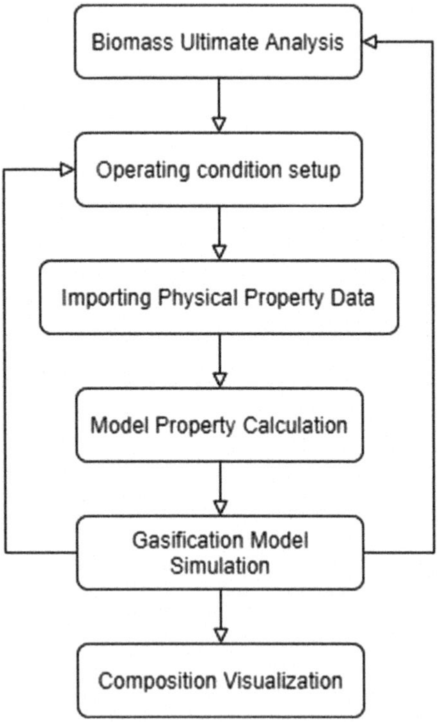

mentioning that the Internet is absolutely required for successful installation of the packages if they don't exist on the system.

The NumPy is a very fast and versatile N-dimensional array packages for matrices' implementations and operations. NumPy is the base foundation for all the three mathematical and scientific packages used in this project. It handles matrix vectorization, indexing, and broadcasting easily which is necessary for standard array computing. The Pandas' package represents a table-like data structure with each row representing a record and the columns a field in a record. Pandas have a function that seamlessly imports Microsoft Excel (Excel), comma-separated value (csv), tab-separated value (tsv), and text file content into a Pandas table. In this project, the physical property data was imported using Pandas. The SciPy package contains several modules for handling algebra (linear and nonlinear), integral, differential, and optimization problems. SciPy makes use of the optimized module to find the optimum composition for the gasification equilibrium model. Lastly, the Matplotlib library is the de facto visualization package in Python. Matplotlib can create static, animated, and interactive visualizations in Python. It easily plots line, bar, pie, and histogram visual chart and gives full control to the user on every aspect of its component such as the line styles, font properties, and axes properties. For this project, it was used to visualize the sensitivity profile of the gasification product with changes with certain operating conditions such as temperature and pressure.

2.2.2 Collection of Biomass Ultimate Analysis

The gasification stoichiometric equilibrium model requires the elemental composition of the biomass feedstock and can be obtained from empirical determination through the ultimate analysis. The biomass feedstock ultimate analysis contains the composition of carbon (C), hydrogen (H_2), oxygen (O_2), nitrogen (N_2), sulfur (S), and sometimes chlorine (Cl_2). The values from the ultimate analysis were assigned to four variables in the next cell of the Gasific Python notebook file (gasific.ipynb) corresponding to carbon, hydrogen, oxygen, and nitrogen in percentages. The values assigned were then normalized for the sum of the four components to be equal to a 100 percent (see Fig. 2) and subsequently used to calculate the fraction between hydrogen-carbon, oxygen-carbon, and nitrogen-carbon in Fig. 3.

2.2.3 Operating Condition Setup

The simulation takes a range of values for the operating condition, the temperature, pressure, steam to feed ratio, and air to feed ratio. The start, step, and end for the four parameters are stored in variables to be called for use during the gasification stoichiometric equilibrium module simulation. The operating condition setup is very important in the overall performance of Gasific result virtualization. The gas compositions of the simulated results are limited to the range of parameters supplied/passed to the model and simulator. It is recommended that the range of the parameters should be well thought ahead of the simulation. However, the app is flexible enough to accommodate changes before the next iteration.

2.2.4 Importing Physical Property Data

In this module, the physical property of each component is imported from a data bank stored in a comma-separated version (csv) file called gasific_data.csv. The gasific_data.csv contains 31 columns representing individual physical properties for each row of components. Some of the properties are components: formula, molecular weight, freezing point, boiling point, critical temperature, critical pressure, critical volume, latent heat of vaporization, the heat of formation, Gibbs energy of formation, specific heat capacity power constants, and Antoine constants. The data are used in computing many intermediate properties required by the model simulation.

2.2.5 Model Property Calculation

The heat of reactions, reaction Gibbs free energy change, change in reaction entropy, and reaction equilibrium constants are calculated using the property data imported in the physical property data module. The calculation of the properties takes into consideration temperature change in the system.

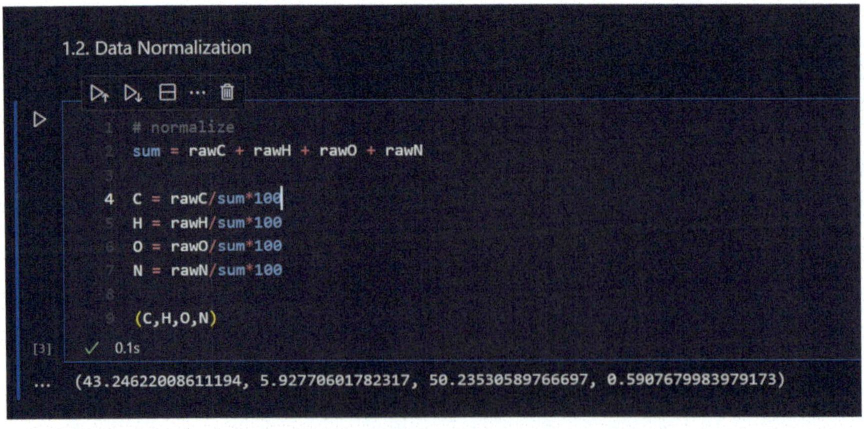

Fig. 2 Normalizing the empirical ultimate analysis data

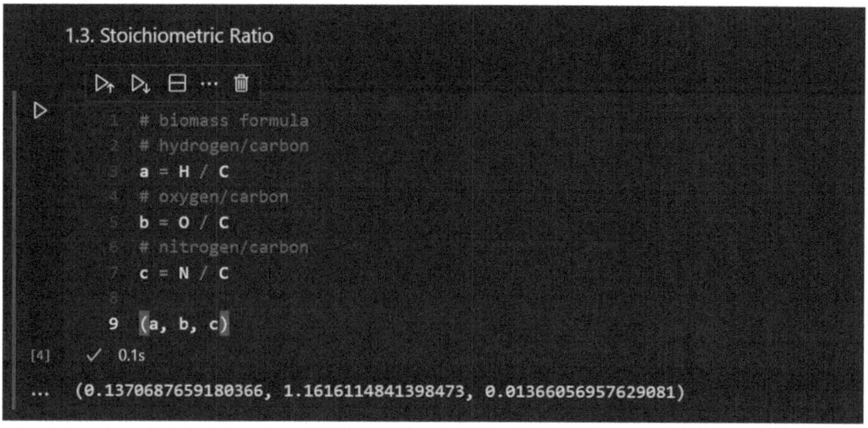

Fig. 3 Fraction of normalized carbon, hydrogen, oxygen, and nitrogen

2.2.6 Equilibrium Model Simulation

The equilibrium model simulation is the heart of Gasific. The module consists of an equilibrium stoichiometric model of the atomic balance of the elemental component of biomass, namely, the carbon (C), hydrogen (H), oxygen (O), and nitrogen (N) in the biomass gasification reaction and the other four reactions: Boudouard reaction, steam gasification, methanation, and water shift reaction. The five chemical reactions and the accompanying atomic equations were developed in earlier work by one of the authors (Oladokun et al., 2016). For the gasification stoichiometric equilibrium model to have a unique solution, three of the reaction's equilibrium equations are needed.

2.2.7 Composition Visualization

After the simulation of the stoichiometric equilibrium model has been simulated, the gas composition results obtained are better represented visually. The visualization of the results is done using the Python Matplotlib package. The python pyplot module inside the Matplotlib package can present data visually using line, bar, pie, and histogram in both 2D and 3D to mention a few. In this study, the data visualization is line plot and 2D; this is since two parameters are mostly compared together such as gas composition-temperature, gas composition-pressure, gas composition-steam, and gas composition-air.

3 Gasific Interface and Example

3.1 Illustrative Examples

In this section, we describe the Gasific application using an example for the gasification of *Imperata cylindrica* a lignocellulose biomass grass. And as highlighted in the software architecture and functionality sections, the application requires the empirical data of the biomass feedstock in the form of ultimate analysis. Oladokun et al. (2016) represented the ultimate analysis of *Imperata cylindrica* and represented in Table 1.

The ultimate analysis values from Table 1 were typed into the ultimate analysis value cell on the gasific.ipynb file (see Fig. 4). The empirical data from the ultimate analysis was normalized as shown in Fig. 2 to give the sum of the composition of the four major components (carbon, hydrogen, oxygen, and nitrogen) as 100 percent, and the normalized compositions for *I. cylindrica* for carbon, hydrogen, oxygen, and nitrogen are 43.2462, 5.9277, 50.2353, and 0.5908, respectively.

The normalized elemental values were subsequently converted to stoichiometric ratio values for a biomass chemical formula of the $CH_aO_bN_c$ in the stoichiometric ratio cell. Figure 3 shows the Python code that implemented the stoichiometric formulae, and the chemical formulae for *I. cylindrical* can be written as $CH_{0.1371}O_{1.1616}N_{0.0137}$.

The next step is to set up the simulation operating condition parameters which include temperature, pressure, steam (steam-feed ratio), and air. The start, end, and the interval step from start to end are set up for temperature as TStart, Tend, and TStep. Similarly, the values are set for pressure (PStart, PEnd, and PStep), steam (SStart, SEnd, and SStep), and air (AStart, AEnd, and AStep). Figure 5 shows the values set for the operating conditions for temperature (100:2000:50), pressure (0:2:0.5), steam (0:2:1), and air (0:2:1).

After the biomass feedstock and operating conditions successfully have been set up, the Gasific app will need to import and load the last set of data needed, and this is the physical property data for the components present in the series of reactions that makes up the biomass gasification reaction. The physical properties are imported from the gasif_data.csv file (gasif_data.csv must be present in the same folder as gasific.ipynb). The code for loading the physical property data is in Fig. 6;

Table 1 *Imperata cylindrica* ultimate analysis

	Weight (%) db.	Typical biomass weight (%) db. (Vassilev et al., 2015)
C	43.19	42–71
H	5.92	3–11
O	50.17	4–36
N	0.59	0.1–12
S	0.14	0.01–2.3

db. dry basis

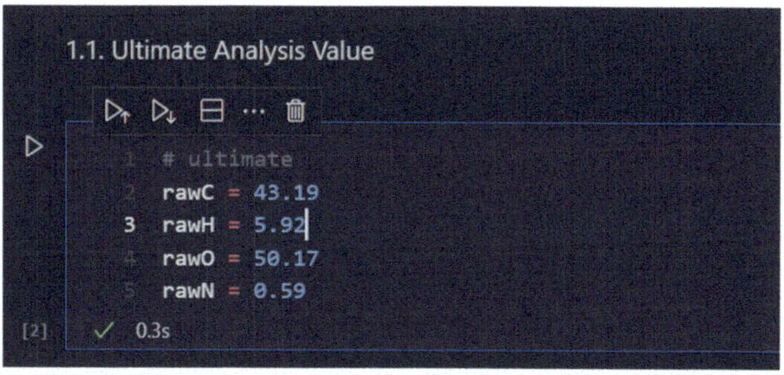

Fig. 4 Ultimate analysis value cell

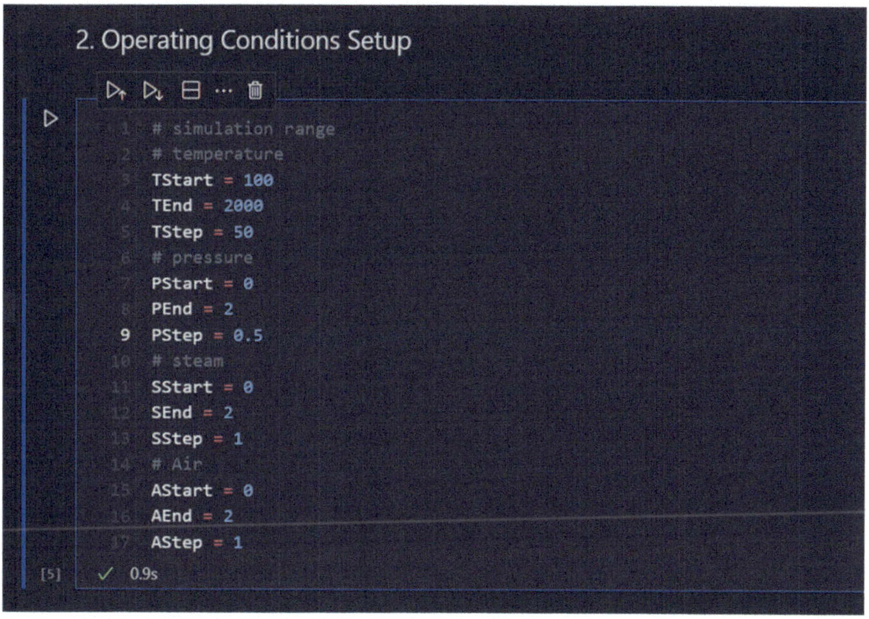

Fig. 5 Operating condition setup

the physical property data are stored in a table-like data structure and contain 33 columns and 9 rows in total.

Figure 6 displays only five components of the nine, and the user can add more components to the gasif_data.csv file; however, the new record must adhere to the structure file and separate the specific number of commas as present in the file record. After the physical property data have been read, a series of Pandas dataframe manipulation operations were carried out in order to make the table-like structure easily searchable for the individual physical data required to compute heat of reaction, standard Gibbs free energy of a reaction, and entropy of a reaction in the model physical property calculation (Sinnott, 2005).

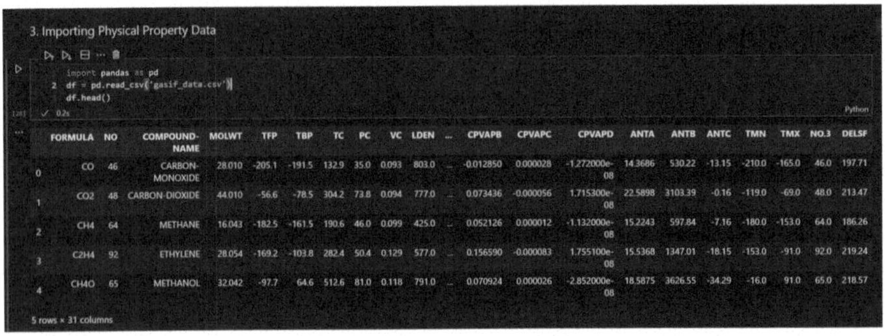

3. Importing Physical Property Data

```
import pandas as pd
df = pd.read_csv('gasif_data.csv')
df.head()
```

✓ 0.2s Python

	FORMULA	NO	COMPOUND-NAME	MOLWT	TFP	TBP	TC	PC	VC	LDEN	...	CPVAPB	CPVAPC	CPVAPD	ANTA	ANTB	ANTC	TMN	TMX	NO.3	DELSF
0	CO	46	CARBON-MONOXIDE	28.010	-205.1	-191.5	132.9	35.0	0.093	803.0	...	-0.012850	0.000028	-1.272000e-08	14.3686	530.22	-13.15	-210.0	-165.0	46.0	197.71
1	CO2	48	CARBON-DIOXIDE	44.010	-56.6	-78.5	304.2	73.8	0.094	777.0	...	0.073436	-0.000056	1.715300e-08	22.5898	3103.39	-0.16	-119.0	-69.0	48.0	213.47
2	CH4	64	METHANE	16.043	-182.5	-161.5	190.6	46.0	0.099	425.0	...	0.052126	0.000012	-1.132000e-08	15.2243	597.84	-7.16	-180.0	-153.0	64.0	186.26
3	C2H4	92	ETHYLENE	28.054	-169.2	-103.8	282.4	50.4	0.129	577.0	...	0.156590	-0.000083	1.755100e-08	15.5368	1347.01	-18.15	-153.0	-91.0	92.0	219.24
4	CH4O	65	METHANOL	32.042	-97.7	64.6	512.6	81.0	0.118	791.0	...	0.070924	0.000026	-2.852000e-08	18.5875	3626.55	-34.29	-16.0	91.0	65.0	218.57

5 rows × 31 columns

Fig. 6 Loading physical property data

Figure 7 is the code for the computation of heat of reactions, Gibbs free energy, and entropy for some of the reactions involved in the gasification of biomass. The three major reactions considered are the Boudouard reaction, water-gas or steam reaction, and methanation reaction (Basu, 2013; Oladokun et al., 2016).

The data generated from Fig. 7 codes are essentially used to determine the equilibrium values of the reactions and subsequently substituted into the gasification stoichiometric-equilibrium model (Nyakuma et al., 2014). The code for the gasification stoichiometric equilibrium (GSE) model is in Fig. 8.

The Python function code for the GSE model is named appropriately gasification (note it could be named differently); it received all the parameters necessary to carry out the simulation and contains an inner function called func. func is the developed system of nonlinear equations for the atomic balance for the GSE model. The nonlinear system of equation (func) was solved using the fsolve function in the optimize module of the Python SciPy package. The fsolve finds the optimized root of the nonlinear equation of the form func(x) = 0 given an initial value for the variables. The initial starting value for all the gas compositions is one (1).

The simulation module put together all the code design and writing from cell one to carry out the gasification calculation for all various operating condition setups in Fig. 9. The simulation results are stored in a list value and subsequently loaded in the popular table-like structure dataframe from Pandas' package.

The simulation code in Fig. 9 starts by importing the Python packages required for the simulation, namely, Pandas and NumPy packages (all other packages used in the Gasific app are scoped in the region of use, hence the advantage of modularizing the application). The early set operation conditions for temperature, pressure, steam, and air are converted into an array or list value from start to the end values at the specified step. For example, for temperature (100:2000:50), the values of temp. in the code range from [100, 150, 200, …, 1900, 1950, 2000], pressure (0:2:0.5) is [0, 0.5, 1.0, 1.5, 2.0], steam (0,2,1) is [0, 1, 2], and air (0:2:1) is [0, 1, 2]. The calculation proceeds in four loops such that gas composition (hydrogen, carbon monoxide, carbon dioxide, methane, and water/steam) is determined from the model for every possible combination of parameters such as when the temperature is 1900 °C,

4.2. Calculate reactions Heat of reaction, Gibbs free energy and Entropy

```
# 2O2

# Carbon Reactions
## Boudouard
## R1: CO2 + C -> 2CO
hR1 = computeHeatOfReaction([hfCO2, hfC, hfCO], [-1, -1, 2])
gfR1 = computeGibbs([gfCO2, gfC, gfCO], [-1, -1, 2])
sfR1 = computeEntropy([sfCO2, sfC, sfCO], [-1, -1, 2])

## water-gas or steam
## R2: C + H2O -> H2 + CO
hR2 = computeHeatOfReaction([hfC, hfH2O, hfH2, hfCO], [-1, -1, 1, 1])
gfR2 = computeGibbs([gfC, gfH2O, gfH2, gfCO], [-1, -1, 1, 1])
sfR2 = computeEntropy([sfC, sfH2O, sfH2, sfCO], [-1, -1, 1, 1])

## hydrogasification
## R3: C + 2H2 -> CH4
hR3 = computeHeatOfReaction([hfC, hfH2, hfCH4], [-1, -2, 1])
gfR3 = computeGibbs([gfC, gfH2, gfCH4], [-1, -2, 1])
sfR3 = computeEntropy([sfC, sfH2, sfCH4], [-1, -2, 1])

## R4: C + 0.5 O2 -> CO
hR4 = computeHeatOfReaction([hfC, hfO2, hfCO], [-1, -.5, 1])
gfR4 = computeGibbs([gfC, gfO2, gfCO], [-1, -.5, 1])
sfR4 = computeEntropy([sfC, sfO2, sfCO], [-1, -.5, 1])

# Oxidation Reactions
## R5: C + O2 -> CO2
hR5 = computeHeatOfReaction([hfC, hfO2, hfCO2], [-1, -1, 1])
gfR5 = computeGibbs([gfC, gfO2, gfCO2], [-1, -1, 1])
sfR5 = computeEntropy([sfC, sfO2, sfCO2], [-1, -1, 1])
```

Fig. 7 Codes for calculating the heat of reaction, Gibbs free energy, and entropy for the gasification model reactions

pressure is 1 atm, steam is 2 kg, and air equals 0 kg. The result of the specified combination is displayed in Fig. 10.

Displaying the gas composition in a tabular format for each combination of operating conditions is tedious and laborious. The tabular format doesn't present the trend easily; therefore, Gasific presents the calculated gas composition results from the model visually. The visualization package used is Matplotlib, a Python package that was installed in the section on external packages at the very beginning of the app. For this example, the visualization codes and the plots are presented in Figs. 11, 12, 13, 14 and 15.

Figures 11 and 12 display the plot of all the gas compositions (H_2, CO, CO_2, and H_2O) in moles and mole fractions, respectively. The trend is similar in both cases and makes use of a twin y-axis plot, where the scale on the left and the right y-axis is different but with a similar x-axis. The twin y-axis plot allows similar components

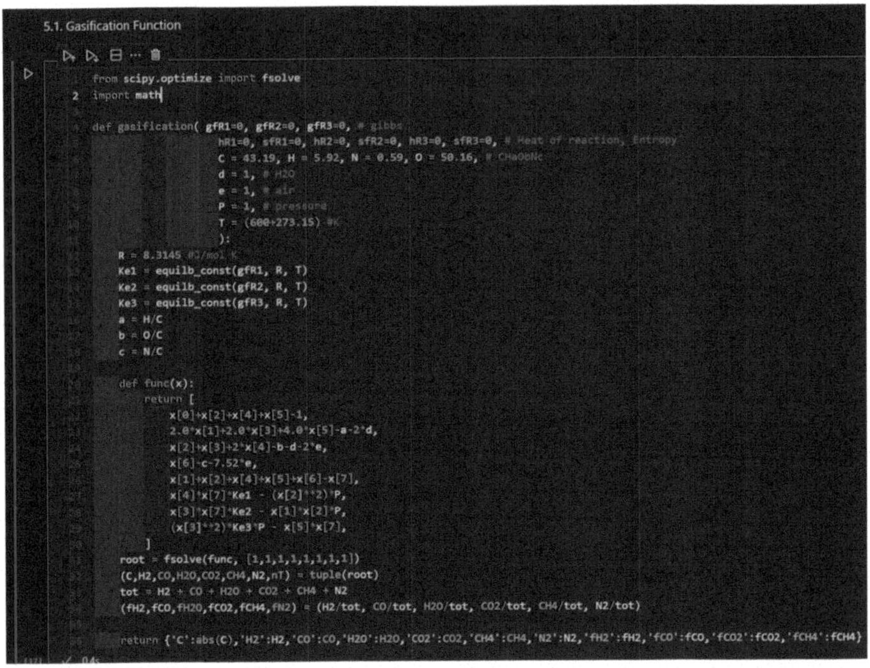

5.1. Gasification Function

```
from scipy.optimize import fsolve
import math

def gasification( gfR1=0, gfR2=0, gfR3=0, # gibbs
                  hR1=0, sfR1=0, hR2=0, sfR2=0, hR3=0, sfR3=0, # Heat of reaction, Entropy
                  C = 43.19, H = 5.92, N = 0.59, O = 50.16, # CHaObNc
                  d = 1, # H2O
                  e = 1, # air
                  P = 1, # pressure
                  T = (600+273.15) #K
                  ):
    R = 8.3145 #J/mol K
    Ke1 = equilb_const(gfR1, R, T)
    Ke2 = equilb_const(gfR2, R, T)
    Ke3 = equilb_const(gfR3, R, T)
    a = H/C
    b = O/C
    c = N/C

    def func(x):
        return [
            x[0]+x[2]+x[4]+x[5]-1,
            2.0*x[1]+2.0*x[3]+4.0*x[5]-a-2*d,
            x[2]+x[3]+2*x[4]-b-d-2*e,
            x[6]-c-7.52*e,
            x[1]+x[2]+x[4]+x[5]+x[6]-x[7],
            x[4]*x[7]*Ke1 - (x[2]**2)*P,
            x[3]*x[7]*Ke2 - x[1]*x[2]*P,
            (x[3]**2)*Ke3*P - x[5]*x[7],
        ]
    root = fsolve(func, [1,1,1,1,1,1,1,1])
    (C,H2,CO,H2O,CO2,CH4,N2,nT) = tuple(root)
    tot = H2 + CO + H2O + CO2 + CH4 + N2
    (fH2,fCO,fH2O,fCO2,fCH4,fN2) = (H2/tot, CO/tot, H2O/tot, CO2/tot, CH4/tot, N2/tot)

    return {'C':abs(C),'H2':H2,'CO':CO,'H2O':H2O,'CO2':CO2,'CH4':CH4,'N2':N2,'fH2':fH2,'fCO':fCO,'fCO2':fCO2,'fCH4':fCH4}
```

Fig. 8 Gasification stoichiometric equilibrium model function code

to be grouped for better visualization since the values of H2 and CO are very small compared to CO_2 and H_2O.

The trend in the plot of gas composition in mole and mole fraction is similar as expected since they are from the same basis. It shows that both hydrogen and carbon monoxide compositions increase with an increase in temperature (100–2000 °C) when all other process parameters are kept constant. Hydrogen and carbon monoxide are produced from water-gas/steam reactions where carbon is steam gasified to hydrogen and carbon monoxide. The compositions of carbon dioxide and methane decrease with an increase in temperature from 100 to 2000 °C. The Boudouard reaction converts most of the available carbon dioxide produced by shift reaction to carbon monoxide. Comparing the hydrogen increasing trend to carbon monoxide, it was observed that the rate of the composition increases sharply at 680 °C for both components and at 800 °C the rate of hydrogen composition increased more than that of carbon monoxide. The further increase in hydrogen over carbon monoxide is due to the shift reaction where carbon monoxide is oxidized by steam to hydrogen and carbon dioxide as well as the reversible reaction of methanation where methane was converted back to carbon and hydrogen gas. Detailed interpretation of the trends can be found in the original papers presented by two authors (Nyakuma et al., 2014; Oladokun et al., 2016).

Figure 13 shows the code and the resulting plot of H_2 composition in mole fraction as a function of temperature at different pressures and constant steam and air

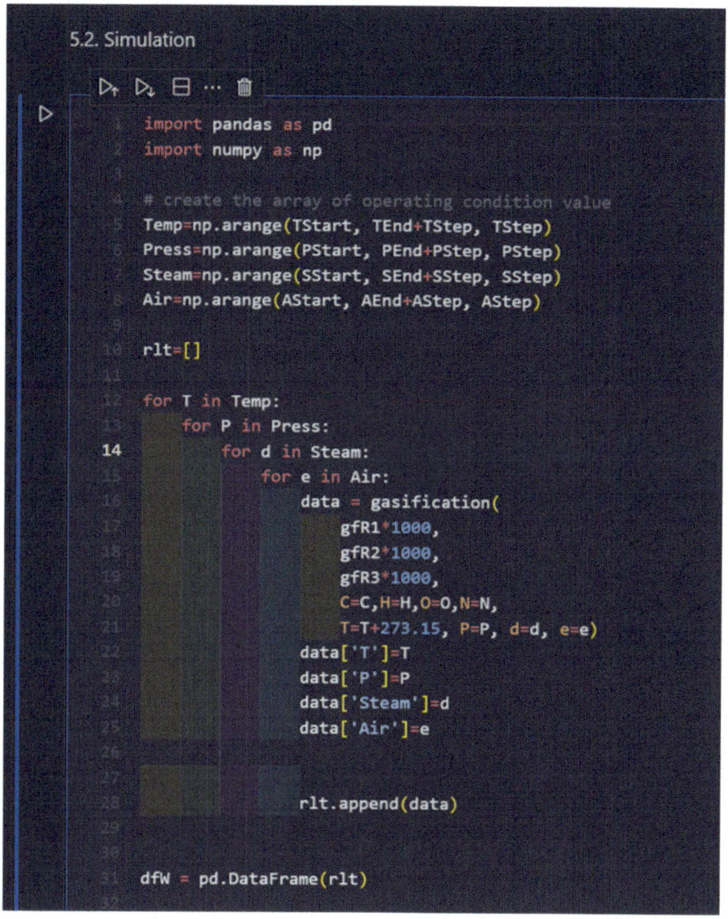

Fig. 9 Biomass gasification simulation code

values. The data was queried from the existing simulated results, and the query results were plotted for each pressure value and varying temperature (T) on the x-axis and hydrogen (H_2) composition. The manuscript is on the design and development of the Gasific app; for interpretation of the visual plot, see the papers co-authored (Nyakuma et al., 2014; Oladokun et al., 2016).

Figure 14 shows the code for querying the data for temperature and hydrogen (H_2) composition when the amount of steam is changing from 0, 1, and 2. The values of air and pressure were held constant at 0 kg and 1 atm, respectively. The effects of steam and air are further discussed in the paper by Oladokun et al. (2016).

Similarly, the effect of temperature on hydrogen (H_2) composition in mole fraction was queried and displayed in Fig. 15, when air is changing and pressure and steam were held constant at 1 atm and 1 kg, respectively.

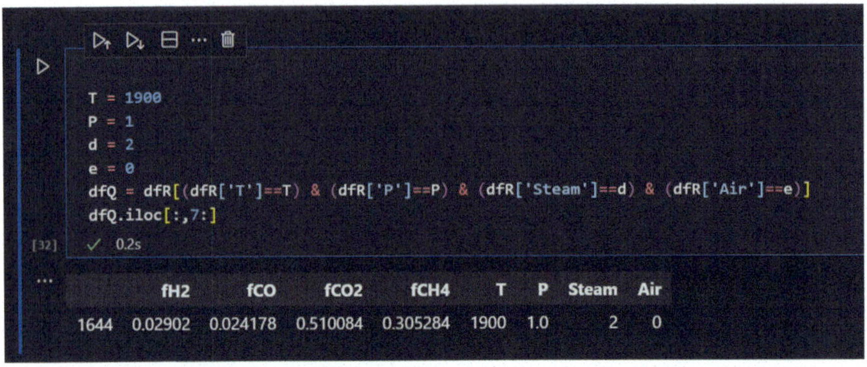

Fig. 10 Gas composition at a specific operating condition

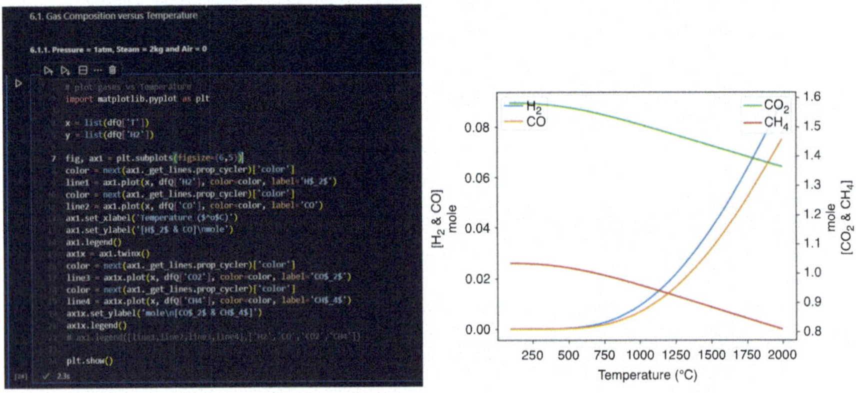

Fig. 11 Gas compositions in moles with temperature

The query code is like for all the composition and variations displayed in Figs. 11, 12, 13, 14 and 15 and can easily be modified to display for other gas compositions such as CO, CO_2, and H_2O. Furthermore, the query is not limited to gas compositions as a function of temperature. The gas compositions can also vary as a function of pressure, steam, and air. All that is required for the query to be modified is the adjustment to the name of the columns and the range for the constants parameters.

4 Impact

In the Gasific app, we provide an easy-to-use and versatile open-source code to model and simulate biomass gasification to produce gas composition products such as hydrogen, carbon monoxide, carbon dioxide, and water/steam. Gasific makes it possible for a researcher to have a free tool for a quick estimate of biomass

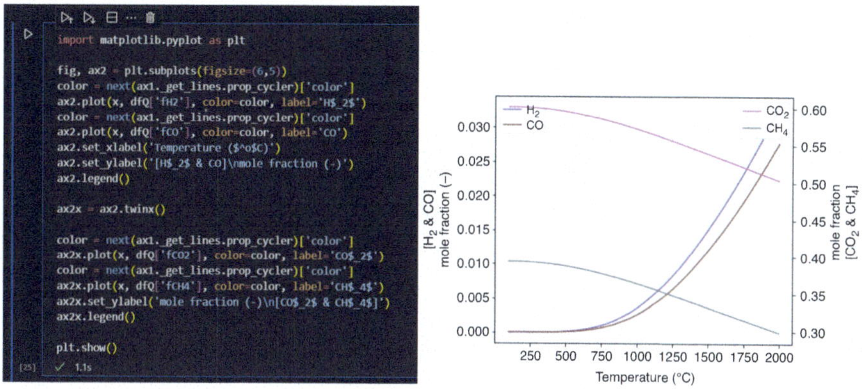

Fig. 12 Gas compositions in mole fraction with temperature

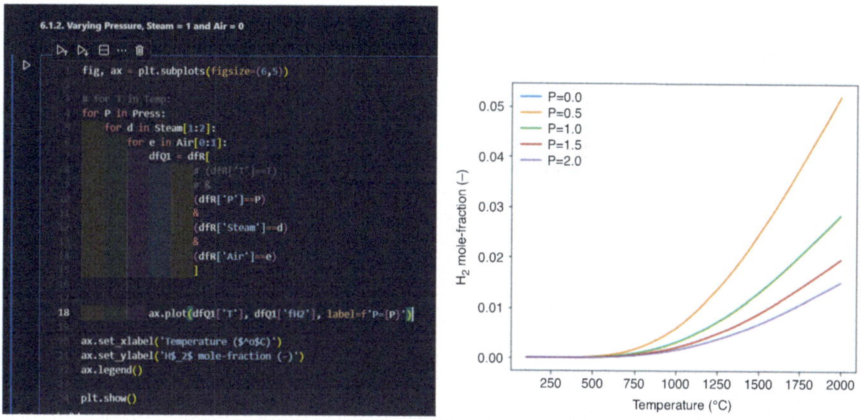

Fig. 13 Hydrogen composition with temperature and varying pressure

gasification products. The standout features of Gasific for biomass gasification include the following:

- It is a self-contain application in a notebook-like-interface that scientists are conversant with, making it extremely user-friendly.
- It doesn't need any third-party license; all the packages used are inherently free Python packages.
- It includes state-of-the-art methods of proven usefulness in many biomasses' gasification models.
- It simplifies a complex biomass gasification model and allows ease of tuning parameters making it usable for researchers with limited programming experience.
- As a modular design app, Gasific allows easy modification and extension for more functionality or integration with other software.

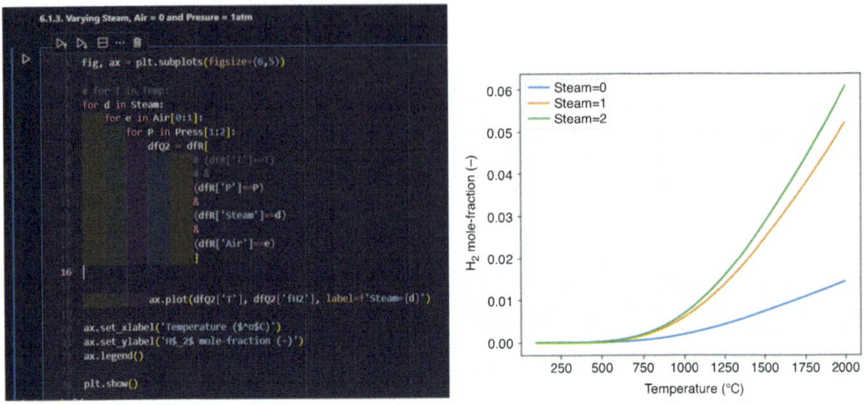

Fig. 14 Hydrogen composition with temperature for varying steam

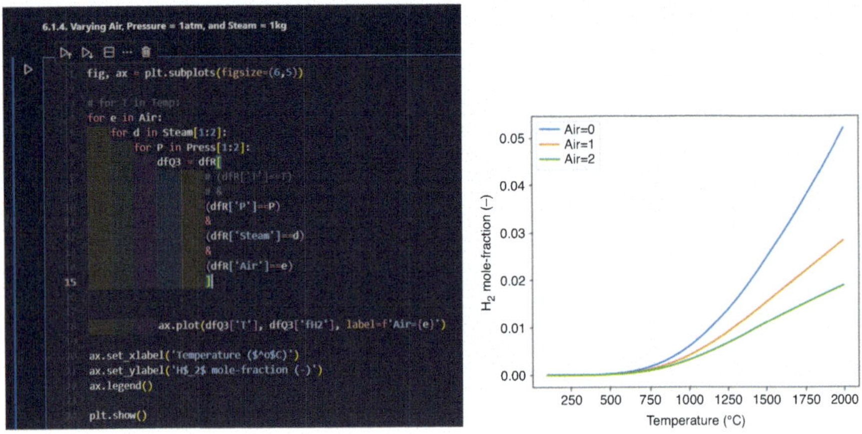

Fig. 15 Hydrogen composition with temperature for varying air

- It can be used as a teaching aid in classrooms for the studying of biomass gasification and as well as some mathematical concepts.

The compactness, clarity, user-friendliness, and versatility of the Gasific app make it extremely usable in the studying of the biomass gasification process, when the empirical data from the ultimate analysis is available. The app/tool is pertinent in renewable energy research and specifically biomass gasification. Furthermore, it can serve as a companion in the teaching and studying of biomass gasification subjects. Just like every scientific model and experimental data been encouraged to be transformed into a usable app, Gasific will contribute significantly to the development of scientific apps and renewable energy production (Barrasa-Fano et al., 2021; Le Houx & Kramer, 2021).

5 Conclusion

Gasific is an open-source Python app and was successfully designed and developed in a Jupyter Notebook for clarity, compactness, and ease of use. Gasific was developed for researchers and renewable energy enthusiasts in biomass gasification, especially for users without strong programming and technical background in biomass gasification. Gasific founds a place in research and classroom for the studying of the gas product compositions of biomass gasification. All the steps required for the gasification of biomass are included in the all-in-one package without any need for a third-party license. It installs any missing packages automatically from the Internet and presents the results from the biomass gasification stoichiometric equilibrium (GSE) model in a visually appealing way for easy understanding. The platform for development was a Jupyter Notebook allowing a familiar background for modification and extension. We expect that researchers and biomass renewable energy engineers will find Gasific useful independent of their programming or technical background. Furthermore, this app design and development lay a good foundation for the conversion of many experimental data and scientific models to open-source software in diverse research areas including the work of Sanni et al. (2021), Ayodeji et al. (2018), Ayeni et al. (2021), Mamudu et al. (2020), Elehinafe et al. (2020), Agboola et al. (2021), and Nyakuma et al. (2021)

Acknowledgments We would like to thank all the contributors that helped over the years in the shaping of Gasific libraries, in particular Prof. Arshad Ahmad, Dr. Tuan Amran Tuan Abdullah, and Dr. Anwar Johari. This work has been partially funded by the Covenant University, Ota, Nigeria.

References

Afanasyev, A., Bianco, M., Mosimann, L., Osuna, C., Thaler, F., Vogt, H., Fuhrer, O., VandeVondele, J., & Schulthess, T. C. (2021). GridTools: A framework for portable weather and climate applications. *SoftwareX*, 15. https://doi.org/10.1016/j.softx.2021.100707

Agboola, O., Sadiku, E., Popoola, P., Fayomi, O., Ayeni, A., Dick, D., Adegbola, A., Moropeng, L., & Ramakhokhovhu, M. (2021). Surface roughness of ternary blends: Polypropylene/chitosan/sisal fiber membranes. *Materials Today: Proceedings, 38*, 2342–2346.

Aspentech. (2021) *Aspen Plus*. https://www.aspentech.com/en/products/engineering/aspen-plus. Accessed 07 Sept 2021

Ayeni, A. O., Agboola, O., Daramola, M. O., Grabner, B., Oni, B. A., Babatunde, D. E., & Evwodere, J. (2021). Kinetic study of activation and deactivation of adsorbed cellulase during enzymatic conversion of alkaline peroxide oxidation-pretreated corn cob to sugar. *Korean Journal of Chemical Engineering, 38*(1), 81–89.

Ayodeji, A. A., Modupe, O. E., Rasheed, B., & Ayodele, J. M. (2018). Data on CaO and eggshell catalysts used for biodiesel production. *Data in Brief, 19*, 1466–1473.

Barrasa-Fano, J., Shapeti, A., Jorge-Peñas, Á., Barzegari, M., Sanz-Herrera, J. A., & Van Oosterwyck, H. (2021). TFMLAB: A MATLAB toolbox for 4D traction force microscopy. *SoftwareX*, 15. https://doi.org/10.1016/j.softx.2021.100723

Basu, P. (2013). *Biomass gasification, pyrolysis and torrefaction: Practical design and theory*. Academic.

Basu, P., & Kaushal, P. (2009). Modeling of pyrolysis and gasification of biomass in fluidized beds: A review. *Chemical Product and Process Modeling, 4*(1). https://doi.org/10.2202/1934-2659.1338

Elehinafe, F. B., Anjorin, S. D., Odunlami, O. A., Mamudu, A. O., Ayeni, A. O., & Okedere, O. B. (2020). Modelling of sweet gas flaring and the resultant gaseous emissions with their emission factors. *Cogent Engineering, 7*(1), 1815931.

Gartner, J. K., Garcia-Perez, M., & van Wie, B. J. (2019). Investigation of biomass char gasification kinetic parameters using a novel miniaturized educational system. *Bioresources, 14*(2), 3594–3614. https://doi.org/10.15376/biores.14.2.3594-3614

JupyterLab. (2021). *JupyterLab documentation — JupyterLab 3.1.9 documentation*. https://jupyterlab.readthedocs.io/en/stable/. Accessed 30 Aug 2021.

Kardani, N., Hedayati Marzbali, M., Shah, K., & Zhou, A. (2021). Machine learning prediction of the conversion of lignocellulosic biomass during hydrothermal carbonization. *Biofuels*, 1–13. https://doi.org/10.1080/17597269.2021.1894780

Le Houx, J., & Kramer, D. (2021). OpenImpala: OPEN source IMage based PArallisable linear algebra solver. *SoftwareX*, 15. https://doi.org/10.1016/j.softx.2021.100729

Mamudu, A., Emetere, M., Okocha, D., Taiwo, S., Ishola, F., Elehinafe, F., & Okoro, E. (2020). Parametric investigation of indigenous Nigeria mineral clay (Kaolin and Bentonite) as a filler in the Fluid Catalytic Cracking Unit (FCCU) of a petroleum refinery. *Alexandria Engineering Journal, 59*(6), 5207–5217.

Matplotlib. (2021). *Matplotlib: Python plotting — Matplotlib 3.4.3 documentation*. https://matplotlib.org/. Accessed 30 Aug 2021.

NumPy. (2021) *NumPy*. https://numpy.org/. Accessed 3 Aug 2021.

Nyakuma, B. B., Oladokun, O. A., Johari, A., Ahmad, A., & Abdullah, T. A. T. (2014). A simplified model for gasification of oil palm empty fruit bunch briquettes. *Jurnal Teknologi, 69*(2), 7–9. https://doi.org/10.11113/jt.v69.3097

Nyakuma, B. B., Oladokun, O., Wong, S. L., & Abdullah, T. A. T. (2021). Torrefaction of oil palm empty fruit bunch pellets: Product yield, distribution and fuel characterisation for enhanced energy recovery. *Biomass Conversion and Biorefinery*, 1–21. https://doi.org/10.1007/s13399-020-01185-z

Oladokun, O., Ahmad, A., Tuan Abdullah, T. A., Nyakuma, B. B., Kamaroddin, M. F. A., Ahmed, M., & Alkali, H. (2016). Sensitivity analysis of biohydrogen production from imperata cylindrica using stoichiometric equilibrium model. *Jurnal Teknologi, 78*(8-3), 137–142. https://doi.org/10.11113/jt.v78.9577

Oladokun, O., Nyakuma, B. B., & Ahmad, A. (2020). Fundamental theories and kinetic models for the pyrolysis of lignocellulosic biomass wastes. In *Handbook of research on resource management for pollution and waste treatment* (pp. 123–151). IGI Global.

Pandas. (2021). *Pandas – Python data analysis library*. https://pandas.pydata.org/. Accessed 30 Aug 2021.

Python. (2021). *Welcome to Python.org*. https://www.python.org/. Accessed 30 Aug 2021.

Safarian, S., Ebrahimi Saryazdi, S. M., Unnthorsson, R., & Richter, C. (2020). Artificial neural network integrated with thermodynamic equilibrium modeling of downdraft biomass gasification-power production plant. *Energy, 213*, 118800. https://doi.org/10.1016/j.energy.2020.118800

Sanni, E., Adefila, S., Anozie, A., Agboola, O., Emetere, E., & Okoro, E. (2021). Ndubuisi a mathematical description of the flow dynamics of sand, crude oil and water in an oil well. In *IOP conference series: Materials science and engineering* (Vol. 1, p. 012107). IOP Publishing.

SciPy. (2021). *SciPy.org – SciPy.org*. https://www.scipy.org/. Accessed 30 Aug 2021.

Sher, F., Iqbal, S. Z., Liu, H., Imran, M., & Snape, C. E. (2020). Thermal and kinetic analysis of diverse biomass fuels under different reaction environment: A way forward to renewable energy sources. *Energy Conversion and Management, 203*, 112266. https://doi.org/10.1016/j.enconman.2019.112266

Sinnott, R. K. (2005). *Chemical engineering chemical engineering design* (Vol. 6). Coulson and Richardson's.

Vassilev, S. V., Vassileva, C. G., & Vassilev, V. S. (2015). Advantages and disadvantages of composition and properties of biomass in comparison with coal: An overview. *Fuel, 158*, 330–350. https://doi.org/10.1016/j.fuel.2015.05.050

Zhong, H., Xiong, Q., Yin, L., Zhang, J., Zhu, Y., Liang, S., Niu, B., & Zhang, X. (2020). CFD-based reduced-order modeling of fluidized-bed biomass fast pyrolysis using artificial neural network. *Renewable Energy, 152*, 613–626. https://doi.org/10.1016/j.renene.2020.01.057

Index

A

AA6061 aluminium alloy, 85
 chemical configuration, 86
 material's properties, 86
 mechanical properties, 90–93
 microhardness tester, 89
 microstructural analysis, AA6061/clay, 93–94
 stir casting process, AA6061/clay composites, 88–89
 ultimate tensile strength, 90
 XRD analysis, 89
  AA6061/clay + RHA composites, 94–96
AA6061/clay composites, 86
ACE (area control error), 302
Activated carbon
 adsorbent, 211
 DBT, 212
 immobilized (*see* Immobilized activated carbon)
 sulfur compound, 212
Additive manufacturing (AM), 320, 327
Additives
 agro-waste adsorbents, 225, 226
Adenosine triphosphate (ATP), 259
Adsorbents
 agro-waste, 225, 226
 agro-waste peel-based adsorbents (*see* Agro-waste peel-based adsorbents)
 packed-bed column, 215–219
 physicochemical characterization, 214–216
AFR λ, 278–282
African continent, 250
Agricultural production, 3, 8
Agricultural waste skins, 226
Agro-industrial wastes (AIW), 224, 232
Agro-resources, 258, 260, 261
Agro-waste adsorbents, 225, 226
 modification technologies, 224
 preparation and modification, 231, 232
Agro-waste peel-based adsorbents
 agricultural waste skins, 226
 banana peels, 229, 230
 bio-agro-wastes, 226
 cassava peels, 230, 231
 citrus waste peels, 227–229
 FVWs, 226
 VFWs, 226
Air pollution, 19
Air to fuel ratio (AFR)
 CO (*see* Carbon monoxide (CO))
 combustion (*see* Combustion)
 emissions (*see* Emissions)
 exhaust emission tests, 279
 HC (*see* Hydrocarbon (HC))
 lambda (λ), 278–282
 lean mixtures, 278
 rich mixtures, 278
Al_2O_3 nanoparticles, 65–69
Alumina, 37, 38, 41–43
Aluminium, 38
Aluminium ingots, 38
Aluminium matrix composites, 37, 38, 42
Aluminium metal matrix composites (AMMC), 38, 39

A. O. Ayeni et al. (eds.), *Advanced Manufacturing in Biological, Petroleum, and Nanotechnology Processing*, Green Energy and Technology,
https://doi.org/10.1007/978-3-030-95820-6

A36 mild steel corrosion resistance, phosphating
  coated mild steels, 190–193
  materials, 187
  metal composition, 189
  pre-treatment process, 187
  reagents concentration, 188
  sample preparation, 187
  surface morphology, 193–195
  time and temperature, 188
  visual inspections and physical appearance, 190
  XRF analysis, 189–190
  XRF on periwinkle particles, 188
Analysis of variance (ANOVA), 170
ANOVA test (F-test), 167
Antifungal agent, 34
Anything as a service (XaaS), 146
Application, nanotechnology
  in agricultural production, 8–9
  in food preservation, 7–9
  in medicine and healthcare, 9–11
  in textiles, 4, 5
  in water treatment, 6, 7
Arch dam, 289, 291
Artificial intelligence (AI), 146, 319
Artificial neural networks (ANN), 151
Artificial/synthetic chelating agents, 273, 274
Assimilation, 259
Atmospheric conditions, 258
Augmented reality (AR), 320, 327
Automated in toilets, 148
Automatic brake disc, 37
Automatic dipstick loading mechanism, 152
Automatic generation control (AGC), 301, 302
Automatic voltage regulator (AVR), 301

B

Banana peels, 229, 230
Batch adsorption mode, 212
Big data analytics (BDA), 318
Bio-agro-wastes, 226
Bio-based plastics, 31, 258
Bio-based polymers, 258
Biodegradability, 31
Biodegradable, 257, 258, 261, 263
Biodegradable plastics, 258
Biodegradation
  assimilation, 259
  atmospheric conditions, 258
  ATP, 259
  biodeterioration, 258
  blends, 261, 262
  cellulose, 261, 262
  cytoplasmic membranes and cell walls, 258
  description, 259
  energy transfer, 258
  enzymatic hydrolysis, 259
  mineralization, 259, 260
  starches, 261, 262
Biodeterioration, 258
Bioengineering, 158
Bioinstrumentation, 159, 162
Biological synthesis, 4
Biomass, 269
Biomass gasification process, 336
Biomass stoichiometric equilibrium model (BSEM), 337
Biomechanics, 159
Biomedical and biological laboratories, 226
Biomedical engineer responsibilities, 161
Biomedical engineering (BME)
  advancements, 159
  analytical and curative applications, 157
  areas of specialization, 159
  bioengineering, 158
  biomedical engineer's responsibilities, 161
  challenges, 162
  countries offering, 158
  definition, 157
  education in Nigeria, 160
  engineering expertise, 159
  freelancers, 162
  healthcare delivery system, 158
  Nigerian universities, 158
  studies, 159
  sustainable development, 157
Biomedical engineers, 161
Biomedical waste management, 254, 255
  and healthcare delivery, 252
  hospital, 251–253
  incinerators, 251, 253
  mitigation strategies, 251, 254
  preservation of public health, 249
  reconnaissance survey, 251
  sub-Saharan Africa, 251
  waste generation in urban centres, 250, 251
Bioplastics, 34
  agro-resources, 260, 261
  applications, 262, 263
  bio-based plastics, 258
  bio-based polymers, 258
  biodegradable plastics, 258
  biodegradation mechanisms (*see* Biodegradation)
  commercial production, 258
  eco-friendly, 257

hydrocarbon plastics, 257
market sectors, 262, 263
production, 260, 261
properties, 257
Biopolymer-based films, 30
Biopolymers, 34, 258, 260, 262
Biotechnology, 258
Biowaste purifier, 20, 21
Blends, 261, 262
Boudouard reaction, 337, 341, 344, 346
BpNp combination, 123
Brake pad
AMMC, 38
brake disc's performance, 37
conventional brake pad, 41
materials, 37
use of aluminium, 42
wear resistance, 41
Bristol stool form scale (BSFS), 152
Brownian motion, 48
Brunauer-Emmett-Teller (BET) analysis, 214
Business opportunities
material-efficient housing, 242, 244
Buttress dam, 289, 291

C

Cadmium selenide nanoparticles, 11
Calcium, 38
Calcium selenide (CaSe) powder, 3
Carbon monoxide (CO), 277–279, 283
Carbon nanotubes (CNTs), 5
Carbon steel degradation, 165
Carbon steels
applications, 165
capacity, 165
structural components and platforms, 165
Cassava peels, 230, 231
Cellulose, 260–262
Celuluzon®, 262
Ceramic nanofluids, 79
Chelating agents, 272–274
amino carboxylic acid chelators, 272
ammonium acetate, 272
artificial/synthetic, 273, 274
biodegradable and non-biodegradable, 272
citranox, 272
EDTA, 272
gluconate, 272
metallic ions, 272
natural, 273
organic phosphonic acid, 272
oxalate, 272
phytoextraction and chelant-enhanced electrokinetic extraction, 272
soil acidity, 272
soil flushing, 272
Chemical precipitation, 223
Chemical wastes, 203
Chemically induced biodegradation, 259
Chlorophenols, 7
Circular economy (CE), 237, 245, 250, 316
greener environment, 238
MSW, 238, 239
negative and positive sides, 241, 242
physical elements, 239
principles, 238
regenerative approach to natural resources management, 238
waste management, 238
waste minimization, 239
Citrus waste peels
FAOSTAT records, 227
orange peels, 227, 228
PPs, 229
Civil engineers, 287
Cleaner environment, 241, 246
Climate issues, 19
Clinoptilolite, 9
Closed-loop human ecosystem, 197
Cloud computing (CC), 148, 319
Cloud platforms, 319, 324
Cloud technology, 319
Coarse soil particles, 269
Coefficient of regression (R^2), 183
Combustion, 278, 283
Communicable water-associated diseases, 202
Composites
aluminium matrix, 37, 38
AMMC, 38
compressive strengths, 41
spent lubricating oil, 39
thermal conductivities, 42
variation and constituent, 38
wear resistances, 41
Compressive strength test, 40
Computational capabilities, 319
Computational fluid dynamics (CFD)
nanofluids, 46
Computer vision flow metering, 152
Containment remedies, 270
Continuous adsorption process, 212
Conventional nano-cutting fluids, 81, 82
Conventional PID control, 302
Conventional plasticizers, 31
Convolutional neural network (CNN), 152
Coolant, 80

Coprecipitation, 4
Core flooding analysis, 126
Core flooding tests, 119
Coronavirus (COVID-19), 9
Corrosion
  definition, 175
  interaction, 175
  metals, 175
Corrosion inhibitor, 165, 175, 176
Corrosion reaction temperature, 181
Corrosion resistance, 194
Cost-effective, 224
Crop residues, 225
Crude palm oil (CPO), 32, 33
Cryogenic cooling, 64
Cupriethylenediamine, 261
Cutting fluids, 62, 76
  coated carbide tool, 65
  effect on cutting forces, 65–66
  effect on cutting temperature, 67–68
  effect on surface roughness, 66–67
  effect on tool wear, 69
  lubricants, 65
  machining with cryogenic cooling system, 64
  machining with solid lubricant, 63
  in machining
    cooling and lubricating properties, 77
    cutting tool and workpiece, 78–79
    lubrication and cooling effects, 77
    lubri-cooling techniques, 77
    MQL machining, 77
  MQL (*see* Minimum quantity lubrication (MQL))
  optimized results, 65
  vegetable oils, 64
Cutting temperature, 62, 65, 67–69
Cytoplasmic membranes and cell walls, 258

D
Dams, 289
Decision-makers, 19
Deep learning (DL), 319
Degradation, 260, 262
Degraded ecosystems, 201
Design
  water retaining structures (*see* Water retaining structures)
Desulfurization, 211, 213–215, 219
Dibenzothiophene (DBT), 212–219
Differential pulse voltammetry (DPV), 7
Digital healthcare services, 146
Digitalization
  IoT, 317
  manufacturing processes, 316
  process digitalization, 322
  and sustainability, 322
Digital technologies (DT), 316
  AI, 319
  BDA, 318
  CC, 319
  deployment in manufacturing sector, 327
  implementation, 316, 322
  IoT, 317
  manufacturing processes, 324
  process control technology and automation, 324
  SACPS, 318
  standard analytics, 323
Digital twin technology, 322
Dipstick method, 152, 153
Dispersion additives/surfactant, 47
Disposal methods, 18
Distribution of energy, 107, 108
Drill point geometry, 57
Drilling, 56, 57
Drug detoxification, 10

E
Eastar Bio™ (polybutylene adipate-co-terephthalate), 263
E-cars, 105
Ecoflex®, 263
Eco-friendly and renewable plastics, 260
Eco-friendly plastic production, 31
Eco-friendly plasticizers, 31, 32
Economic capacity, 205
Economic development, 201, 277
Economic dispatch function, 302
Ecovio®, 263
Effluent treatment, 223–225, 228, 231
Electric vehicles (EVs), 106, 112
Electrokinetic separation, 271
Electronic medical records (EMR), 146
Elevated surface reservoirs, 287
Elvanol®, 263
Embankment dam, 289, 290
Emerging trend, 316
Emissions
  automotive, 278
  CO and hydrocarbon, 278
  CO, CO_2 and HC, 277
  exhaust emission tests, 279
  traffic, 277
  vehicular, 277–282

End milling, 59, 61
Energy
types, 17
Energy applications, 112
Energy consumption pattern, 24
Energy demand, 100, 112
Energy efficient processes
development, 17
Energy generation/management, 24
Energy sources, 101
fossil fuels, 101
nanotechnology
distribution of energy, 107, 108
in electricity generated by
windmill, 101
electricity generation from waste
heat, 102–103
energy storage, 108, 109
generation of steam from
sunlight, 103–105
hydrogen storage for fuel cell-powered
cars, 105–107
Energy storage, 105, 108, 109
Energy transfer, 258
Energy usage, 101, 105, 110
nanotechnology
for high-efficiency light bulbs, 110–111
for transportation energy usage, 111
for vehicle battery, 112
Energy value-added chain, 101
Engineered nanomaterials (ENMs), 6
Engineering infrastructural degeneration, 138
Engineering infrastructures, 198
Enhanced oil recovery (EOR)
biopolymer coated with nanoparticles, 126
BpNp rheological properties, 120
core flooding, 119
core samples characterization, 118
deployment, 118
materials, 118
NPs and polymer characterization,
119, 120
polymer flooding, 117
results, 121, 125
Environment, 249, 251, 253–255, 257–259
Environmental compliance, 201
Environmental engineering
chemical and environmental applications,
198, 199
closed-loop human ecosystem, 197
control, 197
design, 197
design on environment, 207
economic capacity, 205
engineering infrastructures, 198
environmental management plan, 198
environmental pollution, 198
environmental process
engineering, 199–201
environmental sustainability, 205
food waste, 204, 205
governance in technology, society,
economy and ecology, 198, 199
materials engineering principles, 198
methodology, 203
multiple tools and methods, 198
operation, 197
optimization, 197
pollution prevention, 205, 207, 208
poor governance, 203, 204
principles and laws of nature, 197
protection and sustainability, 200–203
public health, 197
resource availability, 205
sustainability, 204, 206
sustainability addresses, 198
sustainability perspectives, 206
sustainable development, 197, 198, 204
systematic computer-based methods, 197
Environmental engineers, 287
Environmental health, 160, 163
Environmental issues, 250
Environmental management plan, 198
Environmental pollution, 138, 198, 203, 267
Environmental Potential Agency (EPA), 211
Environmental process engineer
(EPE), 199–201
efficient processes and products, 26
energy consumption pattern, 24
energy generation/management, 24
and management of processes, 18–20
pollution reduction, 20–23
public health, 17, 25
recycling of plastic waste, 24, 25
and safety, 18
and society, 24
sustainable development, 17, 20–23, 25
technical fundamentals, 18
Environmental Protection Agency (EPA), 224
Environmental safety, 32
Environmental sustainability, 205
Enzymatic hydrolysis, 259
Epoxidized triglycerides, 32
Equipment procurement, 163
Escherichia coli, 34
Essential oils extracts, 165
Ethylenediaminetetraacetic acid (EDTA), 272
European bioplastics, 263

Exhaust emission tests, 279
External plasticization, 30

F

Face milling, 59–61
Fatty acids, 31
Ferrous metals, 185
Field electronic scanning electron microscope (FESEM), 213
Filtration membrane, 223
First law of thermodynamics, 17
Flexural cracking, 293
Fluid thermal conductivity, 46
Fluidized bed separation, 270
FO fuzzy PID, 306–309, 311
Food added substances, 7
Food preservation, 3, 12
Food waste, 204, 205
Food waste on dumping site, 20, 22
Fossil-based plastics, 257
Fossil fuels, 20, 32
Fourier transform infrared, 213
Fractional-order proportional integral derivative (FOPID) controller, 305–308
Froth flotation, 272
Frothing agent, 270
Fruit and vegetable wastes (FVWs), 225, 226
FTIR spectra, 214, 216
FTIR spectrum, 215
Fuel cell electric vehicle (FCEV), 105
Fuzzy FOPID controller, 307, 308
Fuzzy PID controller, 306, 307

G

Gasific application, 342
  BSEM, 337
  in Jupyter Notebook, 337
  schematic flow, 338
  software functionalities
    collection, biomass ultimate analysis, 339, 340
    composition visualization, 341
    equilibrium model simulation, 341
    external packages, 338–339
    gas composition, 345, 346, 348, 349
    GSE model, 344, 346, 347
    H_2 composition, 346, 347, 349, 350
    loading physical property data, 343
    model property calculation, 340
    operating condition, 339, 342
    physical property, 340
    ultimate analysis, 342, 343
Gasification
  biomass process, 336
  empirical analysis, 336
  self-contain biomass gasification model, 336
Gasification stoichiometric equilibrium (GSE) model, 339, 341, 344, 346, 347
Geo-environmental impact assessment, 288
Global climate, 200
Global community, 201
Global environmental impact, 292, 294
Global warming, 19
Glycerol, 33–35
Granular starch, 261
Grapefruit and lemongrass essential oils (GFLG)
  ANOVA data, 170
  concentration, 169, 170
  data assessment, 171
  gravimetric analysis, 167–169
  inhibition efficiency, 168–170
  inhibition values, 169
  intermediary, 166
  LCS, 166–168
Graphite powder, 112
Grasshopper optimization algorithm (GOA), 306–309, 311
Gravity dam, 289, 290
Gravity retaining wall, 289
Green plasticizers, 32
Greenhouse gas (GHG) emissions, 200–201, 204
Gross domestic product (GDP), 99, 100
Groundwater, 250, 288
Gutbuster, 9

H

HCl formulation, 169
Health cloud, 148
Health monitoring systems
  aim, 147
  MHMS, 147
  non-invasive data collection, 153
  organ analysis, 147
  RHMS, 147
  WHMS, 148
Healthcare, 146
Healthcare delivery, 252
Healthcare expenditure, 157
Healthcare services, 146

Healthy ecosystems, 201
Heat transfer applications, 46
Heavy metals, 186, 267
Heavy oil reservoirs, 126
HLPC pump, 213
Hospital, 249–254
Human-induced activities, 267
Hybrid nanofluids, 46
Hydraulic structure, 288
Hydrocarbon (HC), 277–279, 283
Hydrocarbon plastics, 257
Hydrocarbon reservoirs, 117
Hydrocyclones, 270
Hydrodesulfurization, 211
Hydroelectric power, 19
Hydrogen fuel cell cars, 105
Hydrogen storage, 106
Hydrothermal technique, 4

I

Immobilization technology, 212, 219
Immobilized activated carbon
  adsorbents (*see* Adsorbents)
  batch adsorption mode, 212
  characterization, 213
  continuous adsorption process, 212
  desulfurization, 213, 214
  EPA, 211
  HDS, 211
  hydrodesulfurization, 211
  immobilization technology, 212
  materials, 212
  preparation, 212
  sulfur oxides, 211
  surface area and porous structure, 211
Industrial revolution, 316
Industry 4.0, 318, 321, 327
Infectious diseases, 250
Information and communication technology (ICT), 145
Infrastructural decadence, 131
Infrastructure development, 131
Inhibitor concentration, 171
Inorganic inhibitors, 176
Insect vectors, 202
Integral absolute error (IAE), 304, 306, 308, 309
Integral square error (ISE), 304–307
Integral time square error (ITSE), 304, 306, 307, 309, 310
Internal combustion (IC), 278
Internal plasticization, 30
Internet of things (IoT), 145, 317, 318
IoT-based smart toilet system, 151

J

Japanese Industrial Standard (JIS), 34
Jupyter Notebook, 351

K

Kryos, 64

L

Lacty (polylactic acid), and Mirel™, 262
Land, 267
Land pollution, 268
Land/soil pollution
  biomass, 269
  characteristics, 269
  chelating agents, 272–274
  clay and silt, 269
  coarse soil particles, 269
  containment remedies, 270
  direct and indirect routes, 269
  ecosystems, 269
  electrokinetic separation, 271
  Froth flotation, 272
  in situ treatment methods, 270
  methodology, 268
  physical separation, 270
  pyrometallurgical separation, 271
  remediation methods, 270
  soil washing, 271
  source control, 270
  stones and pebbles, 269
Langmuir adsorption, 182, 183
Langmuir isotherm, 183
Laser ablation, 4
Lean mixtures, 278
Life cycle performance, 294–297
Linear and circular waste disposal system, 237, 238
Linear economy
  and CE, 241, 243
Linear low-density polyethylene (LLDPE), 32
Linoleic acid, 31
Lithium-ion batteries, 109, 112
Livestock waste, 225
Load frequency control (LFC), 301, 302
Long-chain carboxylic acid, 33
Low carbon steel (LCS), 166, 171
Low-carbon technologies, 203

Low-cost adsorbents, 225
Lubrication, 50

M

Machine operation settings, 75
Machining
  common operations, 75
  cutting fluids (*see* Cutting fluids)
  drilling operation, 56, 57
  end milling, 59
  fabrication method, 75
  face milling, 59, 60
  metalworking, 55
  milling operation, 58, 59
  operations, 55
  peripheral milling, 60, 61
  roles of cutting fluids, 76–79
  traditional techniques, 56
  turning, 60, 61
Machining operations, 76
Magnetic separation, 270
Maintenance, 157, 158, 160, 161, 163
Management of processes
  and EPE, 18–20
Manufacturing, 62
Manufacturing industry, 56
Manufacturing organizations, 316, 320, 326
Manufacturing supply chain processes, 318
Mater-Bi®, 263
Material efficiency strategies
  reduction of emissions, 241, 243
Material-efficient housing, 242, 244
Materials engineering principles, 198
MATLAB ("MATrix LABoratory") software, 303
Matplotlib library, 339
Matplotlib package, 341
Mechanical separation, 270
Medicine industries, 33
Medium-chain triglycerides (MCT), 11
Metallic alloys, 165
Metallic nanofluids, 46, 49, 76
Metals, 185, 186
Metalworking operations, 78
Microbial activity, 261
Microbial degradation, 223
Microorganisms, 258
Microwave-assisted method, 4
Mild steels, 195
Milling, 58, 59, 75
Milling cutter, 58
Mineralization, 259, 260
Minimum quantity lubrication (MQL), 62, 63, 65–69
Mitigation strategies, 251, 254
Mobile health monitoring system (MHMS), 147
Modern automated manufacturing systems, 55
Modern toilets, 148
Moringa Oleifera corrosion inhibitive behaviour
  corrosion rate, 178
  electrochemical parameters, 181
  Langmuir adsorption, 182, 183
  mild steel chemical composition, 176
  mild steel composition, 178
  phytochemical analysis, 177
  phytochemical screening, 176
  potentiodynamic polarisation measurement, 177
  potentiodynamic polarisation result, 180
  preparation, 176
  weight loss, 177, 179
Moringa oleifera inhibitor concentration, 178
Motorcars, 278, 283
MQL machining, 77
Municipal solid waste (MSW), 238, 239

N

Nano-agro-composite additives
  AIW, 224
  cost-effective, 224
  effluent (*see* Effluent treatment)
  EPA, 224
  freshwater, 223
  O&M costs, 223
  wastewater (*see* Wastewater treatment)
  water pollution, 223
  WHO, 224
Nanoantimicrobial, 7
Nanoclays, 6, 8
Nanocochleates, 7
Nano-cutting fluids, 45, 79, 80
  in machining operations, 80–81
Nanofluids, 45, 79, 117, 118
  applications, 46
  biomedical applications, 50
  ceramic, 79
  CFD analysis, 46
  direct evaporation method, 47
  factors, 48
  heat transfer fluids, 50
  heat transfer properties, 51
  industrial applications, 50

lubrication, 50
metallic, 46
micro-sized solid particles, 51
nano-sized particles, 47
non-metallic, 46
organic/inorganic fluid, 46
physiochemical characteristics, 48
practical application, 51
preparation, 48
properties and characterization, 48–50
silica nanofluids for cooling applications, 81
single-step preparation method, 47
sonication process, 47
stability, 48
techniques, 47
thermal conductivities, 79
tri-nano-additive, 47
two-phase fluid suspensions, 51
Nanofluids lubricating properties, 50
Nanofluids viscosity, 49
Nanoform zeolites, 8
Nano-liquid, 119
Nanomagnets, 8
Nanomaterials, 9, 45
for finishing in textile industry, 5
methods of synthesis, 4
Nanomedicine, 10
Nanomicelle, 7
Nanoparticles, 49
for machining operations, 80
Nanoscale gadgets, 9
Nanoscale scattered eye balm (NDEO), 11
Nanoseeds, 8
Nanosensing, 7
Nanosensors, 8
Nanostructured materials, 10
Nanosystems, 12
Nanotechnology, 3, 79
areas, 3
challenges of applications, 11–12
energy sequence, 100
in energy production, 100
nanofluids (*see* Nanofluids)
in storage of energy, 108–109
value-added chains of the energy sector, 100
Nanotechnology-based agriculture products, 8
Nanowires, 112
Nanozeolites, 9
Nanozeolite science, 13
Natural-based oil, 32, 33
Natural-based plasticizers, 32, 33
Natural chelating agents, 273
Natural fibres, 38
Natural learning processing (NLP), 319
Natural Rubber and Styrene Butadiene Rubber (NR-SBR), 33
NatureFlex™, 262
NatureWorks® PLA, 262
Nauclea diderrichii (ND), 231
Nigerian Institute for Biomedical Engineering (NIBE), 158
Nodax™ (polyhydroxyalkanoates), 262
Non-biodegradable bio-based plastics, 263
Non-environmentally friendly compounds, 176
Non-ferrous metals, 185
Non-metallic nanofluids, 46, 76
Non-oil extract-based inhibitors, 166
Nonrenewable fossil fuels, 101
Nonrenewable resources, 17–27
Nonstoichiometric equilibrium model, 336
NumPy packages, 338, 339, 344

O

Occupational accidents, 202
Oil-based fluids, 77
Oil recovery strategies, 117
Operation and maintenance (O&M) costs, 223
Orange peels, 227, 228
Organic matter, 241
Organization flexibility, 325
Oxidation of wastes, 202

P

Packed bed column, 214–219
Palladium-graphene nanocomposite, 7
Palm oil (PMO), 32
Pandas' package, 338, 339, 344
Particulate matters, 267
Peripheral milling, 60, 61
Persistent organic pollutants, 267
Personal health monitoring, 146
Personalised health monitoring device, 152
Petroleum distillates, 211, 212
Petroleum industries, 33
Petroleum sources, 258
Pharmaceutical industries, 33
Pharmaceutical nanosystem, 10
Pharmaceutical nanotechnology, 10
Phosphate coatings, 186
Phosphating technique
chemical coating process, 185
chemical oxidation reduction, 186

Phosphating technique (*cont.*)
 control metal corrosion, 185
 heavy metals, 186
 steps, 186
Photodynamic treatment, 11
Photosynthesis, 202
Physical disintegration biodegradation, 259
Physical separation, 270
Physiology, 161
PLA fibres, 262
Placcel® (polycaprolactone), 263
Plain milling, 60
Plant oils, 32
Plasticization
 absorption, 30
 discharge, 30
 external, 30
 internal, 30
 mechanical and thermal features, 30
 mixture, 30
 polar group, 31
 re-establishing, 31
Plasticizer
 biopolymer-based films, 30
 classification, 30
 continuous growth, 29
 development, 29
 eco-friendly, 31, 32
 esters, 30
 glycerides, 30
 glycerol, 33–35
 natural-based oil, 32, 33
 oils, 30
 plasticization, 30–31
 polyhydric alcohols, 30
 primary, 30
 properties, 29
 secondary, 30
Policy and regulatory advocacy, 245
Policy makers, 19
Pollution prevention, 205, 207, 208
Pollution reduction, 20–23
Polycyclic aromatic hydrocarbons, 267
Polyethylene terephthalate (PET), 263
Polyhydroxyalkanoates, 263
Polymer matrix, 29
Polysaccharides, 31
Polytrimethylene terephthalate, 258
Pomelo peels (PPs), 229
Poor governance, 203, 204
POVAL, 263
Primary plasticizers, 30
Production efficiency, SM, 323
Proteins, 31
Public health, 17, 20, 25–27, 197, 198, 202, 205, 208, 241, 249, 252
Public health policy-making, 202
Public-private initiatives, 132, 134
Public-private partnerships (PPSs)
 achievement of construction works, 132
 adoption, PPP approach in Nigeria, 132
 benefits for state and taxpayers, 134–135
 challenges facing PPP in Nigeria, 136–137
 civil infrastructure degeneration in Nigeria, 138
 implementation, development and procurement in Nigeria, 135
 initiatives and concession activities, 134
 portfolio of funding choices, 133
 poverty level in Nigeria, 132
 private establishment, 131, 132
 procurement method and implementation process, 132
 project financial options and portfolio representation for government, 132, 133
 projects in Nigeria, 134, 135
 recommendations, 140
 share of the transport sector in Nigeria's public sector, 134
 sustainable development, 132
Pyrometallurgical separation, 271

R
Reaeration, 202
Rechargeable battery, 109
Reconnaissance survey, 251
Recycling, 18–20, 237, 241, 242, 245, 246, 250
Reduction-oxidation reaction, 168
Remediation methods, 270
Remote health monitoring system (RHMS), 147
Renewable energy, 100, 108
Renewable resources, 17–27
Reservoir permeability analyzer (RPT), 119
Reservoir Permeability Tester, 118
Resource availability, 205
Resource efficiency, 250
Respiration, 202
Retaining walls, 289
Reuse, 237
Rice husk ash (RHA), 38
Rich mixtures, 278

S

Sanitation, 149
SbSeI nanowires, 103
Scanning electron microscopy (SEM), 213
Scanning electron microscopy with energy-dispersive X-ray analysis (SEM-EDX), 186
Schistosomiasis, 202
Schweizer's reagent, 261
SciPy package, 339
Secondary plasticizers, 30
Seed germination, 9
SEM analysis, 195
SEM images, 215
SEM morphology, 195
Semiconductor ceramics, 6
Sensitivity analysis, 126
Sensitization, 250, 254
Sensor technologies, 147
Serviceability, 288
Silica nanofluids, 81
Silicon carbide (SiC), 38, 39, 41–43
Silicon nanoparticle, 126
Simulation tools, 322
Slab milling, 60
Slag, 271
Small-scale tests, 270
Smart, 145
Smart cities
  attainment, 145
  business and career, 143
  components, 144
  data collection and data analysis, 153
  development, 144
  ICT, 145
  innovation, 144
  IoT, 145
  replacement, 144
  smart-oriented governance, 144
Smart furniture, 6
Smart healthcare services, 148, 153
Smart personal health monitoring system, 153
Smart services
  automation, 145
  components, 145
  healthcare, 146
  sectors, 146
  smart city indicators, 145
  systems and infrastructures, 146
  XaaS, 146
Smart textiles, 6
Smart toilets
  job opportunities, 150
  luxury and automated services, 148
  personal health monitoring
    Benedict's quantitative solution, 151
    blood sugar level estimation, 150
    clean water use, 151
    healthcare services, 150
    medical status, 152
    personalised health monitoring device, 152
    sensors, 150
    solid waste analysis, 152
    technical alternatives, 151
    urinary tract infection prediction, 151
    urine storage and separation, 151
    vacuum suction pump, 151
Smart toilets, SDGs
  clean water and sanitation, 149, 150
  good health and well-being, 149
  healthy lifestyle, 149
  sustainable cities and communities, 150
Social and economic development, 202
Social equity, 202
Social, educational and environmental conditions, 202
Socially just systems, 201
Socio-economic conditions, 250
Sodium alginate, 212, 214, 219, 220
Soil, 268
Soil acidity, 272
Soil pollutants, 267
Soil pollution, 19, 268
Soil washing method, 271
Solar energy, 19
Sol-gel, 4
Solid lubricants, 63
Solid waste management (SWM), 246
  CE (*see* Circular economy (CE))
  cleaner environment, 241
  defined, 239
  organic matter, 241
  and SDGs, 239, 240
Sonication process, 47
Sorona® (polytetramethylene adipate-co-terephthalate), 263
Source reduction, 205
Soxhlet extractor, 118
Spacer reinforcement, 293
Stand-alone cyber-physical systems (SACPS), 318
Standard deviation (SD), 169
Starch-cellulose composites, 261
Starches, 261, 262
Sterilisation technologies, 150
Stir casting, 38, 39, 42, 88–89
Stoichiometric equilibrium model, 336

Sub-Saharan Africa, 250–253, 255
Substantial healthcare management system, 149
Sulfur oxides, 211
Surface roughness, 66
Suspending nanoparticles, 45
Sustainability, 204, 206, 287, 288, 292, 294, 297, 298
  PPP, 132 (*see also* Public-private partnerships (PPSs))
  water retaining structures, 288
Sustainability addresses, 198
Sustainability perspectives, 206
Sustainable city, 143, 150
Sustainable development, 17, 18, 20, 25–27, 32, 132, 197, 198, 201, 202, 204, 208
  and water retaining structures, 292, 294
Sustainable development agenda, 143
Sustainable development goals (SDGs), 143
  and SWM, 239, 240
Sustainable environment methods, 20–23
Sustainable manufacturing (SM), 62, 69–70, 316
  approach, 321
  challenges, 326
  collaborative manufacturing eco-friendly system, 316
  cost-effective process, 316
  decision-making, 325
  environmental performance, 325
  future, 327
  optimization process, 324
  organization flexibility, 325
  perspectives, 321
  process digitalization, 322
  process innovation and sustainability, 323–324
  production efficiency, 323
  quality control tools, 322–323
  sustainable method, 320
  technological revolution, 316
Sustainable processes, 131
Sustainable waste management
  CE, 245
  circular economy, 237
  linear and circular waste disposal system, 237, 238
  public health, 241
  recycling, 237
  reuse, 237
Synergistic combination effect, 166
Synthesis
  glycerol, 33
  plasticizers, 32
  plastics and bio-based plastics, 35
Synthetic organic compounds, 165
Synthetic plastics, 31
Synthetic polymers, 261
Systematic computer-based methods, 197

T
Template synthesis, 4
Textile businesses, 12
Textiles, 4, 5
Thermal conductivity, 37, 39, 40, 42
Thermal energy storage, 109
Thermal power plant operation, 99
Thermodynamics, 198
Thermoplastic polyester elastomers, 258
Thermoplastic starch, 261
Time-weighted absolute error (ITAE), 304, 306–308, 311
Titanium aluminium nitride (TiAlN), 38
Titanium oxide, 38
TONE™, 263
Tool wear, 61, 64, 66, 68, 69
Traffic emissions, 277
Transport services, 250
Treatment
  additive-based (*see* Additives)
  effluent (*see* Effluent treatment)
  wastewater (*see* Wastewater treatment)
Tri-nano-additive, 47
Turning, 60, 61
Two-area interconnected microgrid network, 303

U
Ultrasound, 4
Underground water reservoirs, 287
Underground water tanks, 287
United States Environmental Protection Agency (US EPA), 268
Urban and rural populations, 18
Urbanization, 250

V
Value chain, 100
Vegetable oils, 64
Vehicular emissions, 277–282
VFWs, 226
Virtual reality (VR), 320

Volatile organic compounds, 267
Volatilization, 271

W
Ward model, 49
Waste generation in urban centres, 250, 251
Waste heat to power (WHP), 102
Waste management
  and inspection of disposal sites
    circular strategies, innovations, life cycle stages and enablers, 242, 244
    framework, 241, 243
    policy, material efficiency and greenhouse impacts, 242, 244
    residential areas in Nigeria, 241
    surveys and data, 241
  challenges, 245
  solid (*see* Solid waste management)
  sustainable (*see* Sustainable waste management)
Waste minimization, 239
Waste to wealth, 245
Wastewater treatment, 223–232
Water and wastewater systems, 287
Water pollution, 223
Water retaining structures
  BS 8007:1987, 288–292
  coated steel sections, 295, 296
  concrete and steel sections for water tanks, 295
  construction site, 289, 292
  hydraulic structure, 288
  life cycle performance, 294–297
  rivers and small streams, 287
  standard building code, 288
  and sustainable development, 292, 294
  sustainability, 288
  swimming pool construction, 295, 296
  types, 289
  water tank walls and rigid base, 297
  world production of cement and steel, 294, 295
Water scarcity, 250
Water treatment, 20, 23
Wear resistance test, 40
Wearable health monitoring system (MHMS), 148
Wind energy, 101
World Health Organization (WHO), 224

X
X-ray diffraction (XRD) analysis
  AA6061/clay composites, 89
  AA6061/clay + RHA composites, 94–96
X-ray fluorescence (XRF) spectroscopy, 186, 189

Z
Zinc, 38
Zirconium oxide, 38
Zn-Ca phosphate layer, 193

The manufacturer's authorised representative in the EU is Springer Nature Customer Service Centre GmbH, Europaplatz 3, 69115 Heidelberg, Germany. If you have any concerns regarding our products, please contact ProductSafety@springernature.com

Printed and bound by CPI Group (UK) Ltd, Croydon, CR0 4YY
30/09/2026
02225233-0006